P9-EMP-248

Colorimetric and Fluorimetric Analysis of Organic Compounds and Drugs

M. PESEZ
Director, Analytical Division
Roussel, UCLAF
Romainville, France

and

J. BARTOS
Director, Analytical
Research Laboratory
Roussel, UCLAF
Romainville, France

MARCEL DEKKER, INC. New York

A la mémoire de

Monsieur Jean-Claud Roussel

MARCEL DEKKER, INC.

270 Madison Avenue, New York, New York 10016

LIBRARY OF CONGRESS CATALOG CARD NUMBER: 73-84815

ISBN: 0-8247-6105-7

Current printing (last digit):
10 9 8 7 6 5 4 3 2 1

PRINTED IN THE UNITED STATES OF AMERICA

INTRODUCTION TO THE SERIES

Analytical biochemistry, particularly as it relates to clinical chemistry, has experienced explosive growth during the past two decades. The availability of new, sensitive techniques such as atomic absorption and flame spectroscopy, immunochemical and radioimmunoassays, chromatographic and electrophoretic separations, and continuous flow and discrete high-speed automated analyzers, has permitted a first look at the biochemistry of some diseases and in other cases, particularly in hormonally related diseases, a reexamination of previously reported biochemical relationships earlier obtained with cruder and less sensitive methods. The analysis of concentrations of materials in the femto and picogram range has become a common everyday procedure.

The purpose of this series in Clinical and Biochemical Analysis is to provide the worker with a collection of critical monographs in specific subjects, each containing not only details for performing assays and their quality control, but also information in depth concerning the significance of the findings from both the cellular and clinical point of view. Our hope is that these monographs will be of value to the worker who needs pragmatic information during his working day, as well as to the student who wishes an in-depth presentation on a specific and well-defined subject.

Morton K. Schwartz, Ph.D.

PREFACE

Colorimetric and fluorimetric analyses of organic compounds and drugs have been used extensively in numerous fields, and it may be stated that they will remain of great value in the future, in spite of the steadily growing resort to purely physical methods which often necessitate very sophisticated and expensive instrumentation. Routine colorimetric and fluorimetric analyses can be performed with very simple instrumentation, resulting nevertheless in sensitive and accurate measurements with the advantages of speed and simplicity.

A few decades ago a new color reaction was often discovered by chance, but today colorimetry and fluorimetry are no longer mere compilations of purely empirical recipes. Many current methods are derived from preliminary theoretical reasoning. This is more particularly the case in functional colorimetry and fluorimetry, which are dealt with in this book. These branches of analysis are based on chemical reactions characteristic of the various functional groups. The molecule bearing such a group reacts with a suitable reagent to give a colored or fluorescent species.

Numerous reactions of this kind have been described in the literature, and a particular reaction has very often given rise to a wholesale lot of sundry procedures. The analyst who finds himself in need of a colorimetric or fluorimetric method may therefore be obliged to survey a tremendous quantity of literature in order to select the one procedure which seems to suit his needs. Then, when he tests the selected procedure, he may ascertain that he cannot reproduce the author's results.

It is the intent of this book to provide practicing analysts with procedures which were carefully tested. All the data given in Results sections were actually obtained in our laboratories.

It is obvious that we could not research and evaluate all the methods described in the literature, and selection was indispensable. Therefore, for each functional group we initially assembled a set of methods covering the major chemical reactions capable of giving rise to colored or fluorescent species. Only methods which did not necessitate the separation of an intermediary compound or of the final dyestuff by techniques other than a mere liquid-liquid extraction were taken into account. Reactions based on halochromism or halofluorism, which were mainly attributable to intramolecular changes caused by an acid or a base, were not dealt with.

The corresponding procedures were then tested. Since a rather great number of them gave immediately satisfactory results they were included in the book without modification, or with only very slight ones.

In some instances, it was observed that some significant modifications led to more accurate, reproducible, or sensitive results. In our text the corresponding references are then preceded by: From...

In a number of cases, only very poor results were obtained, while in others, it was found that a suitable combination of various procedures based on the same principle, or the development of an almost completely recast procedure gave satisfactory results. The corresponding references are thence preceded by the abbreviation: Cf. This same abbreviation is used when a procedure was derived from a paper in which the reaction was studied only from a qualitative, or nonanalytical, standpoint. Obviously, we were obliged to discard some methods which did not yield satisfactory results by any means.

Some methods have no references. These were developed in our laboratories and, to our knowledge, have not been previously published.

A problem arose when it was necessary to select a limited number of compounds which should exemplify each procedure. Instead of selecting arbitrarily some complicated molecules which might concern only a few specialists, we thought that it would be of more value to present results given by simple compounds. Direct comparisons between methods are thereby made easier, since miscellaneous procedures were often tested with the same products.

In structuring our text, we tried to arrange each chapter in a systematic fashion, proceeding from the most general to the most specific reaction. For instance, in the chapter devoted to aliphatic amines, the analyst will at first find the colorimetric reactions which can be applied to all classes of alkylamines, then those which are limited to primary and secondary amines, secondary and teriary amines, primary amines, secondary amines, tertiary amines, and quaternary ammonium compounds. The chapter ends with the fluorimetric methods of determination, which are classified according to the same rule, and at the top of each of these methods, the class of compounds which can be determined is labeled with an (F). Each chapter is preceded by an introduction in which all the reactions are critically summarized.

For each method, as a rule, only the reagents which should be especially prepared are listed under the Reagents heading. Solvents, mineral acids and bases, and their standardized aqueous solutions, which are in common use in all laboratories, are not included.

Unless otherwise stated, solutions of acids and bases are always aqueous.

For all compounds in solution, and unless otherwise stated, the concentration is always expressed in $^{w}/v$ if the solute is a solid, and $^V/v$ if it is a liquid. There are, however, two exceptions: Solutions of hydrogen peroxide and of formaldehyde are both expressed in $^W/v$, since this corresponds to a commonly accepted practice.

When it is said that a reaction is performed at room temperature, it means that this temperature is within the range 18°-24°C.

As far as possible, the names of all compounds are those in use in The Merck Index (8th Edition).

The abbreviations of the titles of periodicals are those recommended by Chemical Abstracts. The titles of two books which are cited several times were also abbreviated:

M.R.A.O. : Méthodes et Réactions de l'Analyse Organique, vol. 3 (1954), by M. PESEZ and P. POIRIER (Masson Ed.).

F.S.A.O.V.: Farbreaktionen in der Spektrophotometrischen Analyse Organische Verbindungen, vol. 1 (1969), by Z.J. VEJDĚLEK and B. KAKÁČ (VEB Gustav Fischer Verlag, Jena).

At the end of the book, we present some short monographs on miscellaneous reactions or reagents which are involved in several determinations of compounds bearing diverse functional groups, and the syntheses of reagents which are not yet commercially available.

The authors wish to thank all of those who contributed to the numerous experimental studies and determinations which allowed the development or checking of methods described in this book, and to the drawing and preparation of developed formulas for camera-ready copy. Without their helpful assistance, this book could not have been written.

M. Pesez
J. Bartos

CONTENTS

Part I

INTRODUCTION

Chapter 1

THEORY AND INSTRUMENTATION

This book was written for practicing analysts who are fundamentally in need of reliable colorimetric and fluorimetric methods of determination of organic compounds. They might therefore consider that spectrophotometers and spectrofluorimeters are but ingenious tools permitting convenient readings from suitable dials, and that the theories of absorption and emission of light are interesting in so far as they confirm that the relationship of concentration to reading should be linear.

On the other hand, theory and instrumentation have already been dealt with in numerous books and treatises, and it might be considered that a chapter devoted to these topics would merely duplicate these works.

We felt, however, that such a chapter could prove fruitful if restricted to the points which we considered as being closely related to the methods proposed in this book. We have tried to present it so that it will not discourage the analyst who has no particular inclination to study mathematics or electronic theories, but who nevertheless wishes to acquire, or to recall to his recollection, a background of knowledge allowing a better comprehension of the measurements and of the instruments with which they are made.

I. GENERAL COMMENTS

It is obvious that the visual comparison of the color intensity of an unknown solution with a series of standards is now obsolete, and modern analysts do not even conceive colorimetric determinations without the use of a photoelectric instrument.

It must thence be mentioned that the terms "colorimetry" and "colorimetric determination," which remain widely used, do not fit the facts when a spectrophotometer is used, since they imply the measurement of a color intensity. Let us take for instance a blue solution: The deepness of the color increases with increasing concentration of the colored species, and the solution is blue through transparency because it does not absorb blue radiations, whereas yellow and orange ones are strongly absorbed. Likewise, a yellow solution absorbs chiefly violet and blue light. It may be said, therefore, that spectrophotometers measure but indirectly the color intensity of the solution, since readings are made at the wavelength of the most strongly absorbed radiation. Strictly speaking, this branch of analysis should hence be referred to only as spectrophotometry in the visible range, because the apparatus measures the absorption of a visible radiation, but it is not the radiation which is seen through transparency, and which corresponds to the color of the solution. Nevertheless, the terms "colorimetry" and "colorimetric determination" will certainly be kept up in the future for the sake of brevity, as we have done in this book.

Contrariwise, the terms "fluorimetry" and "fluorimetric determination" are entirely adequate, since the apparatus measures the intensity of the emitted light.

In order to understand the laws of absorption and emission of light, some mathematical demonstrations are necessary. We tried to reduce them to a minimum in the following paragraphs, and the purely theoretical principles upon which the apparatus is built up were likewise limited, since these topics are beyond the scope of this practical book and are dealt with in detail in various works (1-7).

II. ABSORPTION OF LIGHT

The energy of a molecule consists of electronic energy afforded by the electrons in the atoms, rotational energy afforded by the rotation of the molecule about its center of gravity, and vibrational energy due to the elastic vibrations of the atoms relative to each other along their internuclear axes.

When a molecule absorbs a visible radiation, its electronic energy is raised from the ground state to a higher level through a quantized transition, but changes in rotational and vibrational energies occur also, since the three energies are interdependent. Hence, instead of discrete lines corresponding to allowed quantized electronic transitions, the absorption curves in the visible range will usually present broad and featureless maxima.

A. Bouguer-Lambert-Beer Laws

The absorption of a photon $h\nu$ by a molecule in its ground state G can be represented by the equation

$$G + h\nu \rightarrow G^* .$$

G^* is the molecule in a higher energy level, from which it reverts rapidly (generally within 10^{-8} sec) to the ground state, and the absorbed radiant energy is usually converted into thermal energy.

This reaction can be considered as an irreversible bimolecular reaction between G and $h\nu$, and its rate is therefore given by

$$v = \frac{d[h\nu]}{dt} = k_1[G][h\nu]; \tag{1}$$

$[h\nu]$ is directly proportional to the intensity of the light: $[h\nu] = k_2 I$. $[G]$ is the concentration c of the colored species, and dt is directly proportional to the distance $d\ell$ covered by the photons during this time: $dt = k_3 d\ell$. Equation (1) can thence be written as follows:

$$-\frac{dI}{d\ell} = k'cI$$

or

$$-\frac{dI}{I} = k'c\,d\ell \tag{2}$$

If I_0 is the incident intensity and I the intensity transmitted through the thickness ℓ of the cell, i.e., the intensity of emergent light, integration of Eq. (2) gives

$$\frac{I}{I_0} = e^{-k'c\ell}, \tag{3}$$

or, upon converting into decadic system ($k = 0.4343k'$),

$$T = \frac{I}{I_0} = 10^{-kc\ell}\ . \tag{4}$$

T is the transmittance of the solution, i.e., the ratio of the radiant energy transmitted by the solution to the incident energy.

Absorbance A is the negative logarithm to the base 10 of the transmittance:

$$A = -\log T = \log\frac{I_0}{I} = kc\ell. \tag{5}$$

Equation (5) shows that absorbance is directly proportional to the thickness of the solution (Bouguer-

Lambert law), and to the concentration of the absorbing species (Beer's law).

This demonstration is valid only under definite conditions. It was implicitly admitted that:

(a) All the molecules G are not brought up simultaneously to the excited state; as a matter of fact, with usual spectrophotometers the fraction of G present in any excited state is very low.

(b) All the photons are identical, i.e., the light is monochromatic.

(c) All the photons follow identical paths through the solution, i.e., the incident light is a parallel beam, and the absorption cell has flat and parallel entrance and exit windows which are perpendicular to the direction of the beam.

The equations $T = 10^{-kc\ell}$ and $A = kc\ell$ show that transmittance and absorbance do not depend on the intensity of the incident light, provided that condition (a) is satisfied.

B. Numerical Value of k

The numerical value of k depends on the units in which c and ℓ are expressed.

When c is expressed in gram moles per liter and ℓ in centimeters, the coefficient k becomes ε and is named "molar absorptivity":

$$\varepsilon = A/c\ell.$$

Molar absorptivity can be considered as a physical constant for a given compound, provided that the solvent and the wavelength are defined.

When c is expressed in grams per liter and ℓ in centimeters, k becomes a and is named "absorptivity."

When c is expressed in grams per 100 ml and ℓ in centimeters, k becomes $E_{1cm}^{1\%}$ and is named "extinction, one percent, one centimeter."

The two latter coefficients do not involve the molecular weight of the colored species and are therefore particularly useful for compounds of unknown constitution.

It must be pointed out that confusion has existed in the literature for a long time in regard to the terms used in absorption spectrophotometry. The terms "transmittance," "absorbance," "molar absorptivity," and "absorptivity," and the corresponding symbols T, A, ε, and a, were recommended in 1968 by the Physical Chemistry Division of IUPAC (8), provided that decadic system is used. Other symbols correspond to the Naperian system. The symbol $E_{1cm}^{1\%}$ and the corresponding term are not mentioned in these recommendations, but are still used.

C. Significance of k

Braude (9) showed that, whereas the wavelength of the absorbed light is determined by the allowed transition energy, the absorption coefficient k is determined by the transition probability and depends upon the chromophore area, i.e., that part of the cross section of the molecule within which a given photon must fall in order that interaction may take place.

The maximum theoretical molar absorptivity for an average organic molecule may be calculated to be of the order of 10^5, and as a matter of fact, with a few exceptions, this is the magnitude of the largest coefficients experimentally determined.

D. Additivity

When two absorbing species are in mixture, additivity of the individual absorbances is observed, provided that no interaction occurs.

E. Deviations from Beer's Law

Whereas there is no known exception to the Bouguer-Lambert law, a deviation from proportionality may be observed when the concentration of the colored species varies. This deviation results from the occurrence of a reaction whose extent is concentration-dependent.

For instance, a solution of a weak acid contains both undissociated molecules and anions. It is obvious that the ratio of the concentration of the undissociated species to the concentration of the anion species depends upon the concentration of the solution as well as upon the pH. If the molar absorptivities of the two species at the selected wavelength are not equal, a deviation from Beer's law will be observed, unless the pH and the ionic strength of the solution be kept constant.

Likewise, dimerization, polymerization, and solvation will be responsible for deviations, as far as they are concentration-dependent. These deviations may, of course, be positive or negative.

If a compound is fluorescent, a part of the absorbed radiation may be reemitted at longer wavelengths. A fraction of this light will impinge on the phototube together with the transmitted light. The absorbance read will, therefore, be theoretically lower than expected. However, comparison between the intensities of transmitted and fluorescent lights shows that the influence of fluorescence may practically be neglected in the field of colorimetric determinations, particularly when absorbances are higher than 0.1, self-quenching and inner-filter effect (Section IV.B) becoming important beyond this value.

III. COLORIMETRIC DETERMINATIONS

A. Functional Colorimetry

This branch of colorimetry is based on chemical reactions characteristic of the various functional groups. A given reaction may therefore be applied to different molecules bearing the same group, but it must be emphasized that the color developed and the yield of the reaction may vary with the compound tested, since these factors depend upon the structure of the whole molecule. Therefore, as a rule, functional colorimetry does not allow absolute measurements, as do, for instance, volumetric methods. The result given by the unknown must be compared with a calibration curve prepared from known amounts of the pure compound under the same conditions.

On the other hand, Beer's law can be obeyed even if the reaction is not quantitative, provided that its yield remains constant within the range of concentrations studied.

B. Reporting Results

There is no general rule for reporting results of colorimetric determinations. Numerous authors give either the minimum amount of compound which can be determined or the range within which Beer's law holds. Sometimes, when a given reaction is applied to various compounds, the authors report the absorbances obtained with an equal weight of each species. In order to express the results in terms of sensitivity, the molar absorptivity ε is also often given.

We felt that these ways of reporting results are not satisfactory in functional colorimetry.

A mere glance at the absorbance (or optical density) scale of a spectrophotometer permits the observation that the accuracy of a reading depends on the value read. Unless the apparatus is equipped with an arrangement controlling the amplification of the signal from the phototube, readings of absorbances higher than 0.8 will be but poorly accurate, and checking the validity of Beer's law above this value will allow only a rough conclusion to be drawn. On the other hand, provided that Beer's law is obeyed, the smallest amount of compound which can be determined with a sufficient precision may depend upon the accuracy of the instrument used.

The reporting of molar absorptivity is also not very convenient. It is generally admitted that ε is a physical constant. When the determination is made through a chemical reaction which develops a colored species, this value becomes arbitrary in so far as it is not established that the reaction is quantitative and that only 1 mole of the compound to be determined is involved in the formation of 1 mole of the dyestuff. On the other hand, if the values calculated for various compounds developed through the same reaction allow a satisfactory comparison between reactivities, the analyst who wants to establish a calibration curve must calculate back from ε and the final volume of solution, the sample size which will afford suitable results.

For these reasons, we decided to report, for each compound tested, the sample weight which gives an absorbance value of 0.3, corresponding to a transmittance of about 50%. It was ascertained (10) that very accurate readings are possible in the vicinity of this absorbance value on any spectrophotometer. In this way, no prior calculation is required for the setting up of a calibration curve. On the other hand, when a compound can be determined through various reactions, the analyst can very readily select the method offering the desired sensitivity, and interferences afforded by other compounds of the same class can also be readily estimated from the given data. It is, of course, implied that Beer's law is obeyed up to this absorbance value, and unless otherwise stated its validity was always checked at least up to $A = 0.7$.

C. Sensitivity of Colorimetric Determinations

It can be easily established that if A is the absorbance value read for m' μg of a dyestuff of molecular weight M', dissolved in v ml of a solvent, ε is given by

$$\varepsilon = \frac{10^3 AvM'}{m'}.$$

If the dyestuff is obtained in a colorimetric determination from m μg of a compound of molecular weight M, and if it is admitted that the reaction is quantitative and that 1 mole of dyestuff is formed from 1 mole of the compound to be determined, it is obvious that

$$\frac{M}{M'} = \frac{m}{m'}.$$

Therefore, ε will be given by

$$\varepsilon = \frac{10^3 AvM}{m}$$

and

$$m = \frac{10^3 AvM}{\varepsilon}. \qquad (6)$$

The methods described in this book always lead to a final volume of at least 3 ml. This is the smallest volume which allows a satisfactory filling of the 1 cm cells commonly used. On the other hand, it was said that, as a rule, the value of ε but seldom exceeds 10^5 (Section II.C). As a result, Eq. (6) shows that under the optimum conditions (i.e., $\varepsilon = 10^5$, $v = 3$ ml, quantitative reaction), $A = 0.3$ will be given by

$$m = \frac{900\ M}{10^5}\ \mu g,$$

i.e., about 10^{-2} M µg of compound.

Therefore, under the most favorable conditions, if the compound to determine has a molecular weight of, say 100, it can be hardly expected to read $A = 0.3$ for a sample weight lower than 1 µg.

It must be pointed out that the ε values of dyestuffs formed in numerous reactions are rather frequently of the order of only 10^4, the final volume often exceeds 3 ml, and the yield of the reaction is not always quantitative. It may be concluded therefrom that in the field of functional colorimetry, when a conventional spectrophotometer with 1 cm cells is used, a method is satisfactory when it affords $A = 0.3$ for sample weights ranging from 1 to 50 µg approximately.

IV. FLUORESCENCE

Fluorescence is the emission of light from a molecule which returns to its normal ground state from the lowest vibrational level of an excited singlet state, the excitation being achieved by the absorption of light. Absorption and fluorescence are thereby closely related, since the absorption must precede fluorescence emission. Because some of the absorbed energy is frequently lost through collision or the dissipation of heat, the remaining energy of fluorescence is less than the absorbed energy and is therefore emitted at longer wavelengths.

A. Quantitative Aspects of Fluorescence

It was shown (Section II.A) that under suitable conditions, if I_0 is the intensity of incident light, ℓ the thickness of the cell, and c the concentration of the studied species, the intensity of emergent light is given by

$$\frac{I}{I_0} = e^{-kc\ell} \quad \text{or} \quad I = I_0 e^{-kc\ell}$$

The amount of light absorbed is then given by

$$I_0 - I = I_0(1 - e^{-kc\ell}).$$

When the compound is fluorescent, a fraction of the absorbed light is converted into fluorescence light. The ratio of quanta of light absorbed to quanta emitted, ϕ, is named the fluorescence efficiency. It so happens that for many substances ϕ is approximately independent of the excitation wavelength and of the concentration in dilute solutions. The fluorescence intensity F is therefore given by

$$F = \phi(I_o - I) = \phi I_o(1 - e^{-kc\ell}) \quad (7)$$

Expanding the exponential term we get

$$e^{-kc\ell} = 1 - kc\ell - \frac{k^2c^2\ell^2}{2!} - \frac{k^3c^3\ell^3}{3!} - \cdots \quad (8)$$

When working with dilute solutions, the latter terms are negligible and may be dropped. Equation (7) can thence be written as follows.

$$F = \phi I_0 kc\ell \quad (9)$$

If c is expressed in gram moles per liter, $k = \varepsilon$, and Eq. (9) becomes

$$F = \phi I_0 \varepsilon c \ell. \qquad (10)$$

The value of ε is, of course, that corresponding to the excitation wavelength.

Equation (10) leads to various conclusions.

(a) The intensity of fluorescence is directly proportional to the molar absorptivity. The excitation spectrum should therefore be a replica of the absorption spectrum, and the highest intensity of fluorescence will be attained with the excitation wavelength corresponding to the absorption maximum.

(b) The intensity of fluorescence is directly proportional to the concentration of the fluorescent species. Under suitable conditions, calibration curves will therefore be rectilinear.

(c) Whereas the absorbance value does not depend on the intensity of the incident light (Section II.A), the intensity of fluorescence is directly proportional to the intensity of the exciting light. It can be concluded therefrom that spectrofluorimetry is much more sensitive than absorption spectrophotometry. Exceedingly low intensities of light can be measured with modern phototubes, and very high intensities of exciting light are afforded by sources such as the xenon arc tube.

On the other hand, whereas any spectrophotometer gives the absorbance value directly, conventional spectrofluorimeters give only relative intensities of fluorescence, which are read from an arbitrary scale usually graduated from 0 to 100 or from 0 to 1 in 100 linear divisions. Suitable adjustments allow the reading of solutions within a very wide range of fluorescence intensities. Another disadvantage arises from the fact that excitation and emission spectra given by conventional instruments are distorted. These matters are dealt with in a more detailed fashion in the section devoted to instrumentation. As a result, comparison of analytical data obtained by authors working with various apparatus is often almost impossible.

B. Deviations from a Linear Relationship between Concentration and Fluorescence

The intensity of fluorescence being directly proportional to absorbance, it is obvious that all the factors responsible for deviations from Beer's law (Section II.E) will lead to similar deviations in fluorimetric determinations.

On the other hand, the intensity of fluorescence is proportional to the concentration only in very dilute solutions. At higher concentrations the intensity falls off owing to self-quenching and inner-filter effects.

Self-quenching or concentration-quenching is due to the fact that, as the concentration increases, the mean distance between molecules becomes less, and the number of collisions becomes higher. Therefore, the probability of energy transfer from the excited molecule to a nonexcited one increases. A fraction of the emitted light can also be absorbed by the fluorescent species itself, if it absorbs light at the emission wavelength.

The inner-filter effect is attributable to the fact that at higher concentrations the solution at the back of the cell is exposed to a lower intensity of exciting light, owing to absorption of a part of this light by the intervening solution. Mathematically, this phenomenon is correlated with the latter terms of Eq. (8), which have been dropped. This simplification is valid indeed only when the absorbance of the solution at the exciting wavelength is less than 0.05.

V. FLUORIMETRIC DETERMINATIONS

A. Functional Fluorimetry

Like functional colorimetry (Section III.A), functional fluorimetry is based on chemical reactions characteristic of the various functional groups. A given reaction may therefore be applied to different molecules bearing the same group. Widely used in colorimetry for many years, this approach was until 1967 but scarcely applied to fluorimetric determinations, probably because it is much easier to predict the development of a color than of a fluorescence. But the knowledge of the compounds already described as being fluorescent, the study of their methods of preparation and the replacement, in the course of the synthesis, of one of the components by another molecule bearing the same functional group, has in the past few years led to an appreciable extension of this branch of fluorimetry, to which a detailed study was recently devoted in a review (11).

B. Reporting Results

As in functional colorimetry, the result given by the unknown must be compared with a calibration curve prepared from known amounts of the pure compound under the same conditions. But the situation is more complicated in fluorimetry. In reports of determinations problems arise from the lack of an absolute expression of the fluorescence intensity and from the fact that the blank almost always fluoresces, particularly when a chemical reaction is involved.

Different methods are used to overcome these problems, but as yet, in the field of routine analysis, comparison of results gathered by authors working with various apparatus is usually a tremendous problem. It often may be solved by the analyst only by reproducing the experiments and making measurements with his own instrument.

In 1967, we proposed a notation (12) which facilitates these comparisons, the numerical values given making results readily reproducible with any spectrofluorimeter.

The meter of the instrument is set at zero with the photomultiplier shutter closed. We give two limit values for the determination, the upper corresponding to a deflection of 100 meter units, the lower to the smallest weight of compound which can be estimated with a reasonable accuracy. It is, of course, implied that the relationship of concentration to reading is linear over this range. We also report the sample weight and the concentration of a suitable standard both corresponding to a deflection of 50 units. Therefore, knowing the sample concentrations which give readings 100 and 50, and since linearity is observed, it is easy to calculate the blank deflection. On the other hand, the knowledge of the standard concentration corresponding to the deflection 50 permits the proper adjustment on any spectrofluorimeter. It must be verified that the cells used do not afford a noticeable interfering fluorescence. If they do, the corresponding readings must be subtracted from the respective deflections given by the blank, the samples, and the standard.

The distortion of spectra (Section VI.B.2) does not interfere, since at given excitation and emission wavelengths, the correction coefficient is the same for the sample, the blank, and the reference.

Comparison of results obtained with two instruments, namely Farrand equipped with xenon lamp and gratings, and Electrosynthèse (manufactured in France) with low-pressure mercury lamp and filters, showed only very slight discrepancies, probably due to the widths of the spectral bands.

Therefore, the numerical values given make our results readily reproducible with other instruments, and it also becomes easy to find out whether unexpected discrepancies are related to an abnormal fluorescence of the blank, caused by impure solvents or reagents. This method for reporting results was thence adopted in this book.

C. Selection of Standards

It is obvious that the best reproducibility in comparisons is attained when the maxima of excitation and emission of the standard correspond to the wavelengths at which the determinations are performed.

This is the reason why we referred at first (12) to standards responding to this condition, and in some instances we even synthesized the fluorescent species formed in the analytical reaction. Although theoretically very satisfactory, this concept would, however, lead to an undue multiplicity of standards. We tried thence to limit their number, and extended the use of some of the compounds we had prepared to miscellaneous determinations, taking into account the fact that their excitation and emission maxima can be slightly modified by a suitable selection of solvent or pH.

The syntheses of these standards are described in Chapter 17.II. They cover a satisfactory range of excitation and emission wavelengths, and we recommend their use.

Feeling, however, that some analysts might be reluctant to synthesize these products, we give also (in Appendix II) tables which permit substitution of compounds commercially available for our standards. It must be emphasized that such a substitution can lead to less accurate comparisons.

It was verified that the linear relationship between concentration and reading holds for all standards within the range of concentrations at which they are used. It is therefore not necessary to prepare the dilution corresponding exactly to that given in the Results sections. For instance, if it is said that reading 50 corresponds to 0.5 μg/ml of the standard, the apparatus can be adjusted to reading 60 if a 0.6 μg/ml solution is available, since in all cases the fluorescence of the solvent blank can be neglected (with the exception of 2-hydroxy-3-methylquinoxaline, Appendix II.D).

D. Sensitivity of Fluorimetric Determinations

It was shown (Section IV.A) that spectrofluorimetry is much more sensitive than absorption spectrophotometry. For instance, it is possible to determine as little as 0.1 ng/ml of quinine sulfate.

However, when chemical reactions are involved, the lower limit of the determination is strictly dependent upon the blank fluorescence, which cannot be avoided. Even if the fluorescent species would allow determination at a level of, say 10 ng, the blank may afford a fluorescence intensity corresponding to, say 1 μg, of the species, and therefore accurate determinations become possible only at levels permitting an appreciable difference between the assay and blank signals. Sometimes, the reaction may be rendered more sensitive by eliminating the excess reagent or by isolating the fluorescent species. But the procedures proposed generally involve chromatographic separations, which are not dealt with in this book, and which are, of course, time-consuming.

It may be stated that when compared with functional colorimetry, functional fluorimetry as we present it in this book afford a sensitivity which is usually 10- to 100-fold greater.

Since measurements are performed in the presence of the excess reagent, it must be pointed out that it is not always worth operating at the wavelengths corresponding to the most intense emission of light from the sample, but at those corresponding to the highest difference between sample and blank. In our procedures, whenever possible, we propose an excitation wavelength accessible to apparatus equipped with a low-pressure mercury lamp and glass cells, i.e., 366, 405, 436, and 546 nm, so that our results can be reproduced even with very simple instrumentation. The study of excitation and emission spectra recorded on a Farrand spectrofluorimeter has shown that, in these instances, the selection of the true optimum excitation wavelength would multiply the sensitivity only by a coefficient lower than 2.

E. Interferences from Colored Impurities

As a rule, slightly turbid or colored solutions do not generally lead to serious interferences in fluorimetric determinations. However, it must be pointed out that when determinations are made on impure and highly colored samples, like those which may be given by biological extracts, serious discrepancies can occur.

It is obvious that if the color absorbs at the excitation or emission wavelength, a fraction of the corresponding light will be lost, and the intensity of fluorescence will fall together. Comparison with a calibration curve established with the pure compound will then lead to erratic and low results. The influence of the color can be checked by adding a known amount of pure compound to an aliquot of the unknown solution and comparing the increase of fluorescence after development with the calibration curve.

It can be demonstrated that at first approximation, if A is the absorbance of the impurity at the excitation or emission wavelength, the relation between the intensities of fluorescence developed in its absence (F) and in its presence (F') is given by

$$F' = 10^{-A}F.$$

Therefore if Δc is the increase in concentration, the corresponding increase in fluorescence, instead of being ΔF, will be roughly given by

$$\Delta F' = 10^{-A}\Delta F.$$

VI. INSTRUMENTATION

Spectrophotometers and spectrofluorimeters were subjected to numerous studies from both the theoretical and practical standpoints. We thought therefore that this section should be limited to some points which might particularly concern the practicing analyst.

A. Spectrophotometers

1. General description

The basic parts of any spectrophotometer are the light source, the focusing system, the monochromator, the entrance and exit slits, the sample compartment, the phototube, the amplifier of the phototube current, and the microammeter equipped with two scales affording the transmittance and the absorbance, the former being linear.

The source used in the visible range is a tungsten lamp. The light emitted is focused and directed in a parallel beam through the entrance slit on the monochromator, which is either a prism or a grating. The light is dispersed into its component wavelengths, and a wavelength selector adjusting the position of the monochromator allows the desired radiation to be sent through the exit slit and the sample. Light transmitted by the sample impinges on the phototube, causing a current gain which is amplified and transmitted to the microammeter, or eventually to a recorder.

2. Instrumental sources of errors

It was shown (Section II.A) that Beer's law holds only if the incident light is a parallel beam of monochromatic radiation and if the absorption cell has flat and parallel entrance and exit windows. Various practical disturbing influences such as the effect of reflection on the windows of the cell, stray radiations which are unavoidable, and the electronic part of the instrument must also be taken into account, and no apparatus can completely satisfy all the requirements leading to a strict observance of Beer's law. Fortunately, some slight deviations from the theoretical conditions often have but a negligible effect.

a. Nonparallel beam. As a matter of fact, there is a moderate divergence from parallelism in almost all spectrophotometers, but it can be demonstrated (1) that such a divergence causes but negligible deviation from Beer's law.

b. Nonmonochromatic radiation. It can be demonstrated (1) that when the light sent through the sample is a mixture of two monochromatic radiations, Beer's law is not obeyed unless the molar absorptivity of the compound is the same at the two wavelengths. The deviation increases with increasing absorbance, and the values found are always lower than expected: The absorbance-concentration curves are concave to the concentration axis.

Practically, there is no conventional spectrophotometer which can afford a monochromatic radiation. The range of wavelengths sent through the sample depends upon the exit slit width, and, for a given width, upon the wavelength if the monochromator is a prism, since its dispersion is nonlinear. Two spectral lines whose wavelengths differ by 10 nm in the violet region are separated in the dispersed beam by a much greater physical distance than a corresponding pair of spectral lines in the red region. The heterochromaticity corresponding to a given slit width will therefore increase from the violet to the red range. Contrariwise, the dispersion of a grating is linear, but filters must then be used to eliminate the second-order scatter (see Section VI.B.3).

In colorimetric determinations, absorption curves generally present broad and featureless maxima. As a rule, the curve is almost flat within a range of 5-10 nm at the maximum. It may therefore be considered that the absorptivity almost does not vary within this range, when Beer's law should hold if the light is not strictly monochromatic, since it was shown that there is no deviation in so far as the molar absorptivity is the same at the wavelengths involved in a nonmonochromatic light.

c. Setting of the slit width. In colorimetric determinations, the sample is read against the reagent blank, which often also absorbs at the selected wavelength. Hence the reading affords directly the absorbance of the colored species. In order to obtain this result, an adjustment is necessary to compensate for the blank absorption. In other words, it is necessary to set the meter needle to 100% transmittance (or zero absorbance) for the blank.

On a spectrophotometer like the Beckman B, this setting is achieved by a suitable opening of the slit, in order to send through the blank a radiation of such an intensity that the intensity of transmitted light impinging on the phototube corresponds to the suitable deflection of the needle. This method presents disadvantages.

(1) The radiant energy emitted by the tungsten lamp varies with wavelength. On the other hand, the spectral response of the phototube is also wavelength-dependent, i.e., the current corresponding to a given intensity of a radiation depends upon its wavelength. It can be inferred therefrom that for a given absorbance of the blank (with respect to air, for instance), the slit width allowing 100% transmittance to be obtained will vary with the wavelength, and when drawing the absorption curve, proper adjustment should be performed for each wavelength.

(2) If the blank absorbs strongly, it can be necessary to open the slit to such a width that the heterochromaticity of the light sent through the sample will result in appreciable deviation from Beer's law. At least, if a part of the curve is linear, it will be of smaller slope than expected.

The Beckman DB is equipped with an automatic reference-standardizing circuit avoiding this adjustment. It allows the selection of a defined slit width, and deviations from 100% transmittance for the blank are compensated electronically. Therefore, the heterochromaticity of the light depends only upon the nonlinear dispersion of the prism, and even this latter inconvenience is overcome when the monochromator is a grating. Results obtained when the reagent blank is strongly colored are therefore more accurate, and recording of absorption spectra becomes easy.

d. Reflections. Upon passage of a beam of radiation from one medium to another of different refractive index, some radiation is reflected. This reflection is minimized when the dividing surface is plane and smooth, and when the incidence is perpendicular. This is the case for the cells commonly used in exact spectrophotometric measurements. It can be demonstrated that under these conditions reflection results in the loss of a small fraction of the incident light, but provided that the refractive indexes of the blank and the sample are the same, Beer's law still holds.

e. *Cells*. Rectangular cells are generally used. It is obvious that all the cells used in a series of determinations must be identical with regard to transmittance and light path length. For a given setting of the instrument, all the cells should afford the same transmittance when empty. The light path length can be checked with a standard solution of known absorptivity. As a matter of fact, some very slight discrepancies cannot be avoided, and their limits are often given by the supplier. When accurate determinations are made, readings should, of course, be corrected for this source of error.

In some instruments, cylindrical cells are also used. It can be shown that under defined conditions an approximate agreement with Beer's law is found, but the absorptivity is less than corresponds to the diameter of cell.

f. *Stray light*. Stray light is extraneous light which impinges directly on the phototube and does not correspond to the selected wavelength. It can be shown that its presence results in the same effect as the use of nonmonochromatic light.

An estimate of stray light can be made by measuring the apparent transmittance at a given wavelength when the cell is replaced by an opaque block. It was shown (1) that the presence of 0.1% of stray light allows measurement of A correct to 1% up to A = 1.6. With 1% of stray light, the deviation attains 2% for A = 0.5.

g. Electronics. It is obviously taken for granted that the response of the phototube is directly proportional to the intensity of incident light at any wavelength. This matter was discussed by Cannon and Butterworth (13).

Amid electronic sources of error, uncertainty in reading the microammeter scale, dark current of the phototube, and unavoidable noises due to the entire electronic part must be mentioned. Precision decreases, of course, with decreasing reading-to-noise ratio, and this ratio itself is the lowest when settings correspond to the highest sensitivity range.

3. Precision of colorimetric determinations

The precision of a colorimetric determination depends upon both the purely instrumental errors and the errors attributable to the chemical manipulation.

The importance of instrumental errors varies, of course, with the apparatus and with the settings. It can be estimated by performing measurements with solutions of known absorptivities. With a conventional spectrophotometer, it can hardly be expected to obtain a precision better than 2% in the vicinity of A = 0.3.

Errors due to chemical manipulation may vary very widely. If the reagent itself absorbs at the wavelength of the determination, a slight difference between the volumes of reagent introduced into two tubes will, of course, lead to a corresponding discrepancy between readings. Factors such as a slight lack of homogeneity in temperature inside the bulk of a water bath or irregular exposure to light when the reaction is photosensitive may also intervene. As a whole, an error of 2-3%, depending on the reaction itself, can be considered as admissible from the manipulative standpoint.

The instrumental errors superimpose, of course, on the deviations due to the manipulation. It can be said therefore that in functional colorimetry, a reaction which leads to deviations of the order of 4-5% from Beer's law is still satisfactory, provided that the dots are dispersed in a random fashion with regard to the mean straight-line concentration-absorbance relationship corresponding to the law.

On the other hand, when the reaction is very sensitive to factors such as light, pH, or temperature, discrepancies of the order of 10%, or even 20%, may sometimes be observed between two runs of determinations. In order to permit suitable comparisons, the standards and the unknown samples must then be treated simultaneously.

In this book, the sample weight given for each compound under Results sections is not a mean value, but the value really found in a single run of measurements. All determinations were performed with a Beckman B spectrophotometer (with sensitivity multiplier at highest gain position), or with a Beckman DB (with medium slit function). Some comparisons showed that, as a rule, discrepancies between readings given by the two instruments can be neglected.

B. Spectrofluorimeters

Detailed descriptions of some commercially available spectrofluorimeters and of some peculiar arrangements can be found in various books (2,3,5,7,14).

1. General description

The basic parts of any spectrofluorimeter are the light source, the focusing system, the monochromator for exciting light, the sample compartment, the monochromator for fluorescence light, the phototube, the amplifier of the phototube current, the microammeter, and, eventually, the recorder.

In the field of functional fluorimetry, the excitation is achieved by absorption of ultraviolet or visible light, and the source should therefore emit these radiations.

Only very simple fluorimeters are equipped with a low-pressure mercury lamp, which produces but the sharp lines of mercury. With high-pressure mercury lamps, the spectral lines are broadened, and a higher intensity of continuum is also present. However, the determination of a complete excitation spectrum requires a lamp which emits highly intense radiations throughout the visible and ultraviolet ranges. The xenon arc lamp responds to this condition and is probably the most widely used source in spectrofluorimetry.

The term "fluorimeter" usually designates an instrument equipped with filters. In spectrofluorimeters, the monochromators are prisms or gratings, most instruments manufactured in the United States being provided with the latter for both excitation and emission.

The radiation selected by the exciter monochromator is sent through the sample. A fraction of the fluorescence light so produced is sent to the analyzer monochromator, and the emitted radiation selected by this latter impinges on the phototube, causing a gain current which is amplified and transmitted to the microammeter or the recorder. The monochromators are generally disposed so that the beam of fluorescence light sent to the analyzer be at a right angle with regard to the exciting beam, thus limiting the effect of scattered exciting light and fluorescence from the cell walls.

The deflection of the needle of the microammeter is proportional to the intensity of fluorescence, provided that the phototube responds linearly to the intensity of incident light, but as explained (Section IV.A), only relative values are so obtained.

2. Distortion of spectra

The monochromator for fluorescence being set at an optimum wavelength, the intensity of emission will depend upon the wavelength of the exciting light (Section IV.A). If the exciter monochromator is then scanned across the wavelength region from the lowest available wavelength (usually 200 nm) to the emission setting, a curve is recorded, which is named the excitation spectrum. It gives the intensity of emitted light as a function of the excitation wavelength.

Inversely, when the monochromator for excitation is set at a given wavelength, an emission spectrum can likewise be obtained. It is a curve giving the intensity of emitted light as a function of the emission wavelength.

These spectra are grossly distorted:

(a) It was shown (Section IV.A) that the excitation spectrum should be a replica of the absorption spectrum. In other words, if a compound has the same molar absorptivity at, say 300 and 400 nm, the excitation spectrum should present the same elongation at these two wavelengths. But it was also shown that the intensity of fluorescence is directly proportional to the intensity of the exciting light. Since the intensity of the radiation source varies with the wavelength, if the light is, say threefold, more intense at 400 nm than at 300 nm, the elongation on the excitation spectrum will be threefold greater at 400 nm than at 300 nm. Some important features of the absorption spectrum may even be almost completely obscured on the excitation spectrum.

(b) Emission spectra are distorted because the response of the phototube varies with the wavelength. Therefore, the same intensity of fluorescence light at, say 500 and 700 nm, will correspond to peaks of different elongations.

Other factors, such as the slit widths and transmission efficiency of the monochromators also intervene in these distortions.

Various procedures and instruments have been proposed for the correction of the spectra, but as far as determinations are made at given excitation and emission wavelengths, these corrections are not necessary.

3. Second-order effects

These effects are inherent in all grating monochromators. If the excitation monochromator is set at 250 nm and the emission monochromator at 500 nm (2 × 250), a signal is observed. It represents the so-called 250 nm scatter which is transmitted in the second order at 500 nm. When recording the emission spectrum, the peak observed at 500 nm will therefore correspond to this scatter. The true emission at 500 nm may be separated from this effect by placing a filter blocking the 250 nm radiation and transparent for the 500 nm radiation between the sample cell and the emission monochromator.

4. Instrumental sources of errors

This study is limited to spectrofluorimetric determinations, in which readings are made at given excitation and emission wavelengths.

Since spectrofluorimeters are designed for the detection and estimation of very low signal levels corresponding to weak fluorescence light, it is obvious that the effects of factors such as scattered and stray light, slight imperfections in the optical system, or undue fluorescence of the cell walls will be critical. On the other hand, fluorescence may vary widely with the sample, and spectrofluorimeters are therefore equipped with a device allowing selection of the appropriate sensitivity range. The noise from the electronic part will vary with setting, and at the most sensitive one, fluctuations of the order of 5% are often observed.

It may be concluded therefrom that the instrumental errors which affect spectrofluorimetric measurements cannot be readily estimated and depend among others on the selected sensitivity. Therefore, it is worth checking experimentally the performance of the instrument with standardized fluorescent solutions of compounds such as quinine sulfate or fluorescein sodium.

The resolution of the spectra, of course, depends upon the widths of the spectral bands for both excitation and emission. Since most apparatus is equipped with gratings, the problem is simplified, a given slit width corresponding to a definite bandpass, whatever be the wavelength. Influence of slit width on the precision of fluorimetric determinations is dealt with in the following section.

5. Precision of fluorimetric determinations

The errors caused by chemical manipulation are the same as in colorimetric determinations (Section VI.A.3), but the influence of light and temperature is more critical.

Many fluorescent species are sensitive to ultraviolet light, and in order to avoid photodecomposition, reasonably rapid measurements are recommended.

Fluorescence intensity usually decreases with increasing temperature, owing to the higher probabilities for the other means of deactivation of the excited molecule. Most fluorimetric measurements are made at room temperature, and it is best to take this temperature into account and to obviate heating of the sample chamber by excitation light by using a suitable thermostatic arrangement whenever necessary.

On the other hand, fluorescence of almost all organic compounds is quenched at least slightly by oxygen. However, none of the procedures described in this book necessitates special precautions against atmosphere, provided that undue shaking of solutions is avoided.

It was said (Section V.D) that it is worth performing measurements at the wavelengths corresponding to the highest difference between sample and reagent blank. They often almost coincide with the maxima of excitation and emission of the sample, but there are exceptions, and readings are then made at wavelengths corresponding to more or less sloped portions of the curves.

Excitation spectra may present very sharp maxima when they are located in the ultraviolet range. An error of a few nanometers in the setting of the excitation wavelength may then lead, at least theoretically, to much less sensitive results, or to a highly fluorescent blank. The adjustment of the exciter monochromator should therefore be checked very carefully, and it is advisable to use a slit width of 5 nm only. Provided that these conditions are satisfied, and within the limits of the reactions described in this book, we have ascertained that, from the practical standpoint, such an error is not of the highest importance (Section V.D).

The emission curves are usually broad, and an error of a few nanometers in the setting of the analyzer monochromator is then of less importance. Slit widths of 10 nm, or even 20 nm, may be used. We ascertained that when the excitation wavelength is very accurately set, only slight discrepancies are observed between results obtained with the Farrand spectrofluorimeter and the Electrosynthèse fluorimeter, whose emission monochromator is a suitable filter allowing measurements at all emission wavelengths in the visible range.

According to our experiments, it may be said that in functional fluorimetry, a reaction which leads to deviations of the order of 6-7% is still satisfactory, provided that the dots are dispersed in a random fashion with regard to the mean straight-line concentration-emission relationship. For the reasons presented (Section V.C), the main error can arise from the reference standard, and deviations of the order of 10% may be expected.

In this book, the sample weights and the standard concentration given for each compound under Results sections are not mean values, but the values really found in a single run of measurements. All determinations were performed with the Electrosynthèse fluorimeter or with the Farrand spectrofluorimeter, with 5 nm slit width for excitation and 10 nm slit width for emission.

REFERENCES

1. E.J. Meehan, in *Treatise on Analytical Chemistry* (I.M. Kolthoff and P.J. Elving, eds.), Part I, Vol. 5, Interscience, New York, 1964, pp. 2753-2803.

2. D.M. Hercules, *Fluorescence and Phosphorescence Analysis*, Interscience, New York, 1966.

3. G.G. Guilbault, *Fluorescence*, Dekker, New York, 1967.

4. R. Bourdon, *Mises au Point de Chimie Analytique*, Masson Ed., Paris, 15e série, 1967, p. 1.

5. C.A. Parker, *Photoluminescence of Solutions*, Elsevier, Amsterdam, 1968.

6. M. Pesez, *Mises au Point de Chimie Analytique*, Masson Ed., Paris, 17e série, 1968, p. 172.

7. C.E. White and R.J. Argauer, *Fluorescence Analysis*, Dekker, New York, 1970.

8. IUPAC Information Bulletin Number 32 (August 1968), pp. 22-23.

9. E.A. Braude, *J. Chem. Soc.*, 379 (1950).

10. N.T. Gridgeman, *Anal. Chem.*, 24, 445 (1952).

11. J. Bartos and M. Pesez, *Talanta*, 19, 93 (1972).

12. M. Pesez and J. Bartos, *Talanta*, 14, 1097 (1967).

13. C.G. Cannon and I.S.C. Butterworth, *Anal. Chem.*, 25, 168 (1953).

14. S. Udenfriend, *Fluorescence Assay in Biology and Medicine*, 3rd printing, Academic Press, New York, 1964.

Part II

PRACTICAL METHODS ACCORDING TO FUNCTIONAL GROUPS

Chapter 2

ALCOHOLS

I. INTRODUCTION

A. Colorimetry

All classes of alcohols react with ammonium hexanitratocerate(IV) in acid medium to give an orange-brown anion, but the reaction is poorly sensitive and determinations are possible only at levels of several milligrams. Primary, secondary, and tertiary alcohols can also be determined through reaction with vanadium oxinate and subsequent stripping of the excess reagent from the organic phase. A fairly good sensitivity is attained with primary and secondary alcohols, but about 1.5 mg of a tertiary alcohol is necessary to reach the absorbance value of 0.3. The structure of the colored species is not known with certainty. It can be hydrolyzed, liberating oxine which is then developed with p-nitrobenzenediazonium ion, thus enhancing the sensitivity and allowing determinations of 5-30 μg of primary and secondary alcohols, and of about 100 μg of tertiary alcohols.

Primary and secondary alcohols are readily esterified with acetic anhydride or 3,5-dinitrobenzoyl chloride. The acetates are developed as the ferric hydroxamates, but determinations are possible only at levels of the order of 0.3-0.7 mg. 3,5-Dinitrobenzoates can be extracted into hexane and developed with propylenediamine (A = 0.3 given by about 0.1 mg of the alcohol), or under suitable conditions with piperazine in dimethylformamide without prior extraction or elimination of the excess reagent, therefore allowing determinations at the level of 15 μg.

Primary alcohols are determined as aldehydes. They can be oxidized with ruthenium tetroxide or with 2,6-dichloro-4-trimethyl-ammoniumbenzenediazonium chloride-fluoborate. In both cases, the aldehyde is then developed with 3-methylbenzothiazolin-2-one hydrazone. With the first oxidant, the determination range is 5.5-22 μg; the second method allows the determination of 24-80 ppm of an alcohol in water.

Secondary alcohols react but very weakly under the above conditions. They can be oxidized with acid potassium dichromate, and the corresponding ketone is revealed with 2,4-dinitrophenylhydrazine in alkaline medium. Primary alcohols do not interfere, since they are oxidized to acids. Determinations are possible with 130 to 240 μg of the alochols tested.

Tertiary alcohols can hardly be esterified, since they are often readily dehydrated under the conventional conditions, and oxidation can eventually afford only degradation products. They are therefore converted to the corresponding iodides, which are extracted and hydrolyzed, and elemental iodine is then released from the alkaline iodide obtained with iodate ion. Depending on the compound tested, determinations are possible in the range 0.3-0.8 mg.

1,2-Diols, hexitols, and α-amino primary alcohols are oxidized with periodate ion, affording formaldehyde, which is selectively developed with phenylhydrazine and ferricyanide ion (Schryver reaction) or with chromotropic acid (Eegriwe reaction). The sensitivity varies with the compound tested.

B. Fluorimetry

All classes of alcohols can be determined through reaction with vanadium oxinate (see preceding Section, Colorimetry). The species formed is hydrolyzed, and the liberated 8-hydroxyquinoline is revealed as the magnesium complex. Determinations are possible with about 10 to 90 μg of a primary or secondary alcohol, and at the level of 200 μg for the tertiary alochols tested.

The aldehydes obtained by oxidizing primary alcohols with 2,6-dichloro-4-trimethylammoniumbenzenediazonium chloride-fluoborate are developed through the Hantzsch reaction as 9-substituted 1,8-dioxodecahydroacridines with cyclohexane-1,3-dione and ammonia. With the exception of ethanol, which reacts weakly, reading 50 is given by 1.6 to 19 μg of the alochol tested.

1,2-Diols, hexitols, and α-amino primary alochols are oxidized with periodate ion, and the formaldehyde formed is revealed as 2,6-dimethyl-3,5-dicarbethoxy-1,4-dihydropyridine with ethyl acetoacetate and ammonia. Determinations are possible at levels of a few micrograms.

II. ALCOHOLS

A. Ammonium Hexanitratocerate (IV)

Principle (1,2)

Reaction with hexanitratocerate(IV) ion: orange-brown color.

$$R{-}OH + [Ce(NO_3)_6]^{2-} \longrightarrow [Ce(NO_3)_5(RO)]^{2-} + NO_3^- + H^+$$

Reagent

A 20% solution of ammonium hexanitratocerate(IV) in 4 N nitric acid.

Procedure

To 5 ml of aqueous solution of the alcohol, add 2 ml of reagent. After 5 min, read at 465 nm.

Results

	A = 0.3 (1 cm cell) Sample, μg
1-Butanol	10,000
2-Butanol	10,000
tert-Butyl alcohol	9,000

Aldehydes, ketones, acids, and esters do not react; phenols give a brownish precipitate.

Note

Ammonium hexanitratocerate(IV) or ammonium ceric nitrate, $[Ce(NO_3)_6](NH_4)_2$, slightly deliquescent, orange monoclinic crystals, very soluble in water, is used as a reference standard in oxidimetry (3).

B. Vanadium Oxinate

Principle (4)

On addition of an alcohol to the blue-violet solution of vanadium oxinate in chloroform, a red color is developed. After selective elimination of the excess reagent, the organic phase is acidified: blue color.

Reagents and Solvent

a. Alcohol-free chloroform: To 500 ml of chloroform, add 100 ml of 1% potassium chromate solution in 1 N sulfuric acid, set aside with occasional shaking for 15 min, discard the aqueous layer, and repeat the operation twice more. Then wash the chloroform successively with 100 ml of water, 100 ml of 1 N sodium hydroxide, and 4 times with 100 ml portions of water. Filter through anhydrous magnesium sulfate, and store in an amber bottle under nitrogen. This solvent is stable for about 2 days.

b. To a boiling suspension of 0.040 g of sodium salt of vanadium oxinate (Chapter 17, Section I.I) in 9 ml of chloroform a, add 1 ml of chloroform a containing 0.026 g of glacial acetic acid, keep boiling under reflux for 1 min, cool, and filter. Prepare fresh before use.

c. A 5% solution of dichloroacetic acid in glacial acetic acid.

Procedure

To 4 ml of sample solution in chloroform a, add 1 ml of reagent b, and let stand at 20-22° for 2 hr 30 min. Add 5 ml of 1 N sodium hydroxide, shake until the blue-violet color of the excess reagent has completely faded, collect the organic layer as soon as it is completely clear, and add to it 0.5 ml of reagent c. Read at 600 nm.

Results

	A = 0.3 (1 cm cell) Sample, μg
Methanol	56
Ethanol	85
1-Butanol	115
Isobutyl alcohol	130
2-Butanol	300
tert-Butyl alcohol	1800
1-Pentanol	155
Isopentyl alcohol	150
2-Pentanol	400
tert-Pentyl alcohol	1600
Benzyl alcohol	210
Cyclohexanol	300

Triphenylcarbinol reacts but weakly (A = 0.04 for 2000 μg). Hydroxysteroids can also be determined (Chapter 15, Section II.A).

Notes

1. Other solvents, such as nitrobenzene (5) or benzene (6), may be used instead of chloroform.

2. Ammonium vanadate(V) reacting with 8-hydroxyquinoline in dilute aqueous sodium hydroxide gives yellow crystals of the salt (I) which, upon acidification, is converted into a blue-violet, almost black complex (II), called vanadium oxinate, and whose proper name is di-8-quinolinol orthovanadic acid.

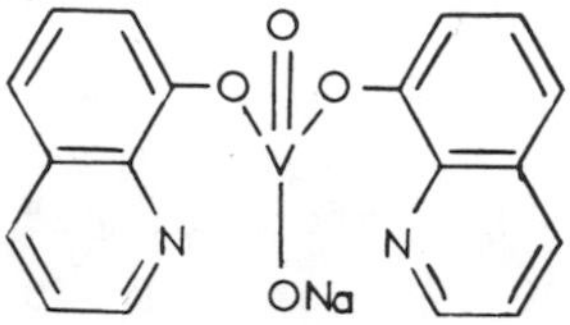

(I) 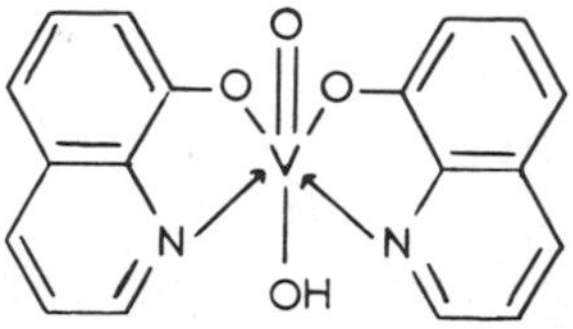 (II)

Acid (II) is soluble in chloroform, giving a deep blue-violet solution which turns to red on addition of an alcohol. The structure of the red species is as yet not well known. According to Bielig and Bayer (7), the alcohol causes the isomerization of the blue-violet trans-syn-syn structure into the red trans-anti-anti form. Feigl (8) and Stiller (9) assume the formation of a complex between one molecule of acid (II) and x molecules of alcohol, whereas Balog and Csaszar (10) admit a true esterification.

3. The unstable red color turns to stable blue upon suitable acidification.

4. The red species can be hydrolyzed in chloroform solution. The 8-hydroxyquinoline liberated reacts with p-nitrobenzenediazonium ion to give a blue azo dye, thus enhancing the sensitivity of the colorimetric determination of alcohols (next Section), or with magnesium ion, thus allowing their fluorimetric determination (Section XII.A).

5. The piperidinium salt of (II) was proposed instead of the sodium salt for the determination of 3-(o-tolyloxy)-

1,2-propanediol-(mephenesin) (11) and of 3-(o-methoxy-phenoxy)-1,2-propanediol (guaiphenesin) (12).

6. Other complexes of vanadium(V) were also proposed for the colorimetric determination of alcohols (6).

7. Based on the same principle, a colorimetric determination of 8-hydroxyquinoline was described (13): Oxine is allowed to react with ammonium vanadate(V) in buffered medium, and the complex is extracted into 1-butanol.

8. Alcohol-free chloroform can also be obtained by shaking 100 ml of chloroform for 2 hr with 10 g of Siliponte NK 20 molecular sieve.

C. Vanadium Oxinate and p-Nitrobenzenediazonium Fluoborate

Principle (14)

On addition of an alcohol to the blue-violet solution of vanadium oxinate in chloroform, a red color is developed. After selective elimination of the excess reagent, the red species is hydrolyzed, thus liberating 8-hydroxyquinoline, which is revealed by coupling with p-nitrobenzenediazonium ion: blue color.

Reagents and Solvent

a. Alcohol-free chloroform: Prepare as described in II.B, under reagent a.

b. Reflux for 10 min a suspension of 0.040 g of finely powdered sodium salt of vanadium oxinate (Chapter 17, Section I.I) in 9 ml of chloroform a, cease heating, add 1 ml of chloroform a containing 0.026 g of glacial acetic acid, and reflux again for 3 min. Cool to 20°, filter, and dilute to 10 ml with chloroform a. Prepare fresh immediately before use.

c. A 1:3 mixture of concentrated hydrochloric acid and ethanol.

d. Dissolve 0.010 g of p-nitrobenzenediazonium fluoborate in 1 ml of 1 N hydrochloric acid and dilute to

20 ml with ethanol. This reagent may be stored at 0° for 1 hr.

Procedure

To 0.5 ml of sample solution in chloroform a, add 0.3 ml of reagent b, plug the tubes with cotton wool, and heat at 50° for 10 min. Cool in ice water, add 1 ml of chloroform a, mix, and transfer the solution to a 10 ml separatory funnel, whose tap should be grease-free. Rinse the tube with 2 ml of 1 N sodium hydroxide, pour the rinsing into the funnel, shake for exactly 20 sec, and let stand until thorough decantation is achieved. Pipet off the aqueous layer, and repeat the washing of the organic phase twice more with 2 ml portions of 1 N sodium hydroxide. Collect the organic layer, pipet a 1 ml aliquot, and add to it 0.1 ml of reagent c, 0.1 ml of reagent d, 0.5 ml of water, and 0.5 ml of 1 N sodium hydroxide. Shake the tube for a few seconds after each addition. Chill in ice water for 2 min, and add 3 ml of dimethylformamide. Read at 580 nm.

Results

	A = 0.3 (1 cm cell) Sample, μg
Methanol	5
Ethanol	8.5
1-Butanol	11
2-Butanol	22.5
tert-Butyl alcohol	95
1-Pentanol	14
2-Pentanol	28
tert-Pentyl alcohol	97
Benzyl alcohol	23
Cyclohexanol	26

Hydroxysteroids can also be determined (Chapter 15, Section II.B).

Notes

1. Structure and properties of vanadium oxinate, see this Chapter, Section II.B, Note 2.

2. Diazo coupling, see Chapter 16, Section I.

III. PRIMARY AND SECONDARY ALCOHOLS

A. Acetic Anhydride, Hydroxylamine Hydrochloride, and Ferric Perchlorate

Principle (15)

Acid-catalyzed acetylation followed by base-catalyzed hydroxamation of the acetate ester, and color development as the ferric hydroxamate chelate: brown-violet color.

$$R{-}OH \xrightarrow{(CH_3CO)_2O} R{-}O{-}CO{-}CH_3 \xrightarrow{NH_2OH} CH_3{-}CO{-}NHOH \xrightarrow{Fe^{3+}} H_3C{-}C({=}O){-}N(OH){-}H \text{ (chelated to Fe/3 via both O)}$$

Reagents

a. Pipet 1 ml of 65% perchloric acid (d = 1.615) into a 50 ml volumetric flask, immerse in ice water, and slowly add 20 ml of pyridine. Allow to come to room temperature, and dilute to 50 ml with acetic anhydride. Prepare fresh daily.

b. Dissolve 195 g of sodium perchlorate in 450 ml of methanol with heating, and pour the solution into a stirred warm solution of 105 g of hydroxylamine hydrochloride in 550 ml of methanol. Add 50 ml of benzene, chill for 1 hr in ice water, filter on a sintered-glass suction filter, and store.

To 20 ml of this stock solution, add 30 ml of 4 N sodium hydroxide and filter. This alkaline solution is stable for 4 hr.

c. Into a 500 ml round-bottomed flask fitted with a reflux condenser, place 7.86 g of iron filings and 168 g of 65% perchloric acid, and heat slowly until reaction starts (slight evolution of chlorine), then reflux for 20 min, cool to room temperature, add 375 ml of ethanol, stir, filter, dilute to 500 ml with ethanol, and store.

Into a 1 liter volumetric flask, pour 800 ml of ethanol, 50 ml of the above stock solution, and 151 g of 65% perchloric acid; allow to cool to room temperature, and dilute to 1 liter with ethanol. This solution should be about 1.03 M in perchloric acid, as checked with a standard solution of 1 N sodium hydroxide in the presence of phenolphthalein. If necessary, it is adjusted to this value by adding perchloric acid or ethanol.

The reagent is obtained by adding, immediately before use, 25 ml of acetone to 350 ml of this solution.

Procedure

To 0.5 ml of sample solution in pyridine, add 0.2 ml of reagent a, and let stand for 10 min. Add 0.3 ml of 20% aqueous pyridine, and after 10 min add 4 ml of reagent b. Let stand for 20 min, and add 18 ml of reagent c. After 5 min, read at 525 nm.

Results

	A = 0.3 (1 cm cell) Sample, μg
Ethanol	300
2-Propanol	410
Cyclohexanol	730

Note

Hydroxamation and hydroxamates, see Chapter 9, Section VIII.A, Notes.

B. 3,5-Dinitrobenzoyl Chloride and Propylenediamine

<u>Principle</u> (16)

Esterification with 3,5-dinitrobenzoyl chloride, extraction of the ester into hexane, and development with propylenediamine: pink color.

<u>Reagent</u>

A 5% solution of 3,5-dinitrobenzoyl chloride in pyridine. Dissolve with gentle warming until clear. Prepare fresh immediately before use.

<u>Procedure</u>

To 1 ml of sample solution in pyridine, add 2 ml of reagent. Let stand at room temperature for 15 min, add 25 ml of 2 N hydrochloric acid, and shake with 10 ml of n-hexane. Collect the organic layer, and repeat the extraction with another 10 ml portion of n-hexane. Combine the hexane extracts, wash with 5 ml of 5% aqueous solution of sodium carbonate, filter the organic layer, and dilute to 20 ml with n-hexane. Pipet a 1 ml aliquot, add to it 5 ml of dimethylformamide and 0.5 ml of propylenediamine. Read within 4 min at 525 nm.

<u>Results</u>

	A = 0.3 (1 cm cell) Sample, μg
Ethanol	105
1-Propanol	150
2-Propanol	140

Glycols react but weakly. The reaction may be applied to phenols, but the color is unstable.

Note

Development of aromatic nitro compounds, see Chapter 16, Section II.

C. 3,5-Dinitrobenzoyl Chloride and Piperazine

Principle (17)

Esterification with 3,5-dinitrobenzoyl chloride and, without prior elimination of the excess reagent, development with piperazine in dimethylformamide: red color.

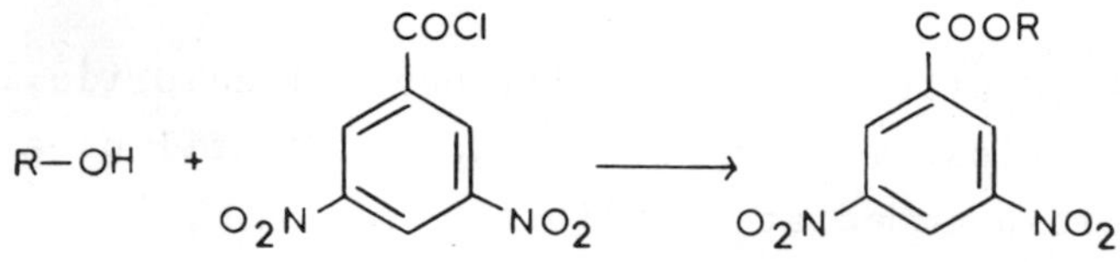

$$\text{R—OH} + 3,5\text{-}(O_2N)_2C_6H_3\text{COCl} \longrightarrow 3,5\text{-}(O_2N)_2C_6H_3\text{COOR}$$

Reagents

a. A 5% solution of 3,5-dinitrobenzoyl chloride in benzene.

b. A 25% solution of piperazine hexahydrate in dimethylformamide.

Procedure

To 0.1 ml of sample solution in a 1:1 mixture of pyridine and benzene, add 0.1 ml of reagent a, and let stand at 18-22° for 20 min. Add 0.1 ml of pyridine, and after 5 min at 18-22°, add 5 ml of reagent b (maintained at 18-22°). Read at 510 nm.

Results

	A = 0.3 (1 cm cell) Sample, μg
Ethanol	9
1-Butanol	15
2-Butanol	17

Hydroxysteroids can also be determined (Chapter 15, Section II.D). Amines and phenols do not react.

Notes

1. 3,5-Dinitrobenzoyl chloride should be very carefully purified, since trace amounts of acid inhibit the reaction. It should be crystallized from carbon tetrachloride, then twice from petroleum ether, and dried over caustic potash and paraffin chips.

2. Based on the same principle, a colorimetric determination of alcohols, amines, and thiols was proposed, the development being achieved with ammonia and ammonium acetate (18).

3. Development of aromatic nitro compounds, see Chapter 16, Section II.

IV. PRIMARY ALCOHOLS

A. Ruthenium Tetroxide and 3-Methylbenzothiazolin-2-one Hydrazone

Principle (19)

Oxidation with ruthenium tetroxide and development of the aldehyde formed with 3-methylbenzothiazolin-2-one hydrazone: blue color.

$$R{-}CH_2OH \xrightarrow{RuO_4} R{-}CHO \longrightarrow$$

See Chapter 8, Section IV.B

Reagents

a. Shake a suspension of 0.010 g of ruthenium dioxide (Note 1) in 5 ml of carbon tetrachloride with 5 ml of 10% aqueous solution of anhydrous sodium metaperiodate until the organic phase turns to yellow-green. The solution of ruthenium tetroxide in carbon tetrachloride so obtained may be used for 2 hr when kept under the aqueous layer. Avoid contact with grease.

b. An 0.5% aqueous solution of 3-methylbenzothiazolin-2-one hydrazone hydrochloride.

c. An 0.3% aqueous solution of hydrogen peroxide (1 volume oxygen).

Procedure

To 3 ml of sample solution in glacial acetic acid, add 0.1 ml of reagent a. Mix, let stand for 5 min, add 0.5 ml of reagent b, and after 10 min add 0.2 ml of reagent c. Let stand for 15 min, and dilute with 5 ml of acetone. After 5 min, read at 660 nm. During the whole procedure, avoid contact with grease.

Results

	A = 0.3 (1 cm cell) Sample, μg
Methanol	22
Ethanol	5.5
1-Butanol	9.5
1-Pentanol	11

Notes

1. The ruthenium dioxide should be amorphous, and the diameter of the particles should not exceed 1 μm. Crystallized samples (as shown by Debye-Scherrer spectrum) are oxidized but poorly or not at all by sodium periodate.

2. Ruthenium tetroxide also oxidizes ethylenic double bonds, yielding carbonyl derivatives (Chapter 12, Section II.A).

3. Properties of 3-methylbenzothiazolin-2-one hydrazone, see Chapter 16, Section III.

B. 2,6-Dichloro-4-trimethylammoniumbenzenediazonium Chloride-fluoborate and 3-Methylbenzothiazolin-2-one Hydrazone

Principle (20,21)

Oxidation with 2,6-dichloro-4-trimethylammonium-benzenediazonium chloride-fluoborate, and development of the aldehyde formed with 3-methylbenzothiazolin-2-one hydrazone: blue-green color.

$$R{-}CH_2OH + \text{[2,6-Cl}_2\text{-4-}\overset{+}{N}(CH_3)_3\,Cl^-\text{-}C_6H_2\text{-}\overset{+}{N}{\equiv}N\ BF_4^-] \longrightarrow \text{1,3,5-}Cl_3C_6H_3 + (CH_3)_3\overset{+}{N}H, BF_4^- + N_2 + R{-}CHO \downarrow$$

See Chapter 8,
Section IV.B

Reagents

a. An 0.5% aqueous solution of 2,6-dichloro-4-trimethylammoniumbenzenediazonium chloride-fluoborate (Chapter 17, Section I.B). Prepare fresh immediately before use.

b. An 0.5% aqueous solution of 3-methylbenzothiazolin-2-one hydrazone hydrochloride.

c. An 0.4% aqueous solution of ferric chloride hexahydrate.

Procedure

To 0.5 ml of aqueous solution of the alcohol, add 0.5 ml of reagent a, and expose to bright light for 10 min at room temperature. Then add 0.5 ml of reagent b, let stand for 20 min, and add 2.5 ml of reagent c. After 10 min, dilute with 10 ml of water, and read.

Results

	λ Max, nm	A = 0.3 (1 cm cell) Ppm of alcohol (v/v) in aqueous solution
Methanol	630	38
Ethanol	610	24
1-Propanol	610	42
1-Butanol	610	64
1-Pentanol	650	86
Ethylene glycol	640	50
Glycerol	640	80

Benzyl alcohol, secondary and tertiary alcohols react but weakly.

Notes

1. When compared with the color developed by the related aldehydes, results obtained with ethanol and 1-butanol establish that the yield of the oxidation step is of about 50%.

2. Properties of diazo compounds, see Chapter 16, Section I.

3. Properties of 3-methylbenzothiazolin-2-one hydrazone, see Chapter 16, Section III.

V. SECONDARY ALCOHOLS

A. Acid Potassium Dichromate and 2,4-Dinitrophenylhydrazine

Principle (22)

Oxidation with acid potassium dichromate and development of the ketone formed with 2,4-dinitrophenylhydrazine in alkaline medium: red color.

Reagents and Solvent

a. Dissolve 15 g of potassium dichromate in 500 ml of water. Slowly add 360 ml of concentrated sulfuric acid with cooling to prevent excessive heating. Dilute to 1 liter with water.

b. Carbonyl-free methanol: Reflux for 4 hr a batch of methanol containing 1% of 2,4-dinitrophenylhydrazine and 0.3% of concentrated hydrochloric acid, then distill.

c. To a suspension of 0.050 g of 2,4-dinitrophenylhydrazine in 25 ml of methanol b, add 2 ml of concentrated hydrochloric acid, and when the crystals are dissolved dilute to 50 ml with methanol b.

d. A 10% solution of potassium hydroxide in methanol, freshly prepared.

Procedure

To 3 ml of reagent a, add 1 ml of sample solution in water or acetonitrile. Let stand at the given temperature for the given length of time (see Results), then while keeping at this temperature add with stirring 0.2 ml of a 50% aqueous solution of hypophosphorous acid (d = 1.274). Let stand at room temperature for 15 min, and dilute to 10 ml with methanol b. Pipet a 3 ml aliquot, chill to 0°, add 3 ml of 4 N potassium hydroxide, and allow to come to room temperature. Add 3 ml of reagent c, and after 30 min add 15 ml of a 4:1 mixture of pyridine and water and 3 ml of reagent d. Let stand for 5 min and filter. After 10 min, read at 440 nm.

Results

	Oxidation time and temperature	A = 0.3 (1 cm cell) Sample, μg
2-Propanol	60 min, room temp.	170
2-Butanol	120 min, room temp.	240
2,5-Hexanediol[a]	60 min, 0°	130

[a]Linear relationship holds down to 80 μg.

Note

Primary alcohols are oxidized to acids and do not interfere.

VI. TERTIARY ALCOHOLS

A. Hydriodic Acid, Potassium Hydroxide, and Potassium Iodate

Principle (23)

Reaction with hydriodic acid, extraction of the tert-alkyl iodide formed into cyclohexane, hydrolysis, and release of elemental iodine by reaction with iodate ion: pink color.

$$\geq C{-}OH \xrightarrow{HI} \geq C{-}I \xrightarrow{OH^-} I^- \xrightarrow[H_2SO_4]{IO_3^-} I_2$$

Reagents

a. A 57% solution of hydriodic acid (d = 1.70).

b. A 2% solution of sodium hydroxide.

c. A 3% aqueous solution of hydrogen peroxide (10 volumes oxygen).

d. A 30% solution of potassium hydroxide.

e. A 10% solution of sulfuric acid.

f. A 4% aqueous solution of potassium iodate.

Procedure

Pipet 2 ml of sample solution in cyclohexane into a 10 ml separatory funnel, add 0.3 ml of reagent *a*, and shake for 5 min. Wash the stopper and the walls of the funnel with 2 ml of water, let stand until thorough decantation, and discard the aqueous layer. To the organic layer, add 1 ml of reagent *b* and 0.15 ml of reagent *c* (to destroy the free iodine), shake for 1 min, allow to separate, and discard the aqueous layer. Wash the organic layer with three 1 ml portions of water, and transfer it into a 10 ml round-bottomed flask. Rinse the funnel with 1 ml of cyclohexane, combine the rinsing with the contents of the flask, add 1 ml of reagent *d*,

1 ml of ethanol, and reflux for 30 min. Allow to cool, transfer into a 10 ml separatory funnel, rinse the condenser and the flask with 2 ml of water, combine the rinsing with the reaction mixture, shake, and allow to separate. Collect the aqueous layer, adjust it to a pH of about 4 with reagent e, add 0.3 ml of reagent f, let stand for 2 min, and extract the released iodine into 5 ml of cyclohexane. Read at 525 nm.

Results

	A = 0.3 (1 cm cell) Sample, μg
tert-Pentyl alcohol	300
2-Ethynyl-2-propanol	330
2-Ethynyl-2-butanol	430
1-Ethynylcyclohexanol	830

Notes

1. The rate of formation of primary and secondary alkyl iodides and the rate of hydrolysis of these iodides are much slower than that of the tertiary ones. Thus, interference from primary and secondary alcohols is negligible: The reaction is negative with 7 mg of both 1- and 2-butanol.

2. Tertiary hydroxyl at position 17 of ketosteroids does not react.

3. 2-Ethynyl-2-butanol (Meparfynol) and its carbamate, and 1-ethynylcyclohexanol, its carbamate (Ethinamate), and its allophanate (Dolcental) are used as hypnotics and sedatives.

VII. 1,2-DIOLS

A. Sodium Metaperiodate, Phenylhydrazine Hydrochloride and Potassium Ferricyanide

Principle (24,25)

Oxidation with periodate ion and development of the

formaldehyde formed with phenylhydrazine and ferricyanide ion: red color.

$$R\text{—}CHOH\text{—}CH_2OH \xrightarrow{IO_4^-} R\text{—}CHO + HCHO$$

$$HCHO + C_6H_5\text{—}NH\text{—}NH_2 \xrightarrow{Fe(CN)_6^{3-}} \text{Formazan dye}$$

Reagents

a. A 6.6% aqueous solution of anhydrous sodium metaperiodate.

b. A saturated aqueous solution of potassium nitrate.

c. A 1% aqueous solution of phenylhydrazine hydrochloride, freshly prepared.

d. A 2% aqueous solution of potassium ferricyanide, freshly prepared.

Procedure

To 0.5 ml of aqueous solution of the diol, add 2.5 ml of reagent a, and let stand at room temperature for 20 min; then chill to 0°, and precipitate out the excess sodium periodate with 0.5 ml of solution b. After 10 min at 0°, pipet a 1 ml aliquot of the clear supernatant, and add to it 2 ml of reagent c and 1 ml of reagent d. Let stand at 0° for 20 min, slowly add with stirring 5 ml of prechilled concentrated hydrochloric acid and 5 ml of ethanol. Let stand at room temperature for 15 min, and read at 520 nm.

Results

	A = 0.3 (1 cm cell) Sample, μg
Ethylene glycol	25
Propylene glycol	50
Glycerol	25
Dihydrostreptomycin (sesquisulfate)	540

Notes

1. The formulas of streptomycin and dihydrostreptomycin are shown by structures (III) and (IV), respectively.

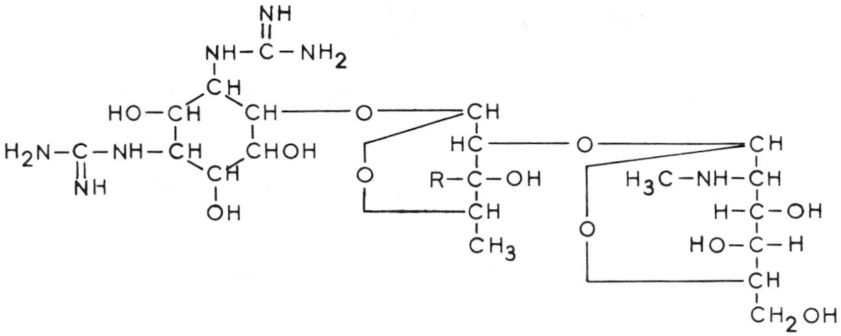

(III) R = CHO

(IV) R = CH_2OH, thus allowing the above colorimetric determination.

2. Periodates and periodate oxidation, see Chapter 16, Section IV.

3. Colorimetric determination of formaldehyde, see Chapter 16, Section V.

VIII. HEXITOLS

A. Potassium Metaperiodate and Sodium Salt of Chromotropic Acid

Principle (26,27)

Oxidation with periodate ion and development of the formaldehyde formed with chromotropic acid: violet color.

$$HOCH_2\text{-}(CHOH)_4\text{-}CH_2OH \xrightarrow{IO_4^-} 2\ HCHO + 4\ HCOOH$$

HCHO + [chromotropic acid: 4,5-dihydroxy-2,7-naphthalenedisulfonic acid, structure with OH, OH, HO_3S, SO_3H] ⟶

Dibenzoxanthylium-type dyestuff

Reagents

a. Dissolve 0.4 g of potassium metaperiodate in 100 ml of 1.6% sulfuric acid.

b. Dissolve 2.35 g of stannous chloride dihydrate in 5 ml of concentrated hydrochloric acid, and dilute to 100 ml with water.

c. Dissolve 5 g of sodium salt of chromotropic acid in water, dilute to 100 ml with water, and keep at 0° for 24 hr before use. The reagent is obtained by diluting 4 ml of this solution to 100 ml with 96% sulfuric acid. Prepare fresh before use.

Procedure

To 2 ml of aqueous solution of the hexitol, add 0.5 ml of reagent *a*. After 10 min, add 0.5 ml of reagent *b*, mix, and add 5 ml of reagent *c*. Heat at 100° for 30 min, cool in a water bath, and dilute to 20 ml with water. Read at 570 nm.

Results

	A = 0.3 (1 cm cell) Sample, μg
Mannitol	37
Sorbitol	37

Notes

1. Periodates and periodate oxidation, see Chapter 16, Section IV.

2. Colorimetric determination of formaldehyde, see Chapter 16, Section V.

IX. α-AMINO PRIMARY ALCOHOLS

A. Sodium Metaperiodate, Phenylhydrazine Hydrochloride, and Potassium Ferricyanide

Principle

Oxidation with periodate ion and development of the formaldehyde formed with phenylhydrazine and ferricyanide ion: red color.

$$R\text{-}CH(NH_2)\text{-}CH_2OH \xrightarrow{IO_4^-} R\text{-}CHO + NH_3 + HCHO$$

$$HCHO \longrightarrow$$

See Section VII.A

Procedure

Follow the procedure for 1,2-diols, Section VII.A.

Results

	A = 0.3 (1 cm cell) Sample, μg
Monoethanolamine	21
Serine	72

X. 2-HYDROXY-1-CARBOXYLIC ACIDS

A. Ferric Chloride

Principle (28)

Reaction with ferric chloride: yellow color.

Reagent

Dissolve 0.10 g of ferric chloride hexahydrate in 10 ml of water, add 0.1 ml of concentrated hydrochloric acid, and dilute to 100 ml with water.

Procedure

To 2 ml of aqueous solution of the acid, add 2 ml of reagent, and read.

	λ Max, nm	A = 0.3 (1 cm cell) Sample, μg
Glycolic acid	360	170
Lactic acid	360	350
Mandelic acid	365	130
Malic acid	360	200
Tartaric acid	360	125
Citric acid	370	360

XI. 2,3-DIHYDROXY-1-CARBOXYLIC ACIDS

A. Sodium Metaperiodate and Resorcinol

Principle (29)

Oxidation with periodate ion and reaction of the glyoxylic acid formed with resorcinol in acid medium to give the lactone of 2,2',4,4'-tetrahydroxydiphenylacetic acid, which is extracted and developed with potassium hydroxide: blue-violet color.

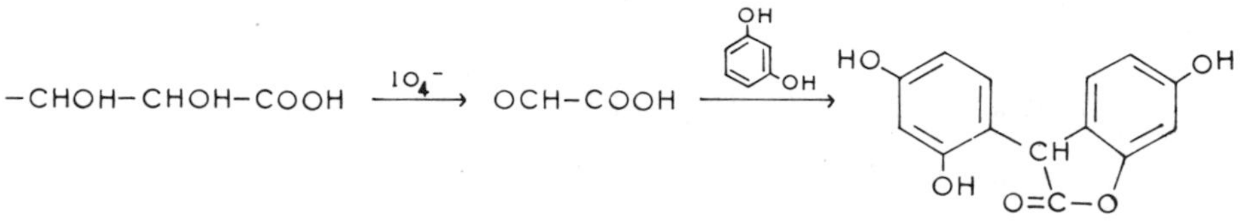

$$-CHOH-CHOH-COOH \xrightarrow{IO_4^-} OCH-COOH \longrightarrow$$

Reagents

a. An 0.05 M aqueous solution of sodium metaperiodate adjusted to pH 10 with 1 N sodium hydroxide.

b. A 10% aqueous solution of ethylene glycol.

c. To 50 ml of 0.1 N aqueous silver nitrate, add 5 ml of 1 N sodium hydroxide. Let the precipitate settle, decant off the supernatant, wash the precipitate with water by decantation, dissolve it in 1 N sulfuric acid, and dilute to 100 ml with the acid. Titrate the silver with potassium thiocyanate.

d. A 1% solution of resorcinol in a 3:7 mixture of concentrated sulfuric acid and water.

e. A 1 N potassium hydroxide solution in ethanol.

Procedure

To 1 ml of neutralized aqueous solution of the acid, add 1 ml of reagent a, and let stand at 20-22° for 30 min. Add 0.1 ml of reagent b to destroy the excess periodate, and let stand for 5 min. Precipitate the iodate by adding the theoretical amount of reagent c (about 1 ml, calculated from the amount of periodate introduced), dilute to 5 ml with water, and filter. Pipet 1 ml of the filtrate, add to it 2 ml of reagent d, heat at 100° for 5 min, and allow to cool. Add 5 ml of water, and extract into 5 ml of ethyl acetate. To 1 ml of the organic layer, add 4 ml of ethanol and 0.5 ml of reagent e. Read at 550 nm within 10 min.

Results

	A = 0.3 (1 cm cell) Sample, μg
Tartaric acid	355
Mucic acid	420

Hexuronic acids can also be determined (Chapter 14, Section XI.B).

Notes

1. Formaldehyde, acetaldehyde, glycolic acid, oxalic acid, formic acid, and malonic dialdehyde do not interfere. Glyoxal reacts weakly, but does not alter the results when present in amounts lower than glyoxylic acid.

2. Glyoxylic acid itself is oxidized by periodate ion, but it was established that in the presence of 0.05 M periodate at 20-22° it is stable at pH 10 (less than 3% oxidized after 1 hr). Between pH 3 and 7, it is completely oxidized within 15 min, whereas the reaction rate is very slow at a pH of about 2 (29).

3. The colorless 2,2',4,4'-tetrahydroxydiphenylacetic acid lactone obtained by condensing glyoxylic

acid with resorcinol is insoluble in water and dilute acids, soluble in alkaline media. The blue-violet solutions so obtained are readily oxidized by atmospheric oxygen, the color rapidly turning to red-purple, then to orange, whereas a yellow-green fluorescence is displayed, thus allowing the fluorimetric determination of glyoxylic acid (this Chapter, Section XVII.A). The quinone structure (V) was tentatively attributed to the oxidized species (30,31).

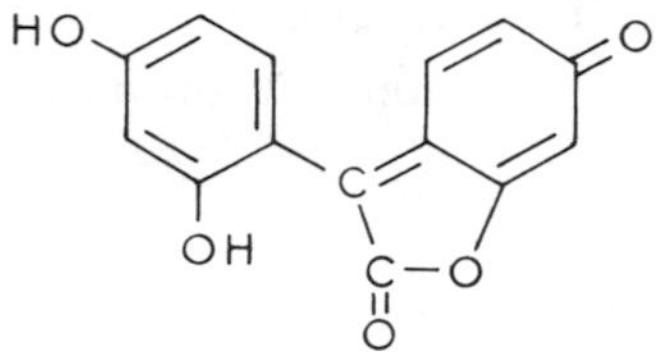

(V)

In a medium of ethanol, the blue-violet color is much more stable, thus permitting spectrophotometric measurements.

4. Periodates and periodate oxidation, see Chapter 16, Section IV.

XII. ALCOHOLS (F)

A. Vanadium Oxinate and Magnesium Acetate

Principle (32)

On addition of an alcohol to the blue-violet solution of vanadium oxinate in chloroform, a red color is developed. After selective elimination of the excess reagent, the red species is hydrolyzed, thus liberating 8-hydroxyquinoline which is revealed as the magnesium complex: blue-green fluorescence.

Reagents and Solvent

a. Alcohol-free chloroform: Prepare as described in Section II.B, under reagent a.

_b_. Vanadium oxinate reagent: Prepare as described in Section II.C, under reagent _b_.

_c_. A 1:3 mixture of concentrated hydrochloric acid and ethanol.

_d_. A 1% aqueous solution of magnesium acetate tetrahydrate.

Procedure

To 0.5 ml of sample solution in chloroform _a_, add 0.3 ml of reagent _b_, plug the tubes with cotton wool, and heat at 50° for 10 min. Cool in ice water, add 1 ml of chloroform _a_, mix, and transfer the solution to a 10 ml separatory funnel, whose tap should be grease-free. Rinse the tube with 2 ml of 1 N sodium hydroxide, pour the rinsing into the funnel, shake for exactly 20 sec, and let stand until thorough decantation is achieved. Pipet off the aqueous layer and repeat the washing of the organic phase twice more with 2 ml portions of 1 N sodium hydroxide. Collect the organic layer, pipet a 1 ml aliquot, and add to it 0.2 ml of reagent _c_, 0.1 ml of reagent _d_, and 0.5 ml of aqueous ammonium hydroxide (d = 0.92, 22° Bé). Shake the tubes for a few seconds after each addition. Dilute with 3 ml of ethanol, and shake until the solution is clear. Read at exc: 366 nm; em: 520 nm.

Standard: Add 0.5 ml of 1% aqueous solution of magnesium acetate tetrahydrate to 5 ml of an ethanol solution containing 10 μg/ml of 8-hydroxyquinoline, and dilute with ethanol.

Results

	Determination limits, μg	Reading 50	
		Sample, μg	Standard, μg/ml of 8-quinolinol
Methanol	5 - 25	11.5	1.49
Ethanol	8 - 40	15	1.33
1-Butanol	10 - 50	18	1.52

	Determination limits, μg	Reading 50	
		Sample, μg	Standard, μg/ml of 8-quinolinol
2-Butanol	20 - 100	38	1.43
tert-Butyl alcohol	100 - 500	210	1.67
1-Pentanol	12 - 60	24	1.52
2-Pentanol	40 - 200	87	2.05
tert-Pentyl alcohol	80 - 400	172	1.45
Benzyl alcohol	25 - 125	47.5	1.67
Cyclohexanol	25 - 125	45	1.52

Hydroxysteroids can also be determined (Chapter 15, Section XV.A).

Note

Structure and properties of vanadium oxinate, see Section II.B, Note 2.

XIII. PRIMARY ALCOHOLS (F)

A. 2,6-Dichloro-4-trimethylammoniumbenzenediazonium Chloride-fluoborate, Cyclohexane-1,3-dione, and Ammonium Acetate

Principle (33,34)

Oxidation of the alcohol with 2,6-dichloro-4-trimethylammoniumbenzenediazonium chloride-fluoborate, and development of the aldehyde formed with cyclohexane-1,3-dione and ammonia: yellow-green fluorescence.

Mechanism of oxidation, see Section IV.B.

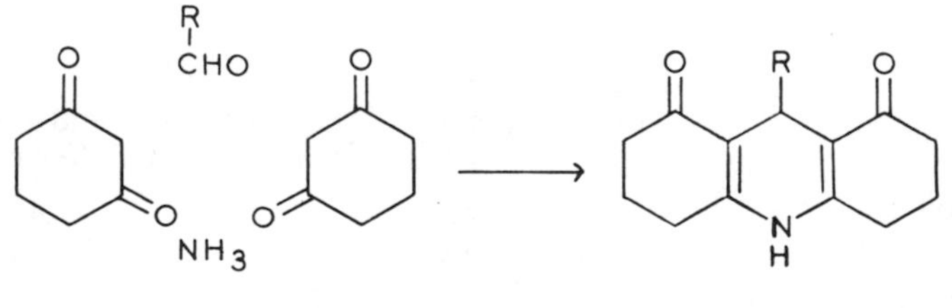

R
CHO
O
O
NH_3
O
O
O R O
N
H

Reagents

a. An 0.1% aqueous solution of 2,6-dichloro-4-trimethylammoniumbenzenediazonium chloride-fluoborate (Chapter 17, Section I.B). Prepare fresh immediately before use.

b. Dissolve 0.025 g of cyclohexane-1,3-dione, 1 g of ammonium acetate, and 0.5 ml of glacial acetic acid in 8 ml of water, and dilute to 10 ml with water.

Procedure

To 0.5 ml of aqueous solution of the alcohol, add 0.5 ml of reagent a, and irradiate for 15 min with a 125 W mercury-vapor lamp, at a distance of about 8 inches. Cool for 1 min in ice water, add 0.5 ml of reagent b, plug the tubes with cotton wool, and heat at 50° for 45 min, protected against light. Cool for 2 min in ice water, and add 3 ml of 2 N sodium hydroxide. Read at exc: 436 nm.

Standard: A solution of 1,8-dioxodecahydroacridine in a 49:1 mixture of 2 N sodium hydroxide and ethanol.

Results

	Em, nm	Determination limits, μg	Reading 50 Sample, μg	Reading 50 Standard, μg/ml
Methanol	525	80 - 400	96	0.15
Ethanol	515	1.2 - 6	1.6	0.17
1-Propanol	515	4.0 - 20	8.6	0.23
1-Butanol	515	4.0 - 20	8.8	0.20
Benzyl alcohol	510	10 - 50	19	0.22

Note

Fluorimetric determination of aldehydes, see Chapter 8, Section XI.A.

XIV. 1,2-DIOLS (F)

A. Sodium Metaperiodate, Ethyl Acetoacetate, and Ammonium Acetate

Principle (34)

Oxidation with periodate ion and development of the formaldehyde formed with ethyl acetoacetate and ammonia: blue fluorescence.

$$R{-}CHOH{-}CH_2OH \xrightarrow{IO_4^-} R{-}CHO + HCHO$$

$$H_5C_2O_2C{-}CH_2{-}CO{-}CH_3 + HCHO + NH_3 + H_2C(CO_2C_2H_5){-}CO{-}CH_3 \longrightarrow$$ 2,6-dimethyl-3,5-dicarbethoxy-1,4-dihydropyridine ($H_5C_2O_2C$, $CO_2C_2H_5$, H_3C, CH_3, N–H)

Reagents

a. An 0.025 M aqueous solution of sodium metaperiodate.

b. Dissolve 2.35 g of stannous chloride dihydrate in 5 ml of concentrated hydrochloric acid and dilute to 100 ml with water.

c. A 2% solution of ethyl acetoacetate in a 20% aqueous solution of ammonium acetate.

Procedure

To 1 ml of aqueous solution of the diol add 0.5 ml of reagent a, and let stand at room temperature for 30 min. Then add 0.5 ml of reagent b, 2 ml of water, and 1 ml of reagent c. Mix well after each addition. Heat at 60° for 20 min, allow to cool, and filter. Read at exc: 366 nm; em: 470 nm.

Standard: A solution of 2,6-dimethyl-3,5-dicarbethoxy-1,4-dihydropyridine in a 49:1 mixture of water and ethanol.

Results

	Determination limits, μg	Reading 50 Sample, μg	Standard, μg/ml
Ethylene glycol	1 - 5	1.8	2.71
Propylene glycol	2 - 10	3.6	2.55
Glycerol	1 - 5	1.6	2.94
Dihydrostreptomycin (sesquisulfate)	20 - 100	35	4.70

Notes

1. Dihydrostreptomycin, see Section VII.A.

2. Periodates and periodate oxidation, see Chapter 16, Section IV.

3. Fluorimetric determination of formaldehyde, see Chapter 16, Section V.

XV. HEXITOLS (F)

A. Sodium Metaperiodate, Ethyl Acetoacetate, and Ammonium Acetate

Principle (34)

Oxidation with periodate ion and development of the formaldehyde formed with ethyl acetoacetate and ammonia: blue fluorescence.

$$HOCH_2-(CHOH)_4-CH_2OH \xrightarrow{IO_4^-} 2\ HCHO + 4\ HCOOH$$

$$HCHO \longrightarrow$$

See preceding method

Procedure

Follow the procedure for 1,2-diols, preceding method.

Results

	Determination limits, μg	Reading 50 Sample, μg	Standard, μg/ml
Mannitol	5 - 25	10	2.63
Sorbitol	3 - 15	6	1.82

XVI. α-AMINO PRIMARY ALCOHOLS (F)

A. Sodium Metaperiodate, Ethyl Acetoacetate, and Ammonium Acetate

Principle (34)

Oxidation with periodate ion and development of the formaldehyde formed with ethyl acetoacetate and ammonia: blue fluorescence.

$$R\text{-}CH(NH_2)\text{-}CH_2OH \xrightarrow{IO_4^-} R\text{-}CHO + NH_3 + HCHO$$

$$HCHO \longrightarrow$$

See Section XIV.A.

Procedure

Follow the procedure for 1,2-diols, Section XIV.A.

Results

	Determination limits, μg	Reading 50 Sample, μg	Reading 50 Standard, μg/ml
Monoethanolamine	1 - 5	2	2.00
Serine	4 - 20	9	2.23

XVII. 2,3-DIHYDROXY-1-CARBOXYLIC ACIDS (F)

A. Sodium Metaperiodate and Resorcinol

Principle (34)

Oxidation with periodate ion and reaction of the glyoxylic acid formed with resorcinol in acid medium to give the lactone of 2,2',4,4'-tetrahydroxydiphenylacetic acid, which is oxidized by atmospheric oxygen to give a quinone derivative: yellow-green fluorescence.

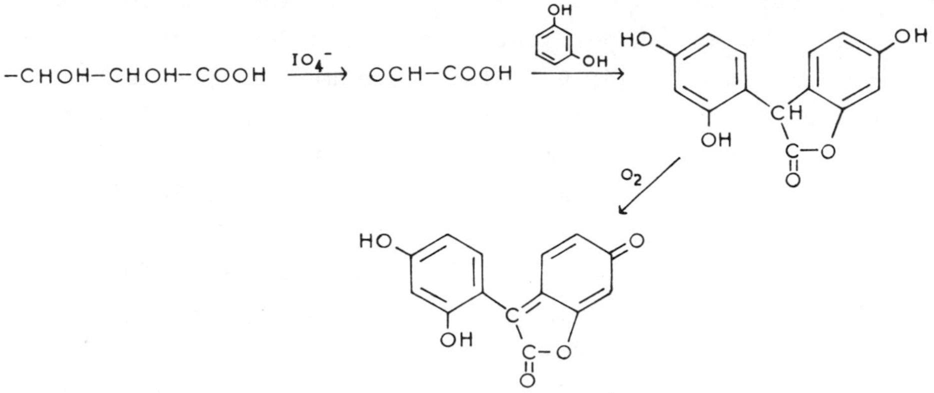

$$-CHOH-CHOH-COOH \xrightarrow{IO_4^-} OCH-COOH \xrightarrow{\text{resorcinol}} \text{lactone} \xrightarrow{O_2} \text{oxidized lactone}$$

Reagents

a. An 0.05 M aqueous solution of sodium metaperiodate adjusted to pH 10 with 1 N sodium hydroxide.

b. An 0.1 N aqueous solution of potassium arsenite (1 liter of solution contains 4.945 g of arsenious oxide, 75 ml of 1 N potassium hydroxide, and 40.0 g of potassium bicarbonate).

c. To 50 ml of 0.1 N silver nitrate, add 5 ml of 1 N sodium hydroxide. Let the precipitate settle, decant off the supernatant, wash the precipitate with water by decantation, dissolve it in 1 N sulfuric acid, and dilute to 150 ml with the acid.

d. A 2% aqueous solution of resorcinol.

e. A saturated aqueous solution of potassium bicarbonate.

Procedure

To 1 ml of neutralized aqueous solution of the acid, add 0.2 ml of reagent a, and let stand at room temperature for 15 min. Add 0.5 ml of reagent b to destroy the excess periodate, let stand for 20 min at room temperature, add 0.3 ml of reagent c to precipitate the iodate, mix, add 1 ml of water, and filter. To 1 ml of the filtrate, add 1 ml of concentrated hydrochloric acid and 0.5 ml of reagent d. Heat at 60° for 20 min, cool to room temperature in a water bath, add slowly while shaking 5 ml of reagent e, and heat at 60° for 15 min. Allow to cool to room temperature, and read at exc: 436 nm; em: 530 nm.

Standard: An aqueous solution of sodium fluorescein.

Results

	Determination limits, μg	Reading 50	
		Sample, μg	Standard, μg/ml
Tartaric acid	1.0 - 5.0	2.1	0.13
Mucic acid	1.0 - 5.0	2.3	0.076

Hexuronic acids can also be determined (Chapter 14, Section XVII.A).

Notes

1. Several compounds afford glyoxylic acid upon suitable chemical reaction. It is selectively and readily obtained from oxalic acid reduced by aluminum amalgam or a zinc-copper mixture (30), from dichloroacetic acid dehalogenated by hydrazine (31), and from acetic acid oxidized by periodate ion and hydrogen peroxide (35).

2. Colorimetric determination of 2,3-dihydroxy-1-carboxylic acids, see Chapter 9, Section V.A.

3. Periodates and periodate oxidation, see Chapter 16, Section IV.

REFERENCES

1. Cf. F.R. Duke and G.F. Smith, *Ind. Eng. Chem., Anal. Ed.*, 12, 201 (1940).

2. V.W. Reid and R.K. Truelove, *Analyst*, 77, 325 (1952); V.W. Reid and D.G. Salmon, *Analyst*, 80, 704 (1955).

3. G.F. Smith and W.H. Fly, *Anal. Chem.*, 21, 1233 (1949).

4. M. Pesez and J. Bartos, *Bull. Soc. Chim. Fr.*, 1930 (1961).

5. R. Amos, *Anal. Chim. Acta*, 40, 401 (1968).

6. M. Tanaka and I. Kojima, *Anal. Chim. Acta*, 41, 75 (1968).

7. H.J. Bielig and E. Bayer, *Justus Liebigs Ann. Chem.*, 584, 96 (1953).

8. F. Feigl, *Spot Tests in Organic Analysis*, Elsevier (Amsterdam), 7th ed., 1966, p. 174.

9. M. Stiller, *Anal. Chim. Acta*, 25, 85 (1961).

10. J. Balog and J. Csaszar, *Magyar Kém. Folyóirat*, 67, 62 (1961).

11. J.R. Plourde, J. Braun, and L. Fernandes, *Can. J. Pharm. Sci.*, 3, 60 (1968).

12. J.R. Plourde and J. Braun, *J. Pharm. Belg.*, 25, 167 (1970).

13. D. Peteri, *Z. Anal. Chem.*, 248, 38 (1969).

14. M. Pesez and J. Bartos, *Bull. Soc. Chim. Fr.*, 340 (1969).

15. From G. Gutnikov and G.H. Schenk, *Anal. Chem.*, 34, 1316 (1962).

16. D.P. Johnson and F.E. Critchfield, *Anal. Chem.*, 32, 865 (1960).

17. M. Pesez and J. Bartos, *Talanta*, 5, 216 (1960).

18. G.R. Umbreit and R.L. Houtman, *J. Pharm. Sci.*, 56, 349 (1967).

19. M. Pesez and J. Bartos, *Ann. Pharm. Fr.*, 22, 609 (1964).

20. M. Pesez and J. Bartos, *Bull. Soc. Chim. Fr.*, 2333 (1963).

21. Cf. H. Meerwein, K. Wunderlich, and K.F. Zenner, *Angew. Chem.*, 74, 807 (1962).

22. F.E. Critchfield and J.A. Hutchinson, *Anal. Chem.*, 32, 862 (1960).

23. From M.P.T. Bradley and G.E. Penketh, *Analyst*, 92, 701 (1967).

24. From P. Desnuelle and M. Naudet, *Bull. Soc. Chim. Fr.* [5], 12, 871 (1945).

25. Cf. S.B. Schryver, *Proc. Roy. Soc.* (London), B 82, 226 (1910).

26. N. Argant, *Bull. Soc. Chim. Biol.*, 31, 485 (1949).

27. Cf. E. Eegriwe, *Z. Anal. Chem.*, 110, 22 (1937).

28. From A. Berg, *Bull. Soc. Chim. Paris* [3], 11, 882 (1894).

29. From M. Pesez and J. Bartos, *Bull. Soc. Chim. Fr.*, 481 (1960).

30. M. Pesez, *Bull. Soc. Chim. Fr.* [5], 3, 676, 2072 (1936); *J. Pharm. Chim.* [9], 2, 325 (1942).

31. M. Pesez, *Ann. Pharm. Fr.*, 9, 187 (1951).

32. J. Bartos, *Ann. Pharm. Fr.*, 27, 323 (1969).

33. See Chapter 8, Section XI.

34. M. Pesez and J. Bartos, *Talanta*, 14, 1097 (1967).

35. M. Pesez and J. Ferrero, *Ann. Pharm. Fr.*, 14, 558 (1956).

Chapter 3

PHENOLS

I. INTRODUCTION

A. Colorimetry

Phenols develop characteristic, although often poorly sensitive, colors with ferric chloride. The wavelength of the absorption maximum depends on the nature and position of the substituents on the phenolic ring, and solvent effect is highly important. In acetic acid medium, only 1,2,3-triphenols react. These compounds also afford a violet color with ferrous ion in the presence of potassium sodium tartrate in buffered medium.

Phenols are reducing compounds and can be determined as such with ferric ferricyanide, phosphomolybdic, or phosphotungstic-phosphomolybdic acids. Determinations are generally possible at levels varying from 2 to about 70 μg (for A = 0.3), depending on the phenol tested and the reagent, phosphomolybdic acid being the less sensitive. It is obvious that these reactions are not selective, and other reducing compounds such as thiols, ascorbic acid, or indole, also react.

The yellow color developed when phenols are reacted with sodium nitrite in acid medium, then with ammonia, is due to ions derived from mono-, di-, or trinitrophenols.

It allows determinations of 50 to 100 μg of a phenol, but the reaction is not a general one.

With Millon reagent, an o-nitrosophenol is at first formed, whose quinone-oxime tautomeric form is then stabilized as a yellow to red mercurial chelate. The sensitivity is strongly dependent on the phenol tested.

The coupling of phenols with diazonium salts obeys the well-known rules (see Chapter 16, Section I) and often allows determinations at the level of a few micrograms. However, with conventional diazo reagents such as diazotized p-sulfanilic acid or p-nitrobenzenediazonium fluoborate, the range of applicability of a given procedure is limited. The sensitivity and color development vary with the phenol tested, and are dependent upon various factors such as the kind and the location of the substituents on the phenolic ring, the pH, and the temperature. These difficulties are overcome with the use of diazotized 2-aminobenzothiazole and of phenitrazole (3-phenyl-5-nitrosamino-1,2,4-thiadiazole), which is dehydrated in acid medium to the corresponding diazo compound. The proposed procedures are independent of the phenol tested and allow very sensitive determinations.

Formation of a benzoquinone anil through oxidative reaction with aniline is of limited scope.

Through an oxidative reaction, phenols react with 4-aminoantipyrine to give a N-substituted quinone-imine. The use of ammonia, potassium ferricyanide, and aminopyrine (4-dimethylaminoantipyrine) leads to the same condensates, but avoids a solvent extraction and enhances the sensitivity (A = 0.3 for 10-20 μg).

Dibromoindophenols are obtained upon reaction of phenols with 2,6-dibromoquinone-chlorimide (Gibbs reaction). The blue to red-violet color developed in alkaline medium affords a sensitivity of 20-50 μg for A = 0.3.

Contrary to some previous statements, this reaction is not specific for p-unsubstituted phenols.

Reduced p-nitroso-N,N-dimethylaniline oxidatively condenses with phenols to give highly colored indophenols. With a few exceptions such as nitrophenols or salicylic acid, determinations are possible at levels of 10-50 μg.

From all these reactions, those involving the formation of indophenols and N-substituted quinone-imines are the most specific for phenols.

The reaction with 1-nitroso-2-naphthol in nitric acid medium is specific for p-substituted phenols only under definite operating conditions. The sensitivity varies widely with the compound tested.

Reactions with molybdate ion or with tungstate and nitrite ions in acidic medium are specific for o-diphenols. The first one affords a molydate ester; the mechanism of the second one is unknown.

According to an unknown mechanism, 3,3-diphenyl-acrolein in hydrochloric acid medium allows the colorimetric determination of m-diphenols within the range 5-30 μg, but other compounds such as various cyclic ketones, dihydroxypyrimidines, or aromatic amines can also react. The condensation with p-dimethylaminocinnamaldehyde is more specific and affords similar sensitivities.

B. Fluorimetry

Only one reaction is available; it is based on the von Pechmann-Duisberg synthesis of coumarins, and the sensitivity varies widely with the compound to be determined.

II. PHENOLS

A. Ferric Chloride

Principle (1)

Reaction with ferric chloride to give a complex of the type $[Fe(OAr)_6]^{3-}$: blue, violet, or red color.

Reagent

A 1 M aqueous solution of ferric chloride.

Procedure

To 5 ml of aqueous solution of the phenol, add 0.5 ml of reagent. Read after 15 min.

Results

	λ Max, nm	A = 0.3 (1 cm cell) Sample, μg
Phenol	560	6,000
Resorcinol	570	3,500
Salicylic acid	530	140

Notes

1. The reaction of phenols with ferric chloride has been extensively studied by Wesp and Brode (1). These authors carried out tests on 60 phenols, and it is therefore recommended that their report be consulted for each particular case.

Usually a blue to violet color (absorption maximum between 540 and 600 nm) develops with phenols substituted with CH_3, OH, NH_2, or halogen. It is quite unstable when these substituents are in the ortho or para position to the hydroxyl group, and much more stable when they are in the meta position. A purple to red color (absorption maximum between 500 and 535 nm) develops in the presence of CHO, COOH, COOR, NO_2, and SO_3H substituents. The carboxyl group stabilizes the color when in the ortho position and makes it unstable when in the meta or para position. Polynitro substitution prevents color reaction.

Solvent effect is very important: The blue-violet color given by phenol completely fades upon acidification or addition of ethanol. With salicylic acid the absorption maximum at 558 nm in methanol solution, shifts to 538 nm with 1-butanol, to 505 nm with secondary alcohols, and to 480 nm with tertiary alcohols. A very close relation is shown between the dielectric constant of the solvent and the intensity of the color produced. Those solvents which have a high dielectric constant produce intense absorption.

2. Various nonphenolic compounds also give a color reaction with ferric ion: formate and acetate ions (red), benzoate ion (pink precipitate), 2-hydroxy-1-carboxylic acids (see Chapter 2, Section X.A), acetylacetone (red), antipyrine (red), camphocarboxylic acid (blue-green), colchiceine (green), and phenylpyruvic acid (green).

3. The enolization of a steroidal carbonyl group may lead to the formation of a colored ferric complex. For instance, a violet color is developed with 2-hydroxymethylene-17α-methyldihydrotestosterone (I). This enolization may depend on the influence of a remote functional group. A blue-green color is developed when ferric chloride is added to a solution of 3α-hydroxy-11,12-diketocholanic acid (IIa) in acetone, whereas the 3α-acetyl derivative (IIb) does not react.

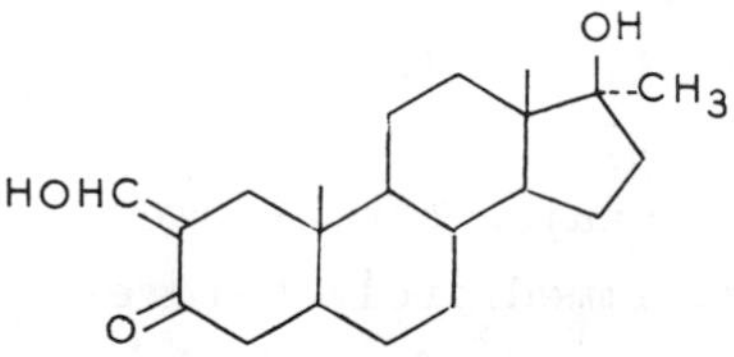

(I)

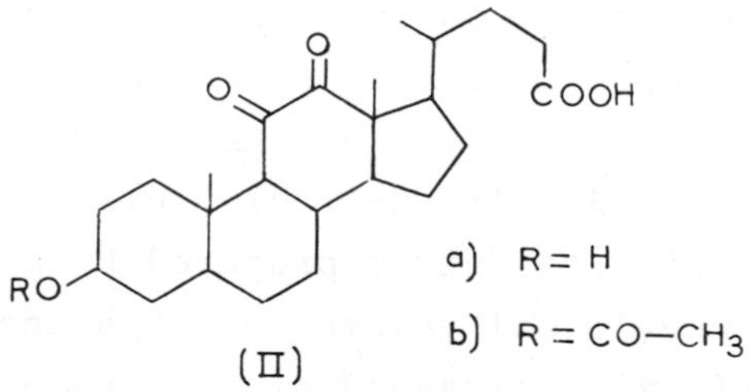

(II)

a) R = H

b) R = CO—CH_3

B. Ferric Ferricyanide

<u>Principle</u> (<u>2</u>)

Reduction of ferric ferricyanide into Prussian Blue.

Reagent

A 1:1 mixture of an 0.6% aqueous solution of potassium ferricyanide and of an 0.9% aqueous solution of ferric chloride hexahydrate. Prepare fresh immediately before use.

Procedure

To 1 ml of sample solution in ethanol, add 0.2 ml of reagent. Let stand in the dark for 5 min, and add 3 ml of 0.1 N hydrochloric acid. After 15 min in the dark, read at 720 nm.

Results

	A = 0.3 (1 cm cell) Sample, μg
Phenol	3.5
Pyrocatechol	6.1
Resorcinol	2.1
Hydroquinone	6.6
Estrone	11
Ethynylestradiol	13

Notes

1. The same reaction is given by aromatic amines (Chapter 5, Section II.A); indole derivatives (Chapter 13, Section III.A); vitamins A, C, and D; and ergosterol.

2. To delay the flocculation of Prussian Blue which may occur during the determination of o-diphenols it was proposed to add, prior to the reagent, a solution of sodium lauryl sulfate as surfactant (3).

3. Alternatively, uranyl ferricyanide can be used. This reagent is prepared by mixing, immediately before use, equal volumes of 1% aqueous solution of potassium ferricyanide and of 1% aqueous solution of uranyl nitrate hexahydrate.

To 1 ml of sample solution in ethanol, add 0.3 ml of reagent. Let stand for 5 min, and add 3 ml of water or 1 ml of water and 2 ml of ethanol. Read the brown-orange color at 450 nm.

Beer's law is followed, and A = 0.3 is given by 8.7 μg of phenol or 40 μg of ethynylestradiol.

C. Phosphomolybdic Acid

Principle (4)

Reduction of Wavelet phosphomolybdic reagent in alkaline medium: blue color.

Reagent

Dissolve 140 g of sodium carbonate decahydrate and 20 g of disodium phosphate dodecahydrate in approximately 500 ml of water, and add 70 g of recently calcined molybdic acid. When dissolved, add 200 ml of nitric acid (d = 1.38, 40° Bé), and dilute to 1 liter with water. Let stand for 24 hr, and filter.

Procedure

To 8 ml of aqueous solution of the phenol, add 1 ml of reagent, and drop by drop (1 drop per second) 1 ml of aqueous ammonium hydroxide (d = 0.92, 22° Bé), with stirring. Read immediately at 700 nm (at 15 ± 1°).

Results

	A = 0.3 (1 cm cell) Sample, μg
Phenol[a]	160
Thymol	65
Guaiacol	32
Eugenol	47
Pyrocatechol	23
Resorcinol	25
Hydroquinone	38
Phloroglucinol	52

[a]Beer's law is not followed.

Esculin and dicoumarol practically do not react.

Notes

1. Other reducing compounds, such as ascorbic acid and thiols, also react.

2. Several heteropoly acids derived from molybdic or tungstic, and silicic or phosphoric acids are used in analysis (5).

Silicomolybdic acid (or 12-silicomolybdic acid)

SiO_2, 12 MoO_3, 2 H_2O, $\underline{n}$ H_2O

It was proposed for the characterization of aldehydes (6, 7). On the other hand, the yellow color of aqueous solutions of the acid and of its salts allows the colorimetric determination of trace amounts of silica.

Phosphomolybdic acid (or 12-phosphomolybic acid)

P_2O_5, 24 MoO_3, 3 H_2O, $\underline{n}$ H_2O

The acid and its salts are yellow. Precipitation of phosphorus as ammonium phosphomolybdate, insoluble in nitric acid medium, is a classical method of gravimetric determination of the phosphate ion.

Silicotungstic acid

SiO_2, 12 WO_3, 2 H_2O, $\underline{n}$ H_2O

It affords highly crystalline salts and is used as a reagent for precipitation and determination of heterocyclic bases and alkaloids. The derivatives obtained (SiO_2, 12 WO_3, 2 H_2O, 4 alkaloid, $\underline{n}$ H_2O) are insoluble in acid medium.

12-phosphotungstic acid

P_2O_5, 24 WO_3, 3 H_2O, $\underline{n}$ H_2O

Its aqueous solutions are colorless. It was used for the precipitation of alkaloids.

Luteophosphotungstic acid (or 18-phosphotungstic acid)

P_2O_5, 18 WO_3, 3 H_2O, $\underline{n}$ H_2O

Its aqueous solutions are yellow.

D. Phosphotungstic-Phosphomolybdic Acid

Principle (8)

Reduction of the Folin-Ciocalteu reagent: blue color.

Reagent

To 100 g of sodium tungstate and 25 g of sodium molybdate, add 700 ml of water, 50 ml of 85% phosphoric acid (d = 1.69), and 100 ml of concentrated hydrochloric acid. Reflux gently for 10 hr. Add 150 g of lithium sulfate, 50 ml of water, and a few drops of liquid bromine. Boil without the condenser attached for 15 min to remove excess bromine. Cool, dilute to 1 liter, and filter.

Procedure

To 1 ml of aqueous solution of the phenol, add 0.5 ml of reagent and 2 ml of 20% aqueous solution of anhydrous sodium carbonate. After 5 min, read at 750 nm.

Results

	A = 0.3 (1 cm cell) Sample, μg
Phenol	6.7
Thymol[a]	16.5
Guaiacol	6.8
Eugenol	5.2
Pyrocatechol	5.2
Resorcinol	5.5
Hydroquinone	10.8
Phloroglucinol	11.5

[a]Dissolved in a mixture of water and ethanol.

Notes

1. Other compounds also reduce the reagent to molybdenum blue: reducing ions, tryptophan, hydroxyproline, 2- and 3-hydroxypyridines, ascorbic acid, uric acid.

2. Various analytical uses of heteropoly acids derived from molybdenum and tungsten have been described (9) (see also Section II.C, Note 2).

E. Sodium Nitrite and Ammonia

Principle (10)

Reaction with sodium nitrite in acid medium, then with ammonia: yellow color.

Reagents

a. Acetate buffer, prepared by mixing 400 ml of glacial acetic acid, 25 ml of water, and 75 ml of 10% solution of potassium hydroxide.

b. A saturated aqueous solution of sodium nitrite.

c. Mix 200 ml of ethanol, 200 ml of aqueous ammonium hydroxide (d = 0.92, 22° Bé), and 100 ml of water.

Procedure

To 1 ml of aqueous solution of the phenol, add 4 ml of buffer a, 0.2 ml of concentrated sulfuric acid, and 0.1 ml of reagent b. Let stand for 1 hr, dilute to 25 ml with reagent c, and read.

Results

	λ Max, nm	A = 0.3 (1 cm cell) Sample, μg
Phenol	400	52
Guaiacol	420	92
Resorcinol	405	48

The reaction is not a general one. For example, it is negative with hydroquinone.

Note

The yellow color is not due to the formation of a p-nitrosophenol, but to ions derived from mono-, di-, or trinitrophenols (11).

F. Millon Reagent

Principle (12)

On reaction with Millon reagent, development of a yellow to red color.

Reagent (13, 14)

In a 200 ml round-bottomed flask immersed in ice and fitted with a reflux condenser, dissolve 20 g of mercury in 15.5 ml of nitric acid (d = 1.39). After dissolution, add 30 ml of water and filter if necessary. Allow to stand for 15 days before use. This reagent is stable for about 2 months.

Procedure

To 3 ml of aqueous solution of the phenol, add 2 ml of reagent. Let stand at the given temperature for the given length of time (see Results), cool or warm to room temperature in a water bath if necessary, and read.

Results

	Oxidation time and temperature	λ Max, nm	A = 0.3 (1 cm cell) Sample, μg
Phenol	20 min, r.t.	495	185
m-Cresol	15 min, 0°	480	140
Pyrocatechol	5 min, r.t.	425	85
Pyrocatechol	10 min, 0°	375	55
Resorcinol	5 min, 70-75°	375	10
Methyl-p-hydroxybenzoate	60 min, 70-75°	510	78

With p-cresol and hydroquinone, Beer's law is not satisfactorily followed.

Notes

1. Millon reagent is a complex acid solution of nitrate, nitrite, mercuric and mercurous ions. It was initially proposed as a reagent for proteins, the color development being due to the phenol group of tyrosine (13,15).

2. The first step of the reaction is assumed to be a nitrosation in the ortho position to the hydroxyl group, yielding an o-nitrosophenol whose quinone-oxime tautomeric form is then stabilized as a strongly colored mercurial chelate.

OH → H O O N ⇌ H O O N → Hg^+ O O N or N O Hg O O O N

3. Vanillin gives only a faint yellow-pink color. However, a violet color (λ Max 530 nm) is developed when 5 ml of aqueous solution of this compound is reacted with 0.5 ml of Millon reagent. This reaction is specific for o-methoxyphenols bearing in the position para to the phenolic hydroxyl a side chain with a carbonyl group in α position to the aromatic ring.

4. Vanillin, as well as other p-substituted o-methoxyphenols can also be determined by a solution of mercuric acetate and sodium nitrite in aqueous acetic acid (16). Only very faint colors are given by usual phenols.

G. Diazotized p-Sulfanilic Acid

Principle

Coupling with the diazonium salt of sulfanilic acid: miscellaneous colors.

$$HO_3S-C_6H_4-\overset{+}{N}\equiv N + C_6H_5-OH \longrightarrow HO_3S-C_6H_4-N=N-C_6H_4-OH$$

Reagent

To 5 ml of 1% solution of p-sulfanilic acid in 1:9 hydrochloric acid, add dropwise 1 ml of 2% aqueous solution of sodium nitrite, and let stand for 5 min before immediate use.

Procedure

To 2 ml of aqueous solution of the phenol, add 5 ml of 1 N sodium hydroxide and 1 ml of reagent. Then, according to the phenol:

A. Dilute to 10 ml with water.

B. Add 1.5 ml of 1 N hydrochloric acid.

C. Add 1 ml of 10% aqueous solution of hydroxylamine hydrochloride.

D. Before adding the reagent, add 1 ml of 10% aqueous solution of hydroxylamine hydrochloride.

E. Add 3 ml of ethanol.

Then, in any case (with the exception of E), dilute to 10 ml with water, and read.

Results

	Procedure	λ Max, nm	A = 0.3 (1 cm cell) Sample, μg
Guaiacol	A	465	17
	B	465	15
	C	465	70
Resorcinol	A	490	12
	C	490	12
Hydroquinone	C	360	32
	D	360	34
	E	370	85
Phloroglucinol	C	425	15
	D	425	62

	Procedure	λ Max, nm	A = 0.3 (1 cm cell) Sample, μg
Phloroglucinol	E	425	21
α-Naphthol	A	510	16.5
β-Naphthol	B	345	90

Notes

1. The sensitivity of the reaction and the color development being dependent upon the phenol tested and on the pH, identification of various phenols therefore becomes possible (17).

2. Application of diazotized p-sulfanilic acid to the determination of imidazole compounds, see Chapter 13, Section V.A.

3. Diazo coupling, see Chapter 16, Section I.

H. p-Nitrobenzenediazonium Fluoborate

Principle

Coupling with p-nitrobenzenediazonium ion in alkaline medium: yellow to red color.

$$O_2N-\bigcirc-\overset{+}{N}\equiv N + \bigcirc-OH \longrightarrow O_2N-\bigcirc-N=N-\bigcirc-OH$$

Reagents

a. An 0.2% solution of p-nitrobenzenediazonium fluoborate in 0.5 N hydrochloric acid.

b. Clark and Lubs buffer for pH 10, see Appendix I.C.

Procedure

To 1 ml of aqueous solution of the phenol, add 0.1 ml of reagent a, cool to 0°, and add 5 ml of buffer b. Read immediately.

Results

	λ Max, nm	A = 0.3 (1 cm cell) Sample, μg
Phenol	470	6
Guaiacol	510	7.5
Resorcinol	440	6
α-Naphthol	440	18
β-Naphthol	460	23

Notes

1. The reaction is highly sensitive to pH.

2. This procedure is applicable only to a limited number of phenols. For example, the color developed by hydroquinone is stable only at pH 9 at 0°, and Beer's law is not followed. Pyrocatechol affords a yellow color at pH 9 after 2 hr at 20-25° in the dark (A = 0.3 for 3.4 μg at 410 nm). With salicylaldehyde, an orange color is developed in a pH 9 medium after 30 min at 0° (A = 0.3 for 10 μg at 450 nm).

3. p-Nitrobenzenediazonium fluoborate also allows the colorimetric determination of primary aliphatic amines (Chapter 4, Section VI.E), aromatic amines (Chapter 5, Section II.C), and α-amino acids (Chapter 9, Section VI.C).

4. Diazo coupling, see Chapter 16, Section I.

I. Diazotized 2-Aminobenzothiazole

Principle (18,19)

Coupling with the diazonium salt of 2-aminobenzothiazole: miscellaneous colors.

Reagent

To 5 ml of 1% solution of 2-aminobenzothiazole in 1:9 hydrochloric acid, add 1 ml of 2% aqueous solution of sodium nitrite, and let stand at room temperature for 30 min. Filter, and use immediately.

Procedure

To 1 ml of aqueous solution of the phenol, add 1 ml of reagent, 1 ml of 2 N sodium hydroxide, mix, and add 1 ml of pyridine. Read immediately.

Results

	Color	λ Max, nm	A = 0.3 (1 cm cell) Sample, μg
Phenol	Yellow-orange	515	3
o-Cresol	Violaceous red	540	3.5
m-Cresol	Violaceous red	520	2.8
p-Cresol	Yellow-brown	555	12.5
Thymol	Orange-red	545	3.1
Guaiacol	Violaceous red	520	7.5
Pyrocatechol	Yellow-brown	380	17
Resorcinol	Yellow-orange	520	5
Hydroquinone	Yellow-brown	380	19
Phloroglucinol	Orange	480	8
Pyrogallol	Yellow-brown	390	37
α-Naphthol	Violaceous red	580	5
β-Naphthol	Red	550	10

Notes

1. The main advantage of diazotized 2-aminobenzothiazole over other diazo reagents rests on the fact that the procedure is independent of the phenol tested.

2. The same procedure allows the determination of 2-methylindole (A = 0.3 for 17 μg at 495 nm) and guanine (A = 0.3 for 15 μg at 570 nm), whereas the reaction is almost negative with adenine (A = 0.08 for 500 μg at 550 nm).

3. Primary aliphatic amines, procaine, p-aminobenzenesulfonamide, ethyl acetoacetate, ergosterol, calciferol, histidine, and histamine do not react.

4. Diazotized 2-aminobenzothiazole also allows the colorimetric determination of aromatic amines (Chapter 5, Section II.E) and nitroaliphatic derivatives (Chapter 7, Section IV.B).

5. Diazo coupling, see Chapter 16, Section I.

J. Phenitrazole

Principle (20)

Coupling with the diazonium salt formed by dehydration of 3-phenyl-5-nitrosamino-1,2,4-thiadiazole (phenitrazole) in the acid reaction medium: yellow, red, or violet color.

$$C_6H_5-\underset{\text{N}\smallsetminus\text{S}\diagup}{\overset{}{C=N-}}C-NH-NO \xrightarrow{H^+} C_6H_5-C=N-\ldots C-\overset{+}{N}\equiv N \xrightarrow{\text{phenol}}$$

Phenitrazole Azo dye

Reagent

An 0.4% solution of phenitrazole (Chapter 17, Section I.F) in ethanol.

Procedure

To 1 ml of sample solution in ethanol, add 1 ml of reagent and 1 ml of 65-70% perchloric acid (d = about 1.62). Let stand at room temperature for 5 min, chill in ice, and add 3 ml of prechilled 5 N sodium hydroxide. Let stand at room temperature for 5 min, and read.

Results

	Color	λ Max, nm	A = 0.3 (1 cm cell) Sample, μg
Phenol	Red	515	4
Guaiacol	Red-violet	530	5
Pyrocatechol	Violet	530	12
Resorcinol	Red-orange	500	5
Hydroquinone	Brown	420	65
Phloroglucinol	Red	500	10
α-Naphthol	Violet	550	7
β-Naphthol	Violet	540	12
o-Aminophenol	Pink	515	21
m-Aminophenol	Red	520	7
p-Aminophenol	Pink	515	45
Salicylaldehyde	Yellow	380	40

Notes

1. Primary aliphatic amines, histidine, histamine, theophylline, caffeine, adenine, thymine, and guanine do not react with phenitrazole, whereas they develop color reactions with various other diazonium salts.

2. Phenitrazole permitted selective determination of phenol and p-cresol in mixtures (21).

3. Phenitrazole also allows the colorimetric determination of aromatic amines (Chapter 5, Section II.F) and of aldehydes developed as formazans (Chapter 8, Section III.A).

4. Diazo coupling, see Chapter 16, Section I.

K. Aniline and Sodium Hypochlorite

Principle (22,23)

Oxidative reaction with aniline, to give a benzoquinoneanil: blue color.

$$C_6H_5\text{-}NH_2 + C_6H_5\text{-}OH \xrightarrow{ClO^-} C_6H_5\text{-}N{=}C_6H_4{=}O$$

Reagents

a. A saturated aqueous solution of aniline.

b. A dilute aqueous solution of sodium hypochlorite: 100 ml contains 0.316 g active chlorine.

Procedure

Mix 4.8 ml of water and 0.2 ml of reagent a. Add 0.3 ml of sample solution in methanol, 0.1 ml of reagent b, and 0.1 ml of 10% aqueous solution of anhydrous sodium carbonate. Let stand for the given length of time (see Results), and read at 600 nm.

Results

	Oxidation time, min	A = 0.3 (1 cm cell) Sample, μg
Phenol	20	98
Thymol	3	18
Carvacrol	3	18
α-Naphthol	3	21

The reaction is not a general one. It is negative with resorcinol, hydroquinone, pyrogallol, gallic acid, salicylic acid, phloroglucinol, and β-naphthol. Pyrocatechol develops a yellow to pink-brown color, guaiacol a yellow, then a greenish color.

Notes

1. According to the same principle, a given phenol (thymol) inversely permits the colorimetric determination of primary aromatic amines (see Chapter 5, Section IV.J).

2. Dimethyl-p-phenylenediamine was proposed instead of aniline, thus affording a highly sensitive reaction (24). The poor stability of this reagent can be overcome by preparing it just before use from p-nitroso-N,N-dimethylaniline (see Section II.O). Aniline can also be replaced by 3,5-dibromo-4-hydroxyaniline (25), and hypochlorite by ferricyanide (26).

3. A nearby similar mechanism was attributed to the Flückiger's reaction (27,28), which involves the action of bromine upon a phenol in the presence of ammonia.

4. Based on the same principle, a determination of ammonium salts by reaction with phenol and hypochlorite in alkaline medium was recently proposed (29).

L. 4-Aminoantipyrine and Potassium Ferricyanide

Principle (30,31)

Condensation with 4-aminoantipyrine in the presence of an alkaline oxidizing agent, to give an N-substituted quinone-imine: red-orange color.

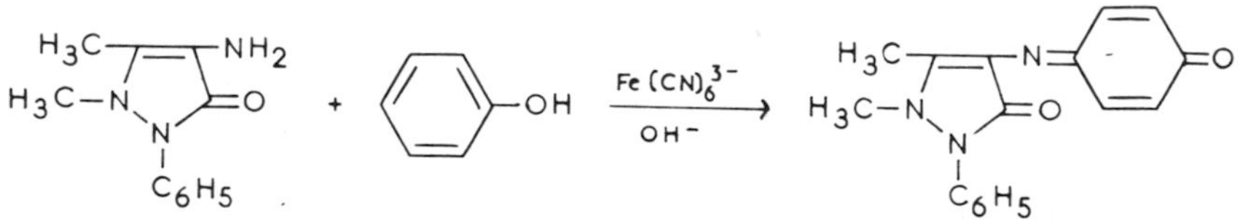

Reagents

a. A solution of 0.5 g of 4-aminoantipyrine in 25 ml of water, freshly prepared and filtered.

b. Buffer solution: Dilute a mixture of 6.75 g of ammonium chloride and 57 ml of aqueous ammonium hydroxide (d = 0.92, 22° Bé) to 100 ml with water, then dilute 2 ml of this stock solution to 1 liter with water.

c. An 8% aqueous solution of potassium ferricyanide, freshly prepared.

Procedure

To the aqueous solution of the phenol, add 1 ml of reagent a, dilute up to about 50 ml with buffer b, add 1 ml of reagent c, mix, and extract with 25 ml, then with two 10 ml portions of chloroform. Combine the organic phases, filter through a small cotton plug, dilute to 50 ml with chloroform, and read at 450-460 nm.

Results

	A = 0.3 (1 cm cell) Sample, μg
Phenol	75
m-Cresol	112
Guaiacol	90
α-Naphthol	240
Methyl salicylate[a]	90
Phenyl salicylate (Salol)	155

[a]Read at 470 nm.

Notes

1. The reaction is given by phenolic compounds with the position para to the phenolic hydroxyl either free or substituted with halogen, OH, OR, SO_3H, or COOH. A nitro or a carboxylic group in the ortho position inhibits the reaction, which is also negative with p-cresol, salicylic acid, p-hydroxybenzoates, vanillin, diethylstilbestrol, hexestrol, and morphine. β-Naphthol affords a green color. A detailed study of the behavior of 26 p-unsubstituted phenols was recently published (32).

2. When atmospheric oxygen acts as the oxidant, only catechols respond to the test. This permits the determination of catechols in the presence of other phenols (33).

3. Application of 4-aminoantipyrine to the determination of ketosteroids with unsaturated ring A, see Chapter 15, Section VII.A.

M. Ammonia, Potassium Ferricyanide, and Aminopyrine

Principle (34,35)

In ammonia buffer in the presence of an oxidant, the phenol is converted into the corresponding quinone-imine, which reacts with aminopyrine (4-dimethylaminoantipyrine) to give an N-substituted quinone-imine: red color.

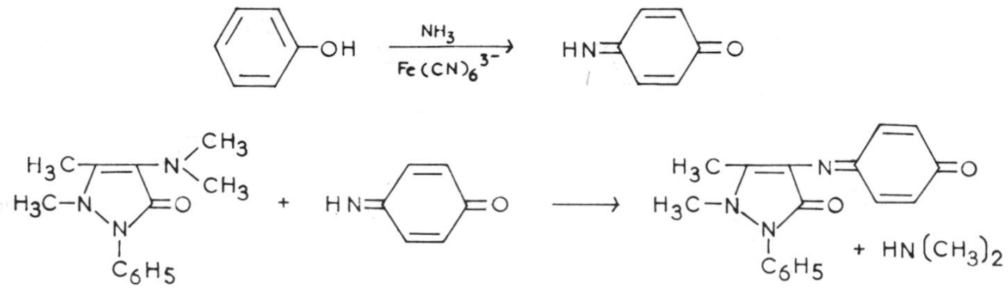

Reagents

a. Buffer for pH 8.0: Dissolve 3.85 g of ammonium acetate in 90 ml of water, adjust the pH to 8.0 with aqueous ammonia, and dilute to 100 ml with water.

b. A 1% aqueous solution of aminopyrine.

c. A 2% aqueous solution of potassium ferricyanide.

Procedure

To 1 ml of aqueous solution of the phenol, add 1 ml of buffer a, 1 ml of reagent b, and 1 ml of reagent c. Let stand at room temperature for 30 min, and read at 500 nm. (When the sample is insoluble in water, it may be dissolved in the minimum amount of 1 N sodium hydroxide, and diluted with water.)

Results

	A = 0.3 (1 cm cell) Sample, μg
Phenol	10
m-Cresol	13.5
p-Chlorophenol	10.8
Guaiacol	19
α-Naphthol	17
Methyl salicylate	17.5
Phenyl salicylate (Salol)	19.2

The reaction is but very slightly positive with p-cresol, negative with β-naphthol.

Notes

1. When compared to the classical determination of phenols with 4-aminoantipyrine (preceding method), the procedure herein described affords a higher sensitivity and avoids any solvent extraction.

2. Aminopyrine offers the advantage of being highly stable, and it may be purchased in a highly pure state, since it is used as a pharmaceutical.

3. According to the same principle, phenol inversely permits the colorimetric determination of 4-dimethylamino- and 4-monomethylaminoantipyrine (*34*), and of ammonia (*36*).

N. 2,6-Dibromoquinone Chlorimide

Principle (*37*)

Reaction with 2,6-dibromoquinone chlorimide in alkaline medium, to give an indophenol: blue to red-violet color.

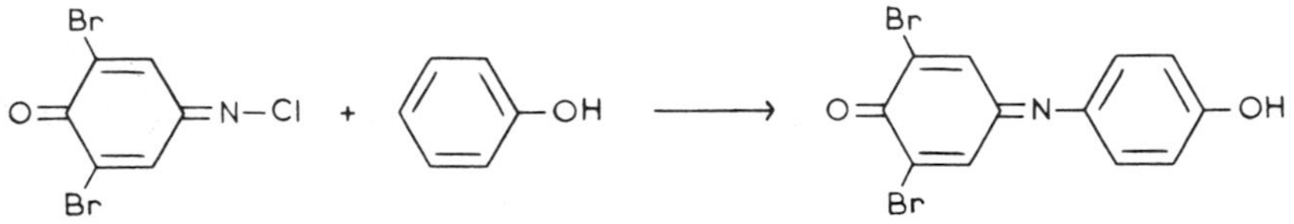

Reagents

a. Borate buffer for pH 8.3: Dissolve 12.37 g of boric acid, 14.9 g of potassium chloride, and 1.60 g of sodium hydroxide in water, and dilute to 1 liter with water.

b. An 0.3% solution of 2,6-dibromoquinone chlorimide in ethanol.

Procedure

To 1 ml of aqueous solution of the phenol, add 4 ml of 0.25 N sodium hydroxide, 10 ml of buffer *a*, and 0.2 ml of reagent *b*. Read after 10 min.

Results

	Color	λ Max, nm	A = 0.3 (1 cm cell) Sample, μg
Phenol	Blue	600	18.5
o-Cresol	Blue	580	26
m-Cresol	Blue	630	36
Thymol	Blue-gray	580	50
Resorcinol	Red-violet	540	46

Pyrocatechol and p-cresol develop a reddish color, but Beer's law is not followed.

Notes

1. The pH of the medium affects the rate of the reaction, as well as the intensity and the stability of the color developed.

2. The reaction of phenolic compounds with 2,6-dibromo- or 2,6-dichloroquinone chlorimide is commonly known as the Gibbs reaction (28,38). When operating with the dichloro reagent, extraction of the color into butanol allows a separation of phenols into four classes (39). This reagent also gives colored species with uric acid (40), theophylline (41), thiouracil (42), and pyridoxal, but not with its phosphoric ester (codecarboxylase).

The dibromo derivative permits the determination of equilenin and equilin in the presence of estrone, which does not react (43).

Initially considered as almost specific for phenolic compounds unsubstituted at the para position, the Gibbs reaction also gives positive results with numerous phenols so substituted (44), and in some instances the molecular extinction coefficient of the colored species formed is higher than that recorded with para-unsubstituted compounds. On the other hand, many highly substituted phenols with the para position unsubstituted do not give a positive test.

4. For other analytical reactions of 2,6-dihalogenoquinone chlorimide, see ref. (45).

5. Colorimetric determination of thiols with 2,6-dichloroquinone chlorimide, see Chapter 11, Section III.D.

O. Reduced p-Nitroso-N,N-dimethylaniline and Potassium Ferricyanide

Principle (46)

Oxidative reaction with dimethyl-p-phenylenediamine, prepared just before use from p-nitroso-N,N-dimethylaniline, to give an indophenol: violet, blue, green, or pink color.

$$(H_3C)_2N-C_6H_4-NH_2 + C_6H_5-OH \xrightarrow{Fe(CN)_6^{3-}} (H_3C)_2N-C_6H_4-N=C_6H_4=O$$

Reagents

a. Dissolve 0.15 g of p-nitroso-N,N-dimethylaniline in 2 ml of ethanol, dilute to 100 ml with water, add 0.2 g of potassium borohydride, and 2 ml of 5% aqueous solution of cupric sulfate pentahydrate. Let stand for 30 min, filter, and acidify to pH 1-2 with a 10% aqueous solution of oxalic acid dihydrate (approximately 6 ml).

b. Clark and Lubs buffer for pH 9.8, see Appendix I.C.

c. A 1% aqueous solution of potassium ferricyanide.

Procedure

To 5 ml of aqueous solution of the phenol, add 0.5 ml of reagent *a*. After 5 min, add 3 ml of buffer *b*, 0.5 ml of reagent *c*, and let stand for 5 min. Extract the color into 5 ml of chloroform, and read.

Results

	Color	λ Max, nm	A = 0.3 (1 cm cell) Sample, μg
Phenol	Blue	590	8
o-Cresol	Blue	590	10
m-Cresol	Blue	590	10
Thymol	Violet	565	17
Pentachlorophenol	Violaceous pink	560	200
Hexachlorophene	Blue-green	700	37
o-Nitrophenol	Green	660	550
m-Nitrophenol	Blue-green	650	335
Guaiacol	Blue	590	10
Pyrocatechol	Blue	590	50
Resorcinol	Violet	550	15
α-Naphthol	Blue	580	20
β-Naphthol	Green	670	50
o-Aminophenol	Blue	590	13
m-Aminophenol	Violet	560	13
Salicylaldehyde	Blue	600	15
p-Hydroxybenzaldehyde	Blue	590	700
Vanillin	Blue	590	450
Salicylic acid	Blue	590	5000
β-Resorcylic acid	Violet	550	40
p-Aminosalicylic acid	Violet	560	136

Phloroglucinol and pyrogallol do not react.

Notes

1. A detailed study was recently devoted to the effects of oxidant and pH on the reaction between phenols and dimethyl-p-phenylenediamine (47).

2. Dimethyl-p-phenylenediamine and most of its common salts are but poorly stable, particularly when

exposed to air or light, which is highly inconvenient when this amine is used as analytical reagent. For this purpose, two stable salts were proposed: the dimethyl-p-phenylenediamine hydrochloride - stannous chloride double salt, and the oxalate. The latter can be prepared in a very pure state (48) and is now commercially available.

3. Application of p-nitroso-N,N-dimethylaniline to the determination of 17-ketolsteroids, see Chapter 15, Section XII.B.

4. Dimethyl-p-phenylenediamine also allows the colorimetric determination of aliphatic aldehydes (Chapter 8, Section IV.D), aromatic aldehydes (Chapter 8, Section VI.A), peroxides (Chapter 10, Section III.A), and ketosteroids (Chapter 15, Sections VIII.A and XI.B).

5. Reduction with borohydride in the presence of a metal salt, see Chapter 7, Section VI.A, Note 2.

III. p-SUBSTITUTED PHENOLS

A. 1-Nitroso-2-naphthol

Principle (49,50)

Reaction with 1-nitroso-2-naphthol in nitric acid medium in the presence of a small amount of sodium nitrite: red color, which turns to stable yellow.

Reagents

a. An 0.05% solution of 1-nitroso-2-naphthol in ethanol.

b. An 0.05% solution of sodium nitrite in 3.5 N nitric acid.

Procedure

To 3 ml of sample solution in ethanol, add 1 ml of reagent *a* and 1 ml of reagent *b*. Glass-stopper the tubes, and heat at 55° for 1 hr. Allow to cool to room temperature, and read.

Results

	λ Max, nm	A = 0.3 (1 cm cell) Sample, μg
p-Cresol	450	36
p-Chlorophenol[a]	440	800
β-Naphthol	485	120
Tyrosine[b]	450	195
Estrone	450	110

[a]Beer's law is not followed.

[b]Dissolve the sample in the minimum amount of 0.1 N sodium hydroxide, and dilute with water. For the determination, add 2.1 ml of ethanol to 0.9 ml of this dilution.

Notes

1. 1-Nitroso-2-naphthol may be purified by washing with cold methanol.

2. Described at first as highly specific for tyrosine and p-substituted phenols (49), this color reaction occurs indeed with all phenols having a positive substituent. It is specific for p-substituted phenols only when the concentration of the reagent is low (51).

3. The reaction was applied to the detection and determination of compounds such as serotonin (52), morphine (53), hexestrol (54).

4. The red color can be stabilized by adding a saturated solution of ammonium ferric sulfate (ferric alum) (55).

5. The mechanism of the reaction does not seem as yet well established.

Anger and Ofri (56) admit that the nitrosonaphthol reacts with p-cresol in its quinone-oxime form. The reaction would involve an oxidation step, finally yielding an acridine derivative, according to the following scheme:

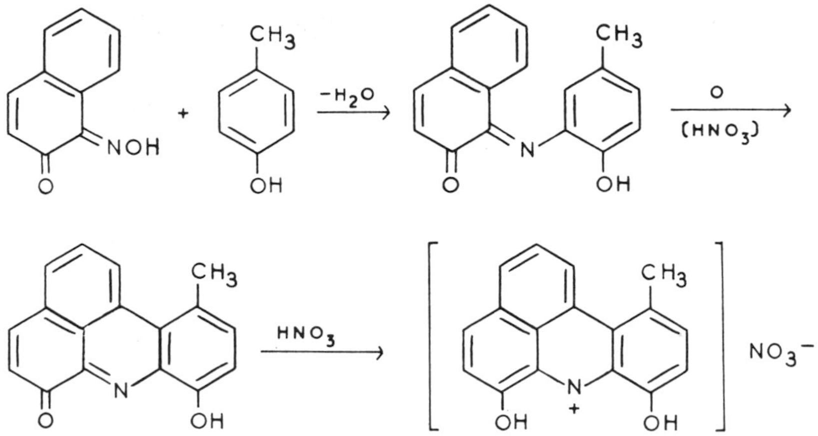

NOH + CH_3 / OH $\xrightarrow{-H_2O}$ N / CH_3 / OH / O $\xrightarrow[(HNO_3)]{O}$

CH_3 / N / OH / O $\xrightarrow{HNO_3}$ [CH_3 / OH / N+ / OH] NO_3^-

Working with p-cresol, Umeda (57) isolated from the reaction mixture a phenothiazine derivative and proposed the following mechanism:

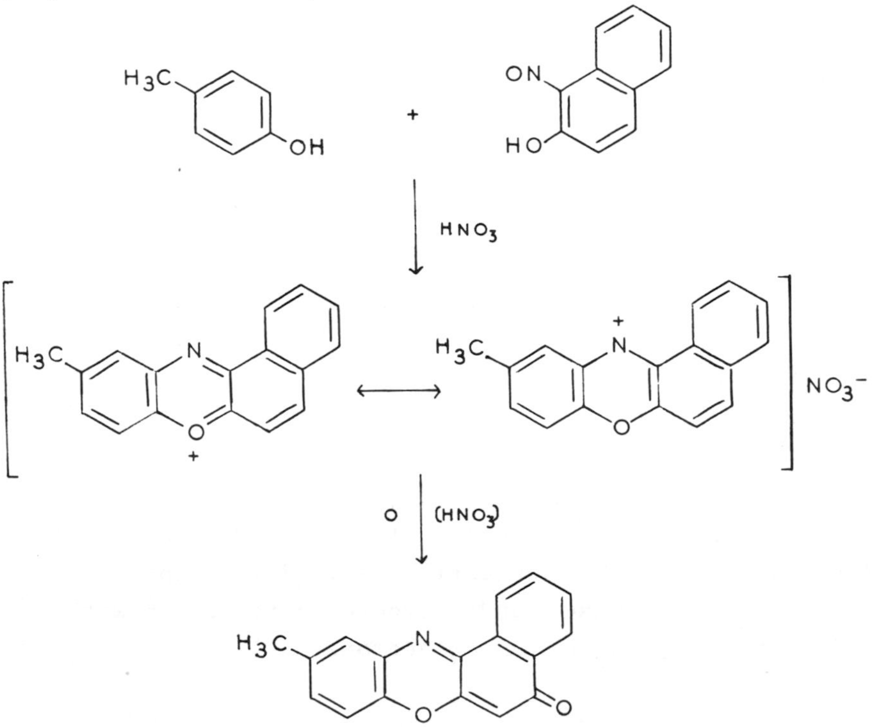

H_3C / OH + ON / HO

$\downarrow HNO_3$

[H_3C / N / O+ $\longleftrightarrow$ H_3C / N+ / O] NO_3^-

$\downarrow$ O (HNO$_3$)

H_3C / N / O / O

The author isolated similar derivatives when working with β-naphthol or p-acetaminophenol.

IV. o-DIPHENOLS

A. Ammonium Molybdate

Principle (58)

Reaction with molybdate in acidic medium: yellow color.

Reagent

A 10% aqueous solution of ammonium molybdate.

Procedure

To 2 ml of aqueous solution of the phenol, add 1 ml of 0.1 N sulfuric acid, 3 ml of reagent, mix, and read.

Results

	λ Max, nm	A = 0.3 (1 cm cell) Sample, μg
Pyrocatechol	370	135
Pyrogallol	360	110
Gallic acid	360	130
Epinephrine	360	183
Esculetin	375	21

Note

Compounds of the general structure, R-CO-NH-NH-R', such as iproniazid (isonicotinic acid 2-isopropylhydrazide) and related compounds, produce a red color with molybdate ion in a medium of acetone (59).

B. Sodium Tungstate and Sodium Nitrite

Principle (60)

Reaction with sodium tungstate and sodium nitrite in acid medium, and development in alkaline medium: pink color.

Reagents

a. A 5% aqueous solution of trichloroacetic acid.

b. A 10% aqueous solution of sodium tungstate.

c. A 1% aqueous solution of sodium nitrite.

Procedure

To 0.5 ml of sample solution in water or 0.002 N sodium hydroxide (when the compound is insoluble in water), add 0.5 ml of reagent a, 1 ml of reagent b, mix, and add 1 ml of 0.5 N hydrochloric acid and 0.5 ml of reagent c. Mix, let stand for 3 min, and add 0.5 ml of 5 N sodium hydroxide. After 5 min, read at 505 nm.

Results

	A = 0.3 (1 cm cell) Sample, μg
Pyrocatechol	13
Protocatechuic acid	40
Epinephrine	38
Norepinephrine	30
Dopa	32

The color is yellow with pyrogallol (A = 0.3 for 40 μg, 330 nm), yellow-orange with o-aminophenol (A = 0.3 for 23 μg, 400 nm).

V. m-DIPHENOLS

A. 3,3-Diphenylacrolein

Principle (61)

Reaction with 3,3-diphenylacrolein in hydrochloric acid medium: red color.

Reagent

A 1% solution of 3,3-diphenylacrolein (Chapter 17, Section I.C) in methanol.

Procedure

To 2 ml of sample solution in methanol, add 2 ml of reagent and 1 ml of concentrated hydrochloric acid. Read after 20 min.

Results

	λ Max, nm	A = 0.3 (1 cm cell) Sample, μg
Resorcinol	485	10
Orcinol	485	17
Pyrogallol	500	30
Phloroglucinol	485	5.5
Naphthoresorcinol	500	7.5

Notes

1. Resorcinol-4,6-disulfonic acid does not react, whereas a color is observed with m-dimethylaminophenol.

2. According to the same procedure, a yellow to yellow-green color is developed with cyclopentanone and cyclohexanone, and a red or orange color is given by the following pyrimidines:

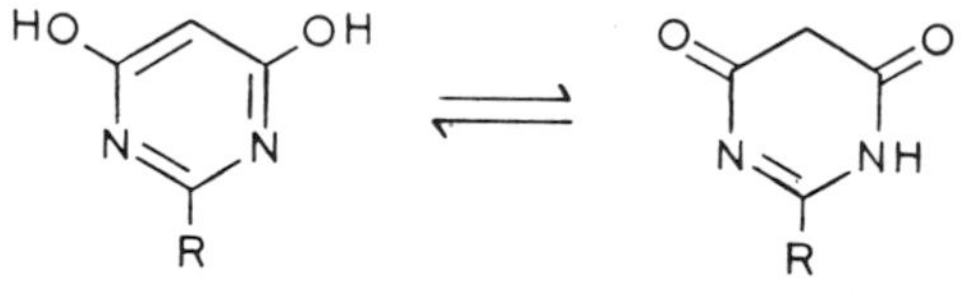

R = C_2H_5	2-Ethyl-4,6-dihydroxypyrimidine
R = H_2N-CO-CH_2	2-Carbamylmethyl-4,6-dihydroxypyrimidine
R = OH	Barbituric acid

3. Application of 3,3-diphenylacrolein to the determination of nitroaromatic compounds, see Chapter 7, Section VI.A.

B. p-Dimethylaminocinnamaldehyde

Principle

Reaction with p-dimethylaminocinnamaldehyde in hydrochloric acid medium: green color.

Reagent

An 0.25% solution of p-dimethylaminocinnamaldehyde in methanol.

Procedure

To 2 ml of sample solution in methanol, add 2 ml of reagent and 1 ml of concentrated hydrochloric acid. Let stand for the given length of time (see Results), and read at 630 nm.

Results

	Reaction time, min	A = 0.3 (1 cm cell) Sample, μg
Resorcinol	25	6
Orcinol	0	8.7
Pyrogallol	25	22.5
Phloroglucinol	0	2.7
Naphthoresorcinol[a]	0	22

[a]Read at 680 nm.

Notes

1. Cyclohexanone and barbituric acid do not react (see preceding Section, Note 2).

2. According to the same principle, catechins develop a blue color, allowing their determination (62).

3. p-Dimethylaminocinnamaldehyde also allows the colorimetric determination of primary aliphatic amines (Chapter 4, Section VI.C), primary aromatic amines (Chapter 5, Section IV.C), and indole derivatives (Chapter 13, Section III.D).

VI. 1,2,3-TRIPHENOLS

A. Ferric Chloride

Principle

Reaction with ferric chloride in acetic acid medium: blue-violet to red-violet color.

Reagents

a. Dissolve 10 g of ferric chloride hexahydrate in 13 ml of water, and dilute 1 ml of this stock solution to 100 ml with glacial acetic acid.

b. A 1% solution of potassium acetate in glacial acetic acid.

Procedure

To 1.5 ml of sample solution in glacial acetic acid, add 1.5 ml of chloroform, 1 ml of reagent a, and 2 ml of reagent b. Let stand for the given length of time (see Results), and read at 570 nm.

Results

	Reaction time, min	A = 0.3 (1 cm cell) Sample, μg
Pyrogallol	10	270
Gallic acid[a]	4 to 8	340
Propyl gallate[b]	0	275

[a]Cool to 4° ± 1° before adding reagent b and keep at this temperature until reading.

[b]Linear relationship holds down to 50 μg.

Triphenols whose hydroxyl groups are not in positions 1,2,3, and other phenols do not react.

B. Potassium Sodium Tartrate and Ferrous Sulfate

Principle (63)

Reaction with tartrate and ferrous ions: violet color.

Reagents

a. A 1.25% aqueous solution of ammonium acetate, used as buffer.

b. An aqueous solution containing 0.1% of ferrous sulfate heptahydrate and 0.5% of potassium sodium tartrate tetrahydrate (Seignette salt).

Procedure

To 1 ml of aqueous solution of the phenol, add 10 ml of buffer *a* and 1 ml of reagent *b*. Read at 540 nm.

Results

	A = 0.3 (1 cm cell) Sample, μg
Pyrogallol	190
Gallic acid	220
Propyl gallate	250

Notes

1. The reaction is also positive, but less sensitive with pyrocatechol, at pH 7-10 (64,65). The violet color turns to yellow-green in a more acidic medium, to orange in a more alkaline one.

2. Ascorbic acid develops a violet color with ferrous fulfate in alkaline medium (66). Uranium salts give a red solution (67).

VII. PHENOLS (F)

A. Ethyl Acetoacetate and Sulfuric Acid

Principle (68)

Condensation with ethyl acetoacetate in sulfuric acid medium, to give a coumarin: blue to blue-green fluorescence.

$$C_6H_5OH + HO{-}C(CH_3){=}CH{-}C({=}O){-}OC_2H_5 \xrightarrow{H_2SO_4} \text{4-methylcoumarin}$$

Reagent

A 2% solution of ethyl acetoacetate in ethanol

Procedure

To 1.5 ml of sample solution in ethanol, add 1 ml of reagent, then add very slowly 2.5 ml of concentrated sulfuric acid, allowing the acid to run down the wall while stirring the tube. Let stand at room temperature for 15 min, and read at exc: 366 nm.

Standard: A solution of quinine sulfate in 0.1 N sulfuric acid.

Results

	Em, nm	Determination limits, μg	Reading 50 Sample, μg	Reading 50 Standard, μg/ml
Phenol[a]	460	100 - 400	120	0.18
Guaiacol	495	10 - 50	20	0.24
Pyrocatechol	495	10 - 45	15	0.16
Resorcinol	430	0.07 - 0.38	0.17	0.34
Hydroquinone[b]	475	6.2 - 19	6.2 (Read. 60)	0.14 (Read. 60)
Phloroglucinol	490	2 - 8	3.5	0.33

[a]After the addition of sulfuric acid, heat at 80° for 15 min, chill in ice water for 3 min, and read.

[b]Linear relationship holds down to reading 60.

Note

This fluorimetric method is based on the von Pechmann-Duisberg synthesis of coumarins: condensation of phenols with β-keto esters in the presence of a dehydrating agent (69).

REFERENCES

1. E.F. Wesp and W.R. Brode, *J. Amer. Chem. Soc.*, 56, 1037 (1934).

2. M. Pesez and J. Bartos, *Ann. Pharm. Fr.*, 23, 281 (1965).

3. J. Michaud, *Bull. Soc. Pharm. Bordeaux,* 105, 219 (1966).

4. Cf. C. Wavelet, *Bull. Soc. Chim. du Nord de la France,* 81 (1898); D. Cristau, Thèse Pharmacie, Marseille (1954).

5. Cf. M. Jean, *Chim. Anal.,* 37, 125, 163 (1955).

6. J.H. Billman, D.B. Borders, J.A. Buehler and A.W. Seiling, *Anal. Chem.,* 37, 264 (1965).

7. J.H. Billman and A.W. Seiling, *Anal. Chim. Acta,* 33, 561 (1965).

8. From O. Folin and V. Ciocalteu, *J. Biol. Chem.,* 73, 627 (1927).

9. L. Vignoli, B. Cristau, and A. Pfister, *Chim. Anal.,* 38, 392 (1956).

10. From L. Lykken, R.S. Treseder, and V. Zahn, *Ind. Eng. Chem., Anal. Ed.,* 18, 103 (1946).

11. J. Gasparic, *Chem. Ind. (London)* (1962), p. 43; *Z. Anal. Chem.,* 199, 276 (1964).

12. Cf. P.C. Plugge, *Z. Anal. Chem.,* 11, 173 (1872); *Arch. Pharm.,* 228, 9 (1890); J. Josselin, Recherches spectrophotométriques sur la réaction de Millon, Thèse Pharmacie Bordeaux, 1961 (published 1963, Imprimerie Drouillard, Bordeaux).

13. E. Millon, *Compt. Rend. Acad. Sci., Paris,* 28, 40 (1849).

14. See: (a) *M.R.A.O.,* vol. 3, p. 138; (b) *Official Methods of Analysis of the A.O.A.C.,* 10th Ed., p. 33. Association of Official Agricultural Chemists, Washington, D.C. (1965).

15. See ref. 14 (b), p. 475.

16. E. Neuzil and H. Hensen, *Bull. Soc. Pharm. Bordeaux*, 105, 137 (1966).

17. L. Légrádi, *Mikrochim. Ichnoanal. Acta*, 865, 870 (1965).

18. Cf. S. Skraup, *Justus Liebigs Ann. Chem.*, 419, 65 (1919); R.F. Hunter, *J. Chem. Soc.*, 1385 (1926).

19. J. Bartos, *Ann. Pharm. Fr.*, 29, 147 (1971).

20. M. Pesez, J. Bartos, and J.F. Burtin, *Talanta*, 5, 213 (1960).

21. P. Zirinis, *Ann. Pharm. Fr.*, 19, 604 (1961).

22. Cf. E. Jacquemin, *Compt. Rend. Acad. Sci.*, *Paris*, 76, 1605 (1873).

23. M. Pesez, *Union Pharm.*, 77, 257 (1936).

24. G.U. Hougton and R.G. Pelly, *Analyst*, 62, 117 (1937).

25. L.V. Levy, *J.S. African Chem. Inst.*, 22, 29 (1939).

26. Y. Shishikura, *J. Japan Biochem. Soc.*, 19, 145 (1947).

27. F.A. Flückiger, *Arch. Pharm.*, 203, 30 (1873).

28. H.D. Gibbs, *Chem. Rev.*, 3, 291 (1927).

29. A.R. Selmer-Olsen, *Analyst*, 96, 565 (1971).

30. Cf. E. Emerson, *J. Org. Chem.*, 8, 417 (1943).

31. C.A. Johnson and R.A. Savidge, *J. Pharm. Pharmacol.*, 10, 171 T (1958).

32. D. Svobodová, J. Gasparic, and L. Nováková, *Coll. Czechoslov. Chem. Communic.*, 35, 31 (1970); see also D. Svobodova and J. Gasparic, *Mikrochim. Acta*, 384 (1971).

33. T.A. LaRue and E.R. Blakley, *Anal. Chim. Acta*, 31, 400 (1964).

34. Cf. S. Ono, R. Onishi, M. Tange, K. Kawamura, and T. Imai, *Yakugaku Zasshi*, 85, 245 (1965).

35. M. Pesez and J. Bartos, *Ann. Pharm. Fr.*, 25, 577 (1967).

36. K. Kawamura, *Chem. Pharm. Bull. Japan*, 16, 626 (1968).

37. From A.J. Singer and E.R. Stern, *Anal. Chem.*, 23, 1511 (1951).

38. H.D. Gibbs, *J. Biol. Chem.*, 72, 649 (1927).

39. E. Boyland, D. Manson, J.B. Solomon, and G.H. Wiltshire, *Biochem. J.*, 53, 420 (1953).

40. W.R. Fearon, *Biochem. J.*, 38, 399 (1944).

41. H.W. Raybin, *J. Amer. Chem. Soc.*, 67, 1621 (1945).

42. R.A. McAllister, *Nature*, 166, 789 (1950).

43. D. Banes, *J. Am. Pharm. Assoc., Sci. Ed.*, 39, 37 (1950).

44. J.C. Dacre, *Anal. Chem.*, 43, 589 (1971).

45. *F.S.A.O.V.*, pp. 255-262.

46. M. Pesez and J.F. Burtin, *Bull. Soc. Chim. Fr.*, 1996 (1959).

47. D.N. Kramer and L.U. Tolentino, *Anal. Chem.*, 43, 834 (1971).

48. M. Pesez and J. Bartos, *Talanta*, 10, 69 (1963).

49. Cf. O. Gerngross, K. Voss and H. Herfeld, *Ber.*, 66, 435 (1933).

50. M. Massin and A.B. Lindenberg, *Bull. Soc. Chim. Biol.*, 39, 1201 (1957).

51. M. Umeda, *Yakugaku Zasshi*, 84, 836 (1964).

52. S. Udenfriend, H. Weissbach, and C.T. Clark, *J. Biol. Chem.*, 215, 337 (1955).

53. H. Sakurai, K. Kato, M. Umeda, and S. Tsubota, *J. Pharm. Soc. Japan*, 83, 811 (1963).

54. A. Carayon-Gentil and J. Cheymol, *Ann. Pharm. Fr.*, 6, 129 (1948).

55. A. Maciag and R. Schoental, *Mikrochemie*, 24, 250 (1938).

56. V. Anger and S. Ofri, *Z. Anal. Chem.*, 203, 350 (1964).

57. M. Umeda, *Yakugaku Zasshi*, 84, 839 (1964).

58. Cf. G.N. Cohen, *Bull. Soc. Chim. Biol.*, 27, 237 (1945).

59. R.J. Colarusso, M. Schmall, E.G. Wollish, and E.G.E. Shafer, *Anal. Chem.*, 30, 62 (1958).

60. From P. Madhusudanan Nair and C.S. Vaidyanathan, *Anal. Biochem.*, 7, 315 (1964).

61. M. Pesez and J. Bartos, *Anal. Chim. Acta*, 20, 187 (1959).

62. M. Thies and R. Fischer, *Mikrochim. Acta*, 9 (1971).

63. From K.F. Mattil and L.J. Filler, Jr., *Ind. Eng. Chem., Anal. Ed.*, 16, 427 (1944).

64. C.A. Mitchell, *Analyst*, 48, 2 (1923).

65. S. Glasstone, *Analyst*, 50, 49 (1925).

66. A. Szent-Györgyi, *Z. Physiol. Chem.*, 225, 168 (1934).

67. M.Z. Barakat, N. Badran, and S.K. Shehab, *J. Pharm. Pharmacol.*, 4, 46 (1952).

68. M. Pesez and J. Bartos, *Talanta*, 14, 1097 (1967).

69. H. von Pechmann and C. Duisberg, *Ber.*, 16, 2119 (1883).

Chapter 4

ALIPHATIC AMINES

I. INTRODUCTION

A. Colorimetry

The basic properties of all classes of aliphatic amines allow their determination through various reactions.

Upon the action of a base, 1,3,5-trinitrobenzene condenses with nitromethane to give a Meisenheimer-like σ complex. The reaction is highly sensitive (5 to 30 μg for $A = 0.3$). Amine or guanidine hydrohalides and quaternary ammonium halides can also be determined, the base being liberated by reacting the salt with silver oxide in the reaction medium. Alkaloids also react.

A mixture of silver nitrate and manganese sulfate reacting with a base affords a precipitate of manganese dioxide and silver. A test paper therefore allows the detection and semiquantitative estimation of amines. A similar reaction is given by a mixture of silver nitrate and ferrous sulfate. The precipitate of ferric oxide and silver is colloidal, thus allowing spectrophotometric determinations within the range 30-220 μg, depending on the amine tested.

The basic properties of secondary and tertiary amines also allow their determination through ion-pair extraction, using picric acid or Tropaeolin 00 as acid dye. The sensitivity varies with the compound tested.

Primary amines do not react, probably because of the higher solubility of their salts with the acid dye in aqueous media, whereas alkaloids can be so determined. Only tertiary amines react with erythrosine.

Aliphatic amines react with a mixture of potassium thiocyanate and cobaltous nitrate in the solid state to give a blue complex soluble in chloroform. The reaction can be applied to various alkaloids, and the sensitivity depends on the compound tested. Low molecular weight amines afford irregular results.

Through an oxidative reaction, aliphatic amines condense with β-naphthol, and the condensate is developed with 2,4-dinitrophenylhydrazine. The green color permits determinations within the range 10-30 μg for A = 0.3.

The hydrogen bonded to the nitrogen atom in primary and secondary amines allows miscellaneous reactions.

Upon reaction with 1-chloro- or 1-fluoro-2,4-dinitrobenzene, N-alkyl-2,4-dinitroanilines are formed, which can be read either directly in the reaction medium or after suitable extraction. Poorly sensitive results are obtained with the first reagent. With the second one, samples are of the order of 5-40 μg for A = 0.3, and the reaction is almost selective for primary amines when operating in nonaqueous medium. In this case, a Janovsky-like reaction is probably involved (sensitivity: 6-17 μg for A = 0.3).

The amino group can be substituted for the sulfonate group of 1,2-naphthoquinone-4-sulfonic acid. The orange N-alkylaminonaphthoquinones formed permit determinations at levels of 25-50 μg. The dye can be extracted into methylene chloride and developed with 2,4-dinitrophenylhydrazine, thus enhancing the sensitivity and making determinations possible within the range 2-8 μg.

Amides obtained by reacting primary and secondary amines with p-nitrophenylazobenzoyl chloride in a non-aqueous medium are developed with benzyltrimethyl-ammonium hydroxide without prior elimination of the excess reagent. The sensitivity varies from 6 to 57 μg.

Upon reaction with carbon disulfide, primary and secondary amines form dithiocarbamates which reduce Blue Tetrazolium (6-37 μg can be so determined). The dithio-carbamates obtained from secondary amines give a cupric salt which is extractable into benzene (30-45 μg of amine for A = 0.3).

The N-alkylvinylamines obtained by condensing primary and secondary amines with acetaldehyde react with chloranil to give blue vinylamino-substituted quinones. Only secondary amines react when 2,3-dichloro-5,6-dicyanobenzoquinone is used instead of chloranil (determination range: 11-33 μg).

In alkaline medium, sodium nitroprusside reacts with primary amines in the presence of acetone and with secondary amines in the presence of acetaldehyde. The mechanism of these reactions is unknown, and they are poorly sensitive.

Only primary amines can be determined through various reactions involving the two hydrogens bonded to the nitrogen atom.

Condensation with p-dimethylaminocinnamaldehyde in nonaqueous medium affords orange Schiff bases (A = 0.3 for 7.5 to 20 μg). The Schiff bases formed with salicyl-aldehyde can be extracted into 1-hexanol as copper complexes, and the copper is then determined with bis(2-hydroxyethyl)dithiocarbamic acid.

With p-nitrobenzenediazonium ion, primary amines give diazoamino derivatives which are developed in alkaline medium (7-16 μg).

Condensation with succinic dialdehyde affords N-substituted pyrroles which can be developed either with p-dimethylaminobenzaldehyde or with flavylium perchlorate. These reactions are highly sensitive; determinations are possible within the range 2-9 μg.

Through the Hantzsch reaction, primary amines can be condensed either with acetylacetone and formaldehyde, but the determinations are poorly sensitive (250-1300 μg for A = 0.3), or with diethyl acetonedicarboxylate and formaldehyde. This latter reaction gives A = 0.3 for 9-50 μg of compound, but the absorption maximum is located in the ultraviolet range.

Primary amines can also be determined with ascorbic acid in dimethylformamide solution. The mechanism of the reaction is not well known. Two absorption maxima are observed, at 390 nm and 530 nm, making determinations possible within the ranges 13-27 and 38-83 μg, respectively.

Secondary aliphatic amines give substituted o-quinones upon reaction with pyrocatechol in the presence of silver oxide. The violet color developed allows determination of 24-95 μg of the amines tested (A = 0.3).

Tertiary amines catalyze the reaction between nitroethane and phenyl isocyanate to give dimethylfuroxane but, as a rule, below a lower limit of concentration the amine does not initiate the reaction. Some alkaloids can also be determined.

Tertiary amines are substituted for the halogen atom in the para position to the diazonium group of 2,4,6-trichlorobenzenediazonium fluoborate, thus giving a yellow quaternary ammonium salt. The yellow color developed allows the determination of 24-180 μg of amine.

Tertiary amines and quaternary ammonium compounds both react with a mixture of cis-aconitic and acetic anhydrides. The red-violet color of the internal salt of aconitic anhydride so formed permits the determination of 14-48 μg of these compounds.

B. Fluorimetry

Fluorescent zinc oxinate is formed when a neutral aqueous solution of 8-hydroxyquinoline and zinc sulfate is exposed to alkaline vapors. All classes of amines react, but the determination is only semiquantitative. It can be performed through microdiffusion or with a test paper.

With 4-chloro-7-nitrobenzofurazan, primary and secondary amines give fluorescent 4-alkyl or dialkyl-amino-7-nitrobenzofurazans. (Reading 50 for 0.4-4.3 μg of amine.)

The N-alkylaminonaphthoquinones obtained from primary or secondary alkylamines and 1,2-naphthoquinone-4-sulfonic acid can be extracted into methylene chloride (see Colorimetry). The quinone is then reduced to the corresponding fluorescent aminodihydroxynaphthalene, allowing determinations within the range 1.3-8.5 μg.

Primary and secondary amines react with dansyl chloride to give highly fluorescent sulfonamides, making determinations possible with 0.7-3 μg samples.

The N-substituted 2,6-dimethyl-3,5-diacetyl-1,4-dihydropyridines obtained by condensing primary amines with acetylacetone and formaldehyde (see Colorimetry) are fluorescent and allow determinations at levels of 1-2 μg.

According to unknown mechanisms, primary amines give fluorescent species with ninhydrin and phenyl-acetaldehyde (0.9-7.6 μg for reading 50) and with o-diacetylbenzene (determination range: 0.21-0.44 μg).

Tertiary amines develop a green fluorescence when treated with a mixture of aconitic acid and acetic anhydride. Reading 50 is given by 2.3-4.5 μg samples.

II. ALIPHATIC AMINES AND QUATERNARY AMMONIUM COMPOUNDS

A. 1,3,5-Trinitrobenzene and Nitromethane

Principle (1)

In the presence of a basic compound, 1,3,5-trinitrobenzene reacts with nitromethane to give a Meisenheimer-like σ complex: red color.

$$O_2N\text{-}C_6H_3(NO_2)_2 + CH_3{-}NO_2 + {\geq}N \longrightarrow \left[H,\ CH_2{-}NO_2;\ O_2N,\ NO_2,\ NO_2\ \sigma\text{-complex}\right]^{-}\ {\geq}\overset{+}{N}H$$

Reagent

An 0.2% solution of 1,3,5-trinitrobenzene in nitromethane. Keep in the dark.

Procedure

A. Free amines. To 0.2 ml of sample solution in ethanol, add 0.2 ml of reagent, mix, let stand at room temperature for 90 sec, and dilute with 3 ml of nitromethane. Read at 565 nm.

B. Hydrohalides and halides. To 0.2 ml of sample solution in ethanol, add approximately 5 mg of silver oxide and 0.2 ml of reagent. Shake all the tubes simultaneously for 2 min with an even oscillating motion, and dilute with 3 ml of nitromethane. Because the solid phase settles out very readily, no filtration is necessary. Decant, and read immediately at 565 nm.

Results

	A = 0.3 (1 cm cell) Sample, μg
A. Free amines	
n-Butylamine	6.8
Di-n-butylamine	9.0
Triethylamine	7.5
B. Hydrochlorides and chlorides	
Methylamine	5.2
Ethylamine	8.0
Di-n-butylamine	12.5
Dibenzylamine	25.0
Trimethylamine	13.0
Triethylamine	10.5
Promethazine	30.0
Acetylcholine	18.0
Benzyltrimethylammonium (iodide)	25.0

Guanidine derivatives also react. A = 0.3 is given by 22 μg of chlorguanide or 7 μg of guanidine hydrochloride. With this latter compound, results are poorly reproducible.

Notes

1. The same procedure allows the colorimetric determination of various alkaloids. For instance, A = 0.3 is given by 45 μg of emetine hydrochloride or 50 μg of morphine hydrochloride.

2. Arylamines are almost unreactive (A = 0.05 for 2 mg of aniline). Pyridine reacts weakly (A = 0.2 for 2 mg), and the color development proceeds for several hours.

3. This method may be bound up with the Janovsky reaction, see Chapter 16, Section VI.

III. ALIPHATIC AMINES

A. Silver Nitrate and Manganese Sulfate

Principle (2)

Practically colorless when neutral, a test paper impregnated with a mixture of silver nitrate and manganese sulfate turns to gray when exposed to vapors of ammonia or of an aliphatic amine.

$$MnSO_4 + 2\,AgNO_3 + 4(\geqslant N) + 2\,H_2O \longrightarrow$$
$$MnO_2 + 2\,Ag + H_2SO_4(\geqslant N)_2 + 2\,HNO_3(\geqslant N)$$

Reagent

Mix 500 ml of 0.1 M aqueous solution of silver nitrate with 500 ml of 0.1 M aqueous solution of manganese sulfate. To the stirred mixture, add dropwise 0.1 N sodium hydroxide. A black precipitate develops which redissolves by stirring. Proceed with the addition until the precipitate persists (about 16 ml of the alkaline solution). Filter, and collect the filtrate, which must be completely limpid and colorless.

Immerse in it Whatman No. 1 paper strips for 15 min, then dry them at room temperature in the dark, protected against acid and alkaline vapors. The test paper so obtained must be colorless.

Procedure

Use a 30-50 ml cylindrical vial with a tight-fitting polyethylene snap cap of about 1 inch in diameter, and insert inside the cap a disk of the test paper cut to the appropriate size. Pipet 1-2 ml of aqueous solution of the amine into the vial, add 0.5 ml of approximately 10 N sodium hydroxide without shaking, allowing the solution to run down the wall of the vial; stopper with the cap, and mix with a circular motion, avoiding projections of droplets on the paper. Heat at 50° for 4 hr. The gray color of the paper can be compared with standards obtained from known amounts of the amine, either visually or spectrophotometrically with the aid of a reflectance accessory.

Results

	Lower limit of detection, μg
Methylamine	4
Ethylamine	6
Isopropylamine	4
n-Butylamine	15
Isobutylamine	15
sec-Butylamine	10
tert-Butylamine	10
Benzylamine	15
Cyclohexylamine	10
Dimethylamine	8
Diethylamine	10
Di-n-butylamine	20
Triethylamine	14

Notes

1. It is obvious that the procedure may be applied only to amines which develop a sufficiently strong alkaline reaction and are at least slightly volatile at 50°.

2. The formation of a black precipitate by the reaction of ammonia or of an inorganic base with a mixture of silver nitrate and a manganese salt was observed as far back as 1837 (3), and this property has been applied to the detection of ammonium salts in drug preparations (4). On the other hand, its extreme sensitivity allows the detection of 0.2 mg of manganese in 1 liter of solution (5).

3. A similar reaction is given by a mixture of silver nitrate and ferrous sulfate (see the following method).

B. Silver Nitrate and Ferrous Sulfate

Principle

Free amines react with a solution of silver nitrate and ferrous sulfate to give a black colloidal precipitate which allows their spectrophotometric determination.

$$2\,FeSO_4 + 2\,AgNO_3 + 6(\geqslant N) + 3\,H_2O \longrightarrow$$
$$Fe_2O_3 + 2\,Ag + 2\,H_2SO_4(\geqslant N)_2 + 2\,HNO_3(\geqslant N)$$

Reagent

Dissolve 0.050 g of ammonium ferrous sulfate hexahydrate (Mohr salt) in 30 ml of water. Immediately before use, add 10 ml of 0.1% aqueous solution of silver nitrate.

Procedure

To 2.5 ml of reagent, add 2 ml of aqueous solution of the amine, mix immediately, and let stand at room temperature for 15 min. Read at 450 nm.

Results

	A = 0.3 (1 cm cell) Sample, μg
Ethylamine	28
n-Propylamine	32
n-Butylamine	42
Benzylamine	70
Diethylamine	45
Di-n-propylamine	67
Di-n-butylamine	74
Triethylamine	57
Tri-n-propylamine	216

A = 0.18 is given by 400 μg of tri-n-butylamine.

Because of the high sensitivity of this reaction, the presence of trace amounts of alkaline vapors in the atmosphere of the laboratory may lead to discrepant results.

Note

The reaction, being due to the basic character of the amine solution, can therefore be extended to all compounds which develop a sufficiently strong alkaline reaction. Aromatic amines do not react.

C. Potassium Thiocyanate and Cobaltous Nitrate

Principle (6,7)

Reaction with a mixture of potassium thiocyanate and cobaltous nitrate in the solid state to give a complex soluble in chloroform: blue color.

$$\text{Amine} + \text{KSCN} + \text{Co(NO}_3)_2 \rightarrow (\text{Amine})_2[\text{Co(SCN)}_4]$$

Procedure

To 5-10 ml of sample solution in chloroform, add approximately 1 g of potassium thiocyanate and 1 g of cobaltous nitrate hexahydrate. Shake for 2 min in a glass-stoppered flask, let stand for 5 min, and filter. Read at 625 nm.

Results

	A = 0.3 (1 cm cell) Sample, μg/ml
Triethylamine	60
Ephedrine	225
1,1-Dimethyl-2-phenylethylamine	120

Beer's law is not followed beyond A = 0.2 with n-propylamine (up to 50 μg/ml), and diethylamine (up to 60 μg/ml).

Notes

1. Low molecular weight amines afford irregular results.

2. The reaction can be applied to various alkaloids. For instance, A = 0.3 is given by 92 µg/ml of sparteine.

3. With some alkaloids, the reaction was also performed in an aqueous solution of ammonium cobaltothiocyanate, the blue color being extracted into chloroform (8). Another related method used a methanol solution of cobaltothiocyanic acid, $H_2[Co(SCN)_4]$. This was prepared by mixing a solution of 1 mole of cobaltous thiocyanate with a solution of 2 moles of thiocyanic acid, obtained by passing a solution of ammonium thiocyanate in methanol through Amberlite IR 120 (9).

4. The structure of some of the complexes formed has been the subject of physical and chemical analysis (10).

5. The above procedure may be applied to the determination of polyethylene glycols (Carbowax), which rather surprisingly also develop a blue color. An absorbance value of 0.3 is obtained with 500 µg/ml of Carbowax 300 and Carbowax 1500. Simple alcohols do not react at the same concentration; for instance, the reaction is negative with a 1% solution of ethanol in chloroform.

6. Polyethylene glycol monooleate can likewise be determined with a mixture of ammonium thiocyanate and cobaltous nitrate in aqueous solution, the blue color being extractable into chloroform (11).

D. β-Naphthol, Hydrogen Peroxide, and 2,4-Dinitrophenylhydrazine

Principle (12)

Oxidative reaction with β-naphthol to give a

substituted o-naphthoquinone, and development with 2,4-dinitrophenylhydrazine: green color.

Reagents

a. A 2% solution of β-naphthol in ethanol.

b. Dilute 0.1 ml of 30% aqueous solution of hydrogen peroxide to 200 ml with ethanol.

c. An 0.1% solution of cupric nitrate trihydrate in ethanol.

d. An 1:9 mixture of concentrated hydrochloric acid and ethanol.

e. A saturated solution of 2,4-dinitrophenylhydrazine in ethanol.

f. A 1:4 mixture of aqueous ammonium hydroxide (d = 0.92, 22° Bé) and ethanol.

Procedure

To 0.6 ml of reagent *a*, add 0.1 ml of reagent *b*, 0.5 ml of sample solution in ethanol, and 0.1 ml of reagent *c*. Heat at 60° for 10 min, cool for 2 min in a water bath, add 0.1 ml of reagent *d*, 1 ml of reagent *e*, and let stand for 2 min. Add 0.2 ml of reagent *f* and 4 ml of dimethylformamide. Read immediately at 630 nm.

Results

	A = 0.3 (1 cm cell) Sample, μg
n-Propylamine	9.5
n-Butylamine	10
Benzylamine	30
Di-n-propylamine	21
Di-n-butylamine	21
Triethylamine	15
Tri-n-butylamine	25

The reaction is negative with 200 μg of benzhydrylamine.

Notes

1. This method was derived from a series of papers published by Brackman and Havinga (13), who established that the oxidation of β-naphthol with molecular oxygen in a solution of morpholine and copper acetate in methanol gives 4-morpholino-1,2-naphthoquinone (I). This condensation is probably accompanied by various side reactions, and it is obvious that the same mechanism cannot be taken into account for tertiary amines.

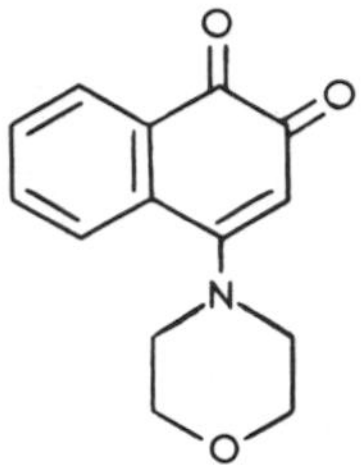

(I)

2. Compare with the determination of secondary aliphatic amines with pyrocatechol and silver oxide (Section VII.C).

3. The same reaction allows the determination of N-heterocyclic derivatives (Chapter 13, Section II.B).

IV. PRIMARY AND SECONDARY ALIPHATIC AMINES

A. 1-Chloro-2,4-dinitrobenzene

Principle (14)

Reaction with 1-chloro-2,4-dinitrobenzene to give a N-alkyl-2,4-dinitroaniline: yellow color.

$$R(R')NH + 1\text{-}Cl\text{-}2,4\text{-}(NO_2)_2C_6H_3 \longrightarrow 1\text{-}(RR'N)\text{-}2,4\text{-}(NO_2)_2C_6H_3$$

Reagent

An 0.5% solution of 1-chloro-2,4-dinitrobenzene in 50% aqueous n-propanol. Keep protected against light.

Procedure

To 5 ml of sample solution in 50% aqueous solution of n-propanol, add 1 ml of reagent, place in a cold water bath, and let stand at room temperature for 30 min. Then gradually raise the bath to the boil over a period of 30 min, and heat at 100° for 30 min. Cool to room temperature, and dilute to 10 ml with 50% aqueous n-propanol. Read at 450 nm.

Results

	A = 0.3 (1 cm cell) Sample, μg
n-Propylamine	560
n-Butylamine	460
Diethylamine	1300
Di-n-propylamine	1600

Ethylamine, diisopropylamine, and diisobutylamine do not react.

Note

1-Chloro-2,4-dinitrobenzene may form ethers with phenols: morphine was gravimetrically determined as 2,4-dinitrophenylmorphine (15).

B. 1-Fluoro-2,4-dinitrobenzene (In Acetone-Water)

Principle (16)

Reaction with 1-fluoro-2,4-dinitrobenzene, to give an N-alkyl-2,4-dinitro aniline: yellow color.

$$R(R')NH + \text{1-fluoro-2,4-dinitrobenzene} \longrightarrow R(R')N\text{-2,4-dinitrobenzene}$$

Reagents

a. Prepare a 1.3% solution of 1-fluoro-2,4-dinitrobenzene in acetone. Immediately before use, dilute 1 ml of this solution to 10 ml with a 2.5% aqueous solution of sodium tetraborate decahydrate.

b. A 1:99 mixture of concentrated hydrochloric acid and dioxane or dimethylformamide.

Procedure

To 0.2 ml of neutral aqueous solution of the amine, add 0.2 ml of reagent a. Heat at 65° for 20 min, cool, add 4 ml of reagent b, and read. When dimethylformamide is used instead of dioxane the absorption maximum shifts about 10 nm towards higher wavelengths.

Results

	λ Max, nm		A = 0.3 (1 cm cell) Sample, μg	
	1	2	1	2
Ethylamine	350	360	5.6	5.7
n-Propylamine	358	366	6.9	5.7
Diethylamine	375	395	10	10.8
Di-n-propylamine	395	395	18.5	14.4
Glycine	355	355	7	6.7

1 = Dioxane

2 = Dimethylformamide

Note

Development of aromatic nitro compounds, see Chapter 16, Section II.

C. 1-Fluoro-2,4-dinitrobenzene (In Ethanol)

Principle (17)

Reaction with 1-fluoro-2,4-dinitrobenzene to give a N-alkyl-2,4-dinitroaniline, which is extracted into cyclohexane after hydrolysis of the excess reagent into 2,4-dinitrophenol: yellow-orange color.

Reagents

a. A 1.2% solution of 1-fluoro-2,4-dinitrobenzene in ethanol.

b. An 0.2 N sodium hydroxide solution in 60% aqueous dioxane.

Procedure

To 0.1 ml of neutral aqueous solution of the amine (if necessary, neutralize with hydrochloric acid), add 0.05 ml of reagent *a*, and 0.1 ml of 0.1 M aqueous solution of sodium bicarbonate. Heat at 60° for 20 min, add 0.4 ml of reagent *b*, and maintain heating at 60° for 60 min more. Dilute to 10 ml with water, extract the color into 10 ml of cyclohexane, and read.

Results

	λ Max, nm	A = 0.3 (1 cm cell) Sample, μg
Ethylamine	330 - 390	12.5 - 30
n-Propylamine	330 - 390	13.5 - 38
Diethylamine	350	17
Di-n-propylamine	360	26

Primary and secondary arylamines can also be determined (Chapter 5, Section III.A).

D. Sodium 1,2-Naphthoquinone-4-sulfonate

Principle (18)

Replacement of the sulfonate group of the naphthoquinone sulfonic acid by an amino group to give a N-alkylaminonaphthoquinone: orange color.

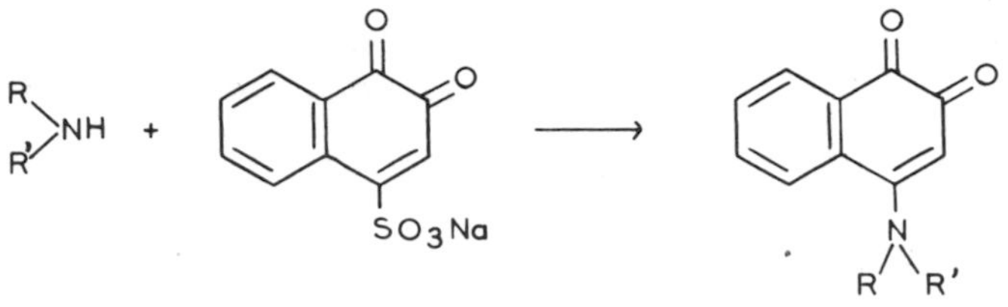

Reagents

a. An 0.5% aqueous solution of sodium 1,2-naphthoquinone-4-sulfonate. Stable for a few hours at 0°, protected against light.

b. A 10% aqueous solution of sodium thiosulfate pentahydrate.

Procedure

To 2 ml of aqueous solution of the amine hydrochloride or of the free amine, neutralized with hydrochloric acid, add 1 ml of 0.01 N sodium hydroxide and 0.5 ml of reagent a. Heat at 50° for 5 min, protected against light, cool for 2 min in ice water, add 0.5 ml of 0.5% acetic acid and 1 ml of reagent b. Mix, and read at 460 nm (primary amines) or 480 nm (secondary amines).

Results

	A = 0.3 (1 cm cell) Sample, μg
Ethylamine (hydrochloride)	36
n-Propylamine	25
n-Butylamine	38
Benzylamine[a]	53
Dimethylamine (hydrochloride)	25
Diethylamine (hydrochloride)	36
Di-n-propylamine	31
Di-n-butylamine	47

[a]Read at 470 nm.

Notes

1. Comparison with absorbances given by the N-alkylaminonaphthoquinones synthesized from n-propylamine and di-n-propylamine evidences an 83% recovery for the analytical reaction with these two amines.

2. The reaction can also be applied to the determination of amino acids, with reading at 470 nm. An absorbance value of 0.3 is given by 36 μg of glycine, 66 μg of phenylalanine, and 42 μg of cysteine. These acids can be determined in the presence of amines in amounts up to 60 μg. Solely the color due to the latter is extractable into methylene chloride. After the condensation and cooling in ice water are performed as described in the procedure, extract the N-alkylamino-naphthoquinones with two 5 ml portions of methylene chloride by shaking moderately for 5 sec each time. Let stand for 2 min, pipet 3 ml of the aqueous layer, and add to it the dilute acetic acid and reagent b as described for amines. Compare with a calibration curve prepared under the same conditions with pure amino acid.

3. The condensate obtained with aniline immediately precipitates out, thus making spectrophotometric determination impossible.

4. Enhancement of the sensitivity, see following method.

5. Reaction between amino compounds and 1,2-naphthoquinone-4-sulfonic acid, see Chapter 16, Section VII.

E. Sodium 1,2-Naphthoquinone-4-sulfonate and 2,4-Dinitrophenylhydrazine

Principle (19)

Replacement of the sulfonate group of the naphthoquinonesulfonic acid by an amino group (preceding method), extraction of the N-alkylaminonaphthoquinone formed into methylene chloride, and development with 2,4-dinitrophenylhydrazine: orange color.

Reagents

a. An 0.5% aqueous solution of sodium 1,2-naphthoquinone-4-sulfonate. Stable for a few hours at 0°, protected against light.

b. To 0.20 g of 2,4-dinitrophenylhydrazine, add 20 ml of ethanol and 0.5 ml of concentrated sulfuric acid. When the crystals are dissolved, dilute to 100 ml with ethanol.

Procedure

To 1 ml of aqueous solution of the amine hydrochloride or of the free amine, neutralized with hydrochloric acid, add 0.2 ml of 0.02 N sodium hydroxide and 0.2 ml of reagent *a*. Heat at 50° for 5 min, and cool for 2 min in ice water. Perform these operations in the dark. Transfer the solution into a 10 ml separatory funnel containing 2 ml of methylene chloride, rinse the tube with 1 ml of water, and pour the rinsing into the funnel. Shake immediately for about 5 sec, let stand for 2 min, and collect the whole bulk of the organic layer. Add to it 3 ml of reagent *b*, fit the tubes with air condensers, and heat at 50° for 45 min for primary amines, 10 min for secondary amines. Chill for a few minutes in ice water, and, still at 0°, add slowly 0.5 ml of concentrated sulfuric acid. Mix, and read at 485-490 nm for primary amines, 490-495 nm for secondary amines.

Results

	A = 0.3 (1 cm cell) Sample, μg
Ethylamine (hydrochloride)	3.2
n-Propylamine	2.0
n-Butylamine	2.3
Benzylamine	7.4
Dimethylamine (hydrochloride)	4.7
Diethylamine (hydrochloride)	5.7
Di-n-propylamine	5.0
Di-n-butylamine	8.0

Primary aromatic amines can also be determined (Chapter 5, Section IV.H).

Notes

1. Since their naphthoquinone derivatives are not extractable into methylene chloride amino acids cannot be determined with this procedure.

2. Reaction between amino compounds and 1,2-naphthoquinone-4-sulfonic acid, see Chapter 16, Section VII.

F. p-Nitrophenylazobenzoyl Chloride and Benzyltrimethylammonium Hydroxide

Principle (20)

Reaction with the acid chloride to give the corresponding amide, and without prior elimination of the excess reagent development in alkaline nonaqueous medium: orange-red color.

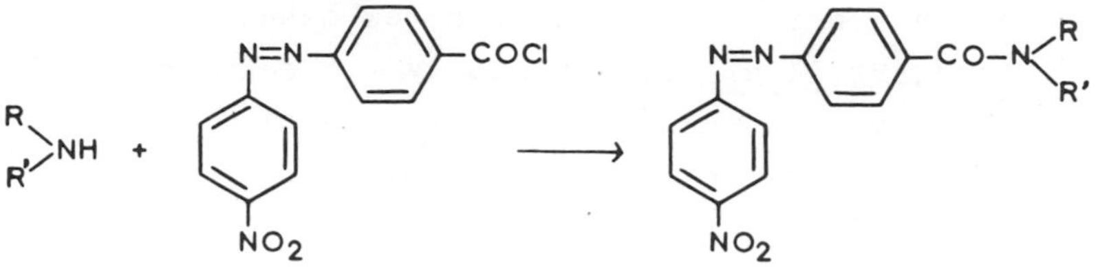

Reagents

a. An 0.01% solution of p-nitrophenylazobenzoyl chloride in dioxane.

b. Dilute 2 ml of 30% aqueous solution of hydrogen peroxide to 100 ml with dimethyl sulfoxide.

c. Immediately before use, dilute 1 ml of 40% solution of benzyltrimethylammonium hydroxide in methanol (see Chapter 17, Section III) to 100 ml with dimethyl sulfoxide.

Procedure

To 2 ml of sample solution in dioxane, add 1 ml of reagent a, and let stand at room temperature for 15 min, protected against light. Add 0.5 ml of reagent b, 2 ml of reagent c, mix, and read.

Results

	λ Max, nm	A = 0.3 (1 cm cell) Sample, μg
Ethylamine	535	6.2
n-Propylamine	530	9.2
n-Butylamine	530	8.0
Benzylamine	540	15.7
Diethylamine	530	21.6
Di-n-propylamine	530	31.0
Di-n-butylamine	530	42.5
Dibenzylamine	530	57.0

Primary and secondary arylamines can also be determined (Chapter 5, Section III.B).

Notes

1. Purification of dimethyl sulfoxide, see Chapter 17, Section IV.

2. Development of aromatic nitro compounds, see Chapter 16, Section II.

G. Carbon Disulfide and Blue Tetrazolium

Principle (21)

Reaction with carbon disulfide, to give a dithiocarbamate which reduces Blue Tetrazolium: pink-violet color.

$$RR'NH + CS_2 \longrightarrow RR'N{-}C({=}S){-}SH$$

$$RR'N{-}C({=}S){-}SH + \text{Blue Tetrazolium} \rightarrow \text{Formazan}$$

(See Chapter 11, Section III.B)

Reagent

An 0.2% solution of Blue Tetrazolium in ethanol.

Procedure

A. Secondary amines. Immerse the tubes in a water bath at 27-28°, and preheat the reagent and solvents to the same temperature.

To 1 ml of aqueous solution of the amine, add 1 ml of carbon disulfide, allowing the solvent to run down the wall of the tube; let stand for 1 min, and add 5 ml of benzene. Shake vigorously, add 1 ml of reagent, and 0.3 ml of 5 N sodium hydroxide. Let stand for 10 min. Perform all these operations in the dark. Filter through a small cotton plug, and let stand at room temperature for 1 hr 30 min. Read at 540 nm.

B. Primary amines. Operate at room temperature and in daylight, otherwise following the above procedure, but let stand for only 10 min before reading at 540 nm.

Results

	A = 0.3 (1 cm cell) Sample, μg
n-Propylamine	6.2
n-Butylamine	10.0
Benzylamine	19.0
Diethylamine	7.3
Di-n-propylamine	8.0
Di-n-butylamine	2.5
Ephedrine	37.0

The reaction is practically negative with tertiary amines.

Notes

1. Colorimetric determination of secondary amines as dithiocarbamates, see Section VII.B.

2. Tetrazolium salts and formazans, see Chapter 16, Section VIII.

H. Acetaldehyde and Chloranil

Principle (22,23)

The N-alkylvinylamine obtained by condensing the amine with acetaldehyde reacts with chloranil to give a vinylamino-substituted quinone: blue color.

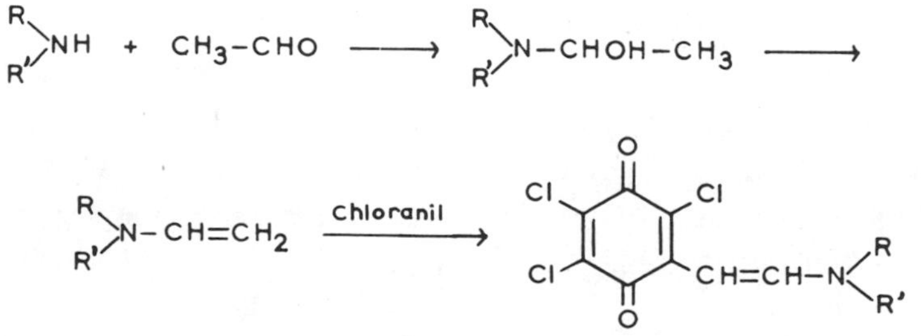

$$RR'NH + CH_3-CHO \longrightarrow RR'N-CHOH-CH_3 \longrightarrow$$

$$RR'N-CH=CH_2 \xrightarrow{\text{Chloranil}} \text{(trichloro-}p\text{-benzoquinone)}-CH=CH-NRR'$$

Reagents

a. A 2% solution of acetaldehyde in dioxane.

b. A 1% solution of chloranil in dioxane.

Procedure

To 1 ml of sample solution in dioxane, add 0.1 ml of reagent a and 0.1 ml of reagent b. Let stand for 5 min at room temperature, dilute with 4 ml of dioxane, and read at 650 nm.

Results

	A = 0.3 (1 cm cell) Sample, μg
n-Propylamine	154
n-Butylamine	180
Benzylamine	245
Diethylamine	31
Di-n-propylamine	40
Di-n-butylamine	49
N-Methylbenzylamine	46
Dibenzylamine	95

Tertiary amines do not react, but enhance the color given by secondary amines.

Notes

1. According to the same principle, only secondary amines react with 2,3-dichloro-5,6-dicyanobenzoquinone instead of chloranil (Section VII.D).

2. Chloranil was also used for the colorimetric determination of primary aromatic amines (24).

V. SECONDARY AND TERTIARY ALIPHATIC AMINES

A. Picric Acid

Principle (25)

Formation of the picrate of the amine, extractable into chloroform: gives a yellow color, but the absorption maximum is located in the ultraviolet range.

Reagents

a. Prepare a buffer by dissolving 300 g of monosodium phosphate dihydrate and 9 g of sodium hydroxide in 750 ml of water.

b. An 0.1% aqueous solution of picric acid.

Procedure

Into a 25 ml separatory funnel, introduce 10 ml of sample solution in chloroform, 2 ml of buffer a, and 1 ml of reagent b. Shake for 1 min, allow the phases to separate, insert a small cotton plug into the stem of the funnel, and collect the organic layer. Read at 346 nm.

Results

	A = 0.3 (1 cm cell) Sample, μg
Di-n-propylamine	33.5
Di-n-butylamine	26.5
Trimethylamine	93
Triethylamine	20.5
Tri-n-propylamine	28.5
Tri-n-butylamine	36.5
N,N-Dimethylbenzylamine	30
Diphenhydramine (hydrochloride)	60

This reaction is also positive with various alkaloids.

	A = 0.3 (1 cm cell) Sample, μg
Ajmaline	69.5
Codeine	60
Pholcodine	79.5
Strychnine	64
Yohimbine	78.5

Notes

1. The absorption curve shows a maximum at 346 nm and a zero slope within the range 385-400 nm. Hence,

readings can also be made at 390 nm, however, with a reduced sensitivity (A = 0.17 for 33.5 μg of tri-n-propylamine, instead of 0.3 at 346 nm).

2. Primary aliphatic amines react but very weakly (A = 0.05 for 570 μg of n-propylamine and 0.18 for 500 μg of benzylamine) or not at all (ethanolamine).

3. For low molecular weight amines, the amount of water has a marked influence on the sensitivity. For example, when the amine or its hydrochloride is dissolved in 2 ml of water, the solution added to 10 ml of chloroform, and the reaction then performed as described under Procedure, A = 0.3 is given by 82 μg of di-n-propylamine or 550 μg of trimethylamine (instead of 33.5 and 93 μg, respectively, under the usual conditions).

4. Using an acid dye as reagent and a chlorinated solvent as extractant, ion-pair extraction has been applied to numerous determinations; among the first examples were the estimation of quinine with eosine (26), and of quaternary compounds with Bromophenol Blue (27). Picric acid, Methyl Orange, Bromothymol Blue, erythrosine, dithizon, Tropaeolin OO, Eriochrome Black, and 3',3",5',5"-tetrabromophenolphthalein ethyl ester (28) are some of the most commonly used dyes.

5. The structure of the species formed may depend upon the operative conditions: concentration of the components, pH of the aqueous phase (29). The color can be modified or intensified upon acidification, or reextracted into a buffer. The presence, on the molecule of the compound to be determined, of hydrophilic substituents such as -OH or -COOH often prevents extraction of the salt into the organic phase.

6. Extraction with benzene may increase the selectivity of the reaction, which then depends upon parameters such as the polarities of the amine and of the dye.

Relatively polar amines, such as arylamines, diamines, and many alkaloids react readily with Bromothymol Blue, poorly or not at all with Bromocresol Purple. Contrariwise, this second dye behaves like the first one with weakly polar amines, such as amine derivatives of diphenylmethane or phenothiazine.

Therefore, compounds may be determined in binary mixtures: The less polar amine is estimated with Bromophenol Blue, the sum of the two amines is given by Bromocresol Purple. Mixtures of diphenhydramine and ephedrine or codeine, prochlorperazine and amphetamine, promethazine and ephedrine were so analyzed (30).

7. According to the same principle, acid compounds can be colorimetrically determined with basic dyes. p-Toluenesulfonic, camphosulfonic, and bromocamphosulfonic acids were estimated with fuchsin (31), chlorotoluenesulfonic acid with Acridine Orange, Rhodamine S, or Chrysoidine (32), and phenylbutazone, which develops a slightly acidic reaction, with Gentian Violet (33).

8. Ion-pair extraction also permits fluorimetric determinations: amitriptyline and allied compounds were so estimated with anthracene-2-sulfonate (34).

9. The same principle has been applied in this book to the determination of secondary and tertiary aliphatic amines with Tropaeolin OO (next Section) and of tertiary aliphatic amines with erythrosine (Section VIII.A).

B. Tropaeolin OO

Principle (35)

Formation of the amine salt of Tropaeolin OO, extractable into methylene chloride: violet color.

Reagents

a. An 0.05% aqueous solution of Tropaeolin OO.

b. A 1% solution of concentrated sulfuric acid in methanol.

Procedure

To 4 ml of aqueous solution of the amine hydrochloride or of the free amine neutralized with hydrochloric acid, add 1 ml of reagent a, mix, let stand for 5 min, and extract the color with 10 ml of methylene chloride. Pipet 5 ml of the organic layer, and add to it 0.5 ml of reagent b. Mix, and read at 545 nm.

Results

	A = 0.3 (1 cm cell) Sample, μg
Di-n-propylamine (hydrochloride)	43
Di-n-butylamine (hydrochloride)	17
Trimethylamine (hydrochloride)	206
Triethylamine (hydrochloride)	24
Tri-n-propylamine (hydrochloride)	18
Tri-n-butylamine (hydrochloride)	21
N,N-Dimethylbenzylamine	18

The reaction is also positive with benzyltrimethylammonium iodide (A = 0.3 for 60 μg) and with alkaloids. Simple primary amines (methylamine, ethylamine, n-propylamine, n-butylamine) as well as dimethylamine and choline chloride do not react. Benzylamine gives a very faint color.

Notes

1. Tropaeolin 00, obtained by coupling diazotized p-sulfanilic acid with diphenylamine, is the sodium salt of p-[(p-anilinophenyl)azo]benzenesulfonic acid.

$$NaO_3S-C_6H_4-N{=}N-C_6H_4-NH-C_6H_5$$

2. Because the reaction depends upon the solubility of the amine salt of Tropaeolin 00 in the organic solvent, the substituents on the amine compound may greatly affect the results. For instance, when extracting with chloroform, the reaction is positive with ephedrine (II), whereas the organic layer remains colorless with phenylephrine (III), which possesses a phenolic hydroxyl (7).

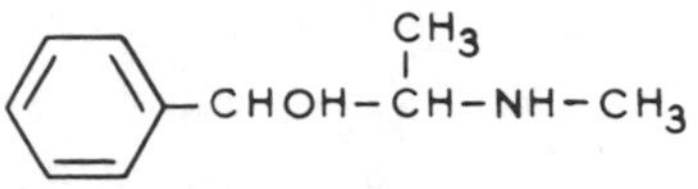

$C_6H_5-CHOH-CH(CH_3)-NH-CH_3$

(II)

$HO-C_6H_4-CHOH-CH_2-NH-CH_3$

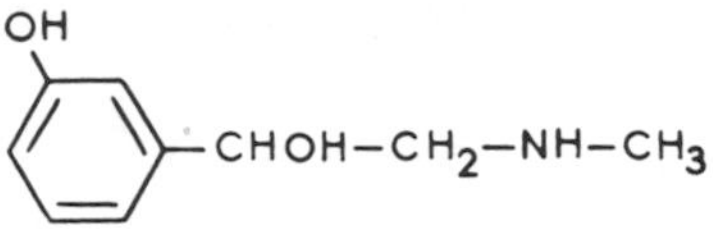

(III)

3. Ion-pair extraction, see preceding Section, Notes 4-9.

VI. PRIMARY ALIPHATIC AMINES

A. Sodium Nitroprusside and Acetone

Principle (36)

Reaction with nitroprusside ion and acetone in alkaline medium: red-violet color.

Reagents

a. A 10% aqueous solution of sodium nitroprusside.

b. A saturated aqueous solution of sodium tetraborate decahydrate.

Procedure

To 3 ml of neutral aqueous solution of the amine, add 0.5 ml of acetone, 0.3 ml of reagent a, and 2 ml of reagent b. Let stand for 1 hr in the dark, and read at 550 nm.

Results

	A = 0.3 (1 cm cell) Sample, μg
Ethylamine	582
n-Propylamine	1750
Benzylamine	530

Notes

1. Under the same conditions, secondary and tertiary amines give only a yellow-orange color. Amino acids behave like primary amines, but the sensitivity is weaker.

2. Nitroprusside also allows the colorimetric determination of secondary aliphatic amines (Section VII.A), α-methylene ketones (Chapter 8, Section VIII.A), imidazoline derivatives (Chapter 13, Section VI.A), and Δ^4-3-ketosteroids (Chapter 15, Section X.A).

3. Relations between nitroprusside and pentacyanoferrates, see Chapter 5, Section IV.A, Note 1.

4. Other analytical uses of nitroprusside, see reference (37).

B. 1-Fluoro-2,4-dinitrobenzene (In Dimethylformamide) and Benzyltrimethylammonium Hydroxide

Principle (38)

Reaction with 1-fluoro-2,4-dinitrobenzene to give a N-alkyl-2,4-dinitroaniline, and without prior elimination of the excess reagent development in alkaline nonaqueous medium: violet color.

$$R{-}NH_2 + F{-}C_6H_3(NO_2)_2 \longrightarrow R{-}NH{-}C_6H_3(NO_2)_2$$

Reagents

a. An 0.025% solution of 1-fluoro-2,4-dinitrobenzene in dimethylformamide.

b. Dilute 1 ml of 40% solution of benzyltrimethylammonium hydroxide in methanol (see Chapter 17, Section III) to 100 ml with dimethylformamide. Prepare freshly.

Procedure

To 1 ml of sample solution in dimethylformamide, add 0.5 ml of reagent a, and let stand at room temperature for 10 min. Add 4.5 ml of reagent b and 0.1 ml of nitromethane. After 10 min, read at 570 nm.

Results

	A = 0.3 (1 cm cell) Sample, μg
Ethylamine	5.7
n-Propylamine	7.6
Benzylamine	15.2
Cyclohexamine	17.0
Monoethanolamine	7.8

Secondary aliphatic amines react weakly. The reaction is negative with tertiary aliphatic amines and aromatic amines.

Notes

1. Upon addition of nitromethane, the color turns to violet and the sensitivity is enhanced. This is probably due to a Janovsky-like reaction (see Chapter 16, Section VI).

2. Development of aromatic nitro compounds, see Chapter 16, Section II.

C. p-Dimethylaminocinnamaldehyde

Principle (39)

Reaction with p-dimethylaminocinnamaldehyde in nonaqueous medium to give a Schiff base: orange color.

$$R{-}NH_2 + OHC{-}CH{=}CH{-}C_6H_4{-}N(CH_3)_2 \longrightarrow R{-}N{=}CH{-}CH{=}CH{-}C_6H_4{-}N(CH_3)_2$$

Reagent

An 0.05% solution of p-dimethylaminocinnamaldehyde in nitromethane.

Procedure

To 2 ml of sample solution in nitromethane (or in 0.1 ml of ethanol diluted with 2 ml of nitromethane), add 2 ml of reagent, and heat at 100° for 25 min. Cool in a water bath, dilute to 10 ml with nitromethane, and read at 475 nm.

Results

	A = 0.3 (1 cm cell) Sample, μg
Ethylamine	12
n-Propylamine	7.5
Benzylamine	20
Cyclohexylamine	18
Monoethanolamine	11

Note

p-Dimethylaminocinnamaldehyde also allows the colorimetric determination of m-diphenols (Chapter 3, Section V.B), primary aromatic amines (Chapter 5, Section IV.C), and indole derivatives (Chapter 13, Section III.D).

D. Salicylaldehyde, Diethanolamine, and Carbon Disulfide

Principle (40)

The Schiff base (IV) formed by reacting salicylaldehyde with the amine is extracted into 1-hexanol as its copper complex. The copper is then developed with bis(2-hydroxyethyl)dithiocarbamic acid (V): yellow color.

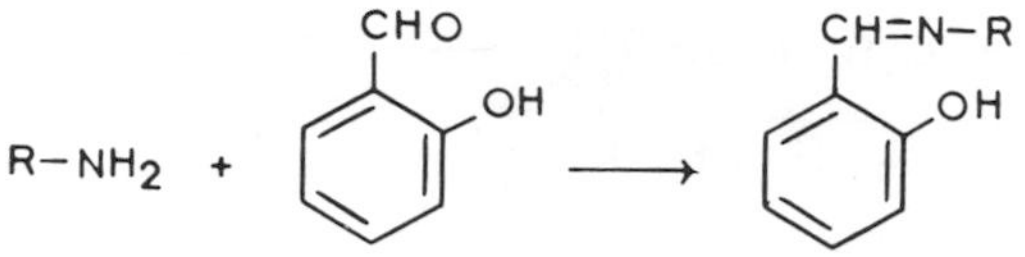

$R{-}NH_2$ + (salicylaldehyde: CHO, OH) $\longrightarrow$ (CH=N−R, OH)

(IV)

$(HOCH_2{-}CH_2)_2N{-}C(=S){-}SH$

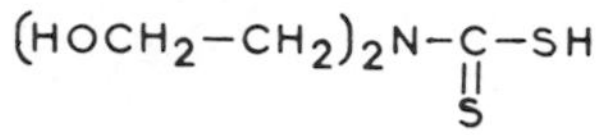

(V)

Reagents

a. Dissolve 15 ml of triethanolamine, 0.5 ml of salicylaldehyde, and 0.25 g of cupric chloride dihydrate in water to give 100 ml of solution.

b. Solution of bis(2-hydroxyethyl)dithiocarbamic acid: Mix equal volumes of 2% solution of carbon disulfide in methanol, and of 5% solution of diethanolamine in methanol.

Procedure

To 2 ml of reagent a, add the aqueous solution of the amine, and dilute to 10 ml with water. Mix thoroughly, let stand for 15 min, add 15 ml of 1-hexanol, shake vigorously for 3 min, and allow the phases to separate. Pipet 5 ml of the organic layer, add to it 5 ml of reagent b, and dilute to 25 ml with methanol. Read at 430 nm.

Results

	A = 0.3 (1 cm cell) Sample, μg
n-Propylamine	270
n-Butylamine	300
Benzylamine	373

Notes

1. The reaction is not quantitative with primary aromatic amines, ethylenediamine, diethylenetriamine, and with primary aliphatic amines substituted at position 2, but propylenediamine and 2-ethylhexylamine react normally.

2. Ammonia interferes, whereas secondary and tertiary amines do not impede the reaction at levels up to 0.5 g.

3. As a matter of fact, this method, which is based on the estimation of the copper extracted as copper-salicylaldehydeimine complex, is an indirect method of determination of the amine.

4. Determination of secondary aliphatic amines as cupric dithiocarbamates, see Section VII.B.

E. p-Nitrobenzenediazonium Fluoborate

Principle (41)

Condensation with p-nitrobenzenediazonium ion to give a diazoamino derivative, developed in alkaline medium: orange color.

$$R{-}NH_2 + \overset{+}{N}{\equiv}N{-}C_6H_4{-}NO_2 \longrightarrow R{-}NH{-}N{=}N{-}C_6H_4{-}NO_2 \longrightarrow R{-}N{=}N{-}NH{-}C_6H_4{-}NO_2 \xrightarrow{OH^-} R{-}N{=}N{-}N{=}C_6H_4{=}NO_2^-$$

Reagents

a. Buffer solution: Dissolve 4.08 g of monopotassium phosphate and 1.60 g of sodium tetraborate decahydrate in 70 ml of water, add 12.7 ml of 2.5 N sodium hydroxide, and dilute to 100 ml with water.

b. An 0.2% solution of p-nitrobenzenediazonium fluoborate in 0.5 N hydrochloric acid.

Procedure

To 1 ml of neutral aqueous solution of the amine, add 5 ml of buffer a, 1 ml of reagent b, and let stand for 20 min. Add 3 ml of 2.5 N sodium hydroxide, let stand for 1 hr, and read at 510 nm.

Results

	A = 0.3 (1 cm cell) Sample, μg
Methylamine (hydrochloride)	7
Ethylamine	7.5
Benzylamine (hydrochloride)	14
Cyclohexylamine (hydrochloride)	9
Histamine (dihydrochloride)	16.5

The reaction is negative with 100 μg of benzhydrylamine.

Notes

1. Proposed at first for the determination of amphetamine (42,43), this reaction also allows the colorimetric determination of amino acids (see Chapter 9, Section VI.C).

2. Other determinations with p-nitrobenzenediazonium fluoborate, see Chapter 3, Section II.H, Note 3.

3. Diazo coupling, see Chapter 16, Section I.

F. 2,5-Diethoxytetrahydrofuran and p-Dimethylaminobenzaldehyde

Principle (44)

The N-substituted pyrrole obtained by reacting the amine with succinic dialdehyde is developed with p-dimethylaminobenzaldehyde: pink-orange to pink-violet color.

Succinic dialdehyde is formed in the reaction medium by hydrolytic cleavage of 2,5-diethoxytetrahydrofuran.

$$H_5C_2O\text{-}(C_4H_6O)\text{-}OC_2H_5 \xrightarrow{CH_3\text{-}CO_2H} OHC\text{-}CH_2\text{-}CH_2\text{-}CHO \rightleftharpoons HO\text{-}HC{=}CH\text{-}CH{=}CH\text{-}OH$$

$$HO\text{-}HC{=}CH\text{-}CH{=}CH\text{-}OH + R\text{-}NH_2 \longrightarrow C_4H_4N\text{-}R \xrightarrow[H^+]{(H_3C)_2N\,C_6H_4\,CHO} C_4H_3N^+(R){=}CH\text{-}C_6H_4\text{-}N(CH_3)_2$$

Reagents

a. Dilute 1 ml of 2,5-diethoxytetrahydrofuran to 100 ml with glacial acetic acid.

b. A 2% solution of p-dimethylaminobenzaldehyde in a 1:19 mixture of concentrated hydrochloric acid and glacial acetic acid.

Procedure

To 1 ml of sample solution in 50% ethanol, add 1 ml of reagent a, and heat at 80° for T_1 min (see Results). Cool to room temperature in a water bath, add 3 ml of reagent b, let stand at room temperature for T_2 min (see Results), and read at 558 nm.

Results

	T_1, min	T_2, min	A = 0.3 (1 cm cell) Sample, μg
n-Propylamine	20	2	1.5
n-Butylamine	20	2	1.9
Benzylamine	2	5	2.8

Notes

1. This method also allows the colorimetric determination of primary aromatic amines (Chapter 5, Section IV.E) and of amino acids (Chapter 9, Section VI.D).

2. According to the same principle, amphetamine salts were determined in various drug preparations (45).

3. The pyrrole ring can also be developed with flavylium perchlorate (see following method).

4. Analytical uses of p-dimethylaminobenzaldehyde, see Chapter 5, Section IV.B, Notes 1 and 2.

G. 2,5-Diethoxytetrahydrofuran and Flavylium Perchlorate

Principle (46,47)

The N-substituted pyrrole obtained by reacting the amine with succinic dialdehyde (see preceding method) is developed with flavylium perchlorate: yellow to orange color.

Reagents

a. Dilute 1 ml of 2,5-diethoxytetrahydrofuran to 100 ml with glacial acetic acid.

b. An 0.05% solution of flavylium perchlorate (see Chapter 17, Section I.D) in glacial acetic acid.

Procedure

To 1 ml of sample solution in 50% ethanol, add 1 ml of reagent *a*, and heat at 80° for T_1 min (see Results). Cool to room temperature in a water bath, add 3 ml of reagent *b*, and heat at 60° for T_2 min (see Results), protected against light. Let cool, and read.

Results

	T_1, min	T_2, min	λ Max, nm	A = 0.3 (1 cm cell) Sample, μg
n-Propylamine	20	5	470	7.5
n-Butylamine	20	5	470	6.0
Benzylamine	5	10	460	9.0

Notes

1. The method also allows the colorimetric determination of primary aromatic amines (Chapter 5, Section IV.F) and of amino acids (Chapter 9, Section VI.E).

2. Analytical uses of flavylium perchlorate, see Chapter 5, Section II.B, Notes 1-5.

H. Acetylacetone and Formaldehyde

Principle (48)

Reaction with acetylacetone and formaldehyde to give

an N-substituted 2,6-dimethyl-3,5-diacetyl-1,4-dihydropyridine: yellow-orange color.

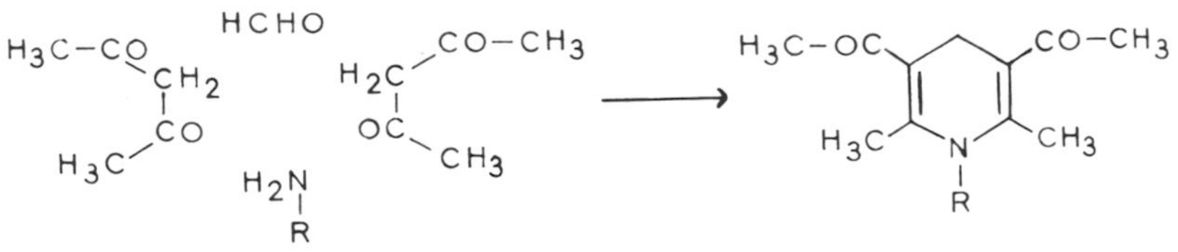

$$H_3C{-}CO{-}CH_2{-}CO{-}CH_3 + HCHO + H_2N{-}R + H_2C(CO{-}CH_3)(CO{-}CH_3) \longrightarrow$$ N-R 2,6-dimethyl-3,5-di($CO{-}CH_3$)-1,4-dihydropyridine

<u>Reagent</u>

Immediately before use, mix 2 ml of acetylacetone and 1 ml of 35% aqueous solution of formaldehyde, and dilute to 50 ml with pyridine.

<u>Procedure</u>

To 5 ml of nearly neutral aqueous solution of the amine, add 1 ml of reagent, heat at 100° for 10 min, cool in a water bath, and dilute to 10 ml with pyridine. Read at 425 nm.

<u>Results</u>

	A = 0.3 (1 cm cell) Sample, μg
Ethylamine	250
n-Propylamine	330
n-Butylamine	460
Benzylamine	1300

The reaction is negative with 2 mg of benzhydrylamine.

<u>Notes</u>

1. This method is based on the Hantzsch reaction, which allows miscellaneous colorimetric and fluorimetric determinations (see Chapter 16, Section IX).

2. Compare with fluorimetric determination, Section XII.A.

I. Diethyl Acetonedicarboxylate and Formaldehyde

Principle

Reaction with diethyl acetonedicarboxylate and formaldehyde. The formation of a N-substituted 3,5-dicarbethoxy-1,4-dihydro-2,6-pyridinediacetic acid diethyl ester may be assumed: yellow color. An absorption maximum is also observed in the ultraviolet range.

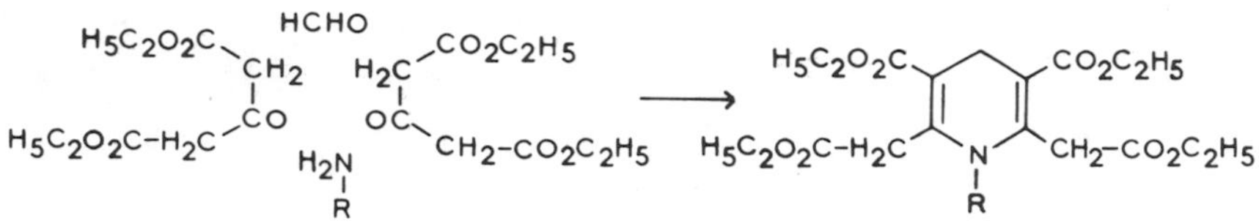

$H_5C_2O_2C$–CH_2–CO–H_2C–$CO_2C_2H_5$ + HCHO + H_2N–R + $H_5C_2O_2C$–H_2C–CO–CH_2–$CO_2C_2H_5$ ⟶ N-R-1,4-dihydropyridine with 3,5-$CO_2C_2H_5$ and 2,6-CH_2–$CO_2C_2H_5$ ($H_5C_2O_2C$–H_2C / CH_2–$CO_2C_2H_5$)

Reagent

Immediately before use, mix 1 ml of diethyl acetonedicarboxylate and 2 ml of 35% aqueous formaldehyde, and dilute to 50 ml with pyridine.

Procedure

To 1 ml of amine hydrochloride solution in pyridine, add 1 ml of reagent, and let stand at 0° for 4 hr, protected against light. Add 3 ml of pyridine, and read at 385 or 308 nm.

When operating with a free amine, neutralize with 1 N or 0.1 N hydrochloric acid, then dilute with pyridine. The sample should be of such a size that after dilution, the amount of water so introduced is lower than or equal to 1%.

Results

	A = 0.3 (1 cm cell) Sample, µg	
	385 nm	308 nm
Ethylamine (hydrochloride)	117	27
n-Propylamine	30.5	8.7
n-Butylamine (hydrochloride)	50	15

	A = 0.3 (1 cm cell) Sample, μg	
	385 nm	308 nm
Benzylamine (hydrochloride)[a]	75	52
Cyclohexylamine	60	17

[a]The sample was dissolved in 3 ml of ethanol, and diluted to 50 ml with pyridine. At 308 nm, Beer's law is not followed, but a linear relationship holds above A = 0.05.

Notes

1. Aniline can also be determined at 385 nm, A = 0.3 for 45 μg. The reaction is almost negative at 308 nm.

2. Glycine, p-aminobenzoic acid, and sulfanilamide also react, but Beer's law is not followed.

3. Inversely, a mixture of diethyl acetonedicarboxylate and ammonium acetate allows the determination of aliphatic aldehydes in the ultraviolet range (Chapter 8, Section IV.E).

4. This method is based on the Hantzsch reaction, which allows miscellaneous colorimetric and fluorimetric determinations (see Chapter 16, Section IX).

J. Ascorbic Acid

Principle (49)

Reaction with ascorbic acid in dimethylformamide solution: red color.

The mechanism of the reaction is as yet unknown, but as a matter of fact the true reagent is dehydroascorbic acid formed by air-oxidation of the ascorbic acid (see Chapter 9, Section VI.H, Note 3).

Reagent

Dissolve 0.050 g of ascorbic acid in 0.5 ml of water, and dilute to 50 ml with dimethylformamide.

Procedure

To 3 ml of sample solution in dimethylformamide, add 2 ml of reagent, heat at 100° for 10 min, and cool to 15-20° in a water bath. Read at 390 nm or 530 nm.

When operating with amine hydrochloride, dissolve in 10 volumes of water, then dilute with dimethylformamide. The sample should be of such a size that after suitable dilution, the amount of water so introduced is negligible (about 1%).

Results

	A = 0.3 (1 cm cell) Sample, μg	
	390 nm	530 nm
Ammonium chloride	22	80
Methylamine (hydrochloride)	14	38
Ethylamine (hydrochloride)	21	67
n-Propylamine	13	44
n-Butylamine	15	50
Benzylamine	27	83

Notes

1. The reaction is selective for ammonia and primary aliphatic amines of the type $R - CH_2 - NH_2$, whereas secondary and tertiary amines do not react. When reading at 530 nm, benzylamine may be determined in the presence of 20-fold larger amounts of di- or tribenzylamine. Di- and tri-n-propylamine interfere in the determination of n-propylamine, when the ratio of the concentrations is higher than 2:1 and 3:1, respectively.

Other amines behave diversely: Benzhydrylamine and glucosamine can be read at 530 nm only, and the reaction is poorly sensitive (A = 0.3 for 1000 μg and 400 μg, respectively). Isopropylamine and α,α-dimethylphenetylamine (phentermine) develop a fluorescent yellow color, and Beer's law is not followed at any wavelength. Aniline gives similar results, but no fluorescence is observed.

2. Application of ascorbic acid to the colorimetric determination of amino acids, see Chapter 9, Section VI.H.

VII. SECONDARY ALIPHATIC AMINES

A. Sodium Nitroprusside and Acetaldehyde

Principle (50,51)

Reaction with nitroprusside ion and acetaldehyde in alkaline medium: blue-violet color.

Reagents

a. Buffer for pH 9.8: Dissolve 9.53 g of sodium tetraborate decahydrate and 5.30 g of anhydrous sodium carbonate in water, and dilute to 1 liter with water.

b. A 1% solution of sodium nitroprusside in a 10% aqueous solution of acetaldehyde.

Procedure

To 20 ml of the buffer a, add 5 ml of reagent b then 1 ml of aqueous solution of the amine. Let stand for the given length of time (see Results), and read.

Results

	Reaction time, min	λ Max, nm	A = 0.3 (1 cm cell) Sample, μg
Dimethylamine	20	565	520
Di-n-propylamine	6 - 8	580	1300
Di-n-butylamine	6 - 8	580	2250
Diisobutylamine[a]	6 - 8	590	600
Di-n-amylamine[b]	6 - 8	580	1950

[a]Linear relationship holds down to 400 μg.

[b]Dissolve in dilute ethanol.

The reaction is also positive with pyrrolidine (A = 0.3 for 190 μg at 575 nm after 15 min), piperidine (A = 0.3 for 750 μg at 565 nm after 60 min) and proline

(A = 0.3 for 860 μg at 585 nm after 10 min). Diisopropylamine, di-sec-butylamine, ammonium salts, primary and tertiary amines do not react.

Notes

1. Nitroprusside also allows the colorimetric determination of primary aliphatic amines (Section VI.A), α-methylene ketones (Chapter 8, Section VIII.A), imidazoline derivatives (Chapter 13, Section VI.A), and Δ^4-3-ketosteroids (Chapter 15, Section X.A).

2. Relations between nitroprusside and pentacyanoferrates, see Chapter 5, Section IV.A, Note 1.

B. Carbon Disulfide and Cupric Chloride

Principle (52)

Reaction with carbon disulfide and cupric ion to give the corresponding cupric dithiocarbamate: yellow color.

$$RR'NH + CS_2 \longrightarrow RR'N{-}C({=}S){-}SH \xrightarrow{Cu^{2+}} [RR'N{-}C({=}S){-}S]_2Cu$$

Reagents

a. A mixture of 35 ml of carbon disulfide, 25 ml of pyridine, and 65 ml of 2-propanol.

b. Dissolve 0.1 g of cupric chloride dihydrate in 250 ml of water, and dilute to 500 ml with pyridine.

Procedure

To 1 ml of aqueous solution of the amine, add 4 ml of reagent a and 2 ml of reagent b, and let stand for 15 min. Add 3 ml of 10% acetic acid and 3 ml of benzene, shake, and let stand for 15 min. Pipet 4 ml of the supernatant phase, add to it 0.5 ml of 2-propanol, and let stand for 30 min. Read at 438 nm.

Results

	A = 0.3 (1 cm cell) Sample, μg
Diethylamine	30
Di-n-propylamine	40
Di-n-butylamine	45

Piperidine also reacts (A = 0.3 for 33 μg).

The reaction is weakly positive with primary amines, negative with tertiary amines.

Notes

1. The dithiocarbamate formed may also be determined with Blue Tetrazolium (Section IV.G).

2. Bis(2-hydroxyethyl)dithiocarbamic acid allows the colorimetric determination of copper, hence an indirect determination of primary amines (Section VI.D).

C. Pyrocatechol and Silver Oxide

Principle (53,54)

Oxidative reaction with pyrocatechol in the presence of silver oxide to give substituted o-benzoquinones (VI) and (VII): violet color.

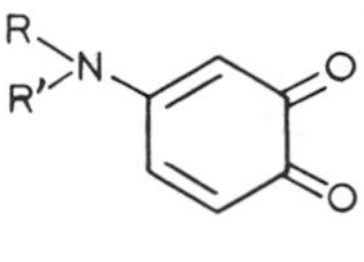

(VI) 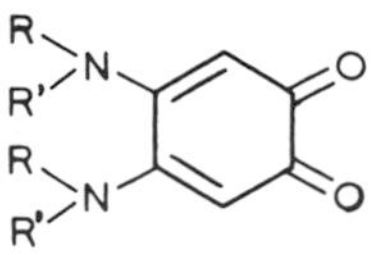 (VII)

Reagent

An 0.1% solution of pyrocatechol in acetone.

Procedure

A. Free amines. To 0.5 ml of sample solution in acetone, add 1 ml of reagent and approximately 2 mg of silver oxide. Shake for a few seconds , and let stand for 10 min. Add 2 ml of acetone, decant (no filtration is necessary), and read at 510 nm.

B. Hydrochlorides. To 0.5 ml of aqueous solution of the amine hydrochloride, add 1 ml of reagent, 2 ml of acetone, and approximately 2 mg of silver oxide. Shake for a few seconds, and let stand at room temperature for 1 hr. Decant (no filtration is necessary), and read at 510 nm.

Results

	A = 0.3 (1 cm cell) Sample, μg
A. Free amines	
Diethylamine	24
Di-n-butylamine	45
Diethanolamine	60
B. Hydrochlorides	
Dimethylamine	30
Diethylamine	28
Di-n-butylamine	53
Dibenzylamine	75
Diethanolamine	95
Adrenalone	70
Ephedrine	90

Contrariwise to appreciable amounts of primary amines, tertiary amines do not interfere.

D. 2,3-Dichloro-5,6-dicyanobenzoquinone

Principle (55)

See the similar reaction with chloranil, Section IV.H: violet color.

Reagents

a. An 0.055% solution of 2,3-dichloro-5,6-dicyanobenzoquinone in dioxane.

b. A 5% solution of acetaldehyde in dioxane.

Procedure

To 1 ml of sample solution in dioxane, add 0.1 ml of reagent a and 0.1 ml of reagent b. Heat at 70° for 3 min, cool for 1 min in ice water, and add 2.5 ml of dioxane. Read at 560 nm.

Results

	A = 0.3 (1 cm cell) Sample, μg
Di-n-propylamine	11
Di-n-butylamine	14
N-Methylbenzylamine	13
Dibenzylamine	33

The reaction is also positive with piperidine (A = 0.3 for 18 μg) and morpholine (A = 0.3 for 11 μg). With pyrrolidine, Beer's law is not followed (A = 0.25 for 22.5 μg).

Primary and tertiary amines practically do not react, and only primary amines interfere.

VIII. TERTIARY ALIPHATIC AMINES

A. Erythrosine Sodium

Principle

Formation of the amine salt of erythrosine, extractable into methylene chloride: pink color.

Reagent

An 0.1% aqueous solution of erythrosine sodium.

Procedure

To 5 ml of aqueous solution of the amine hydrochloride, add 1 ml of reagent, and extract the color with 5 ml of methylene chloride. As soon as the decantation is achieved, read at 530 nm.

Results

	A = 0.3 (1 cm cell) Sample, μg
Hydrochlorides	
Trimethylamine	3500
Triethylamine	185
Tri-n-propylamine	18.8
Tri-n-butylamine	12.5

Notes

1. Erythrosine is the sodium (or potassium) salt of 2',4',5',7',-tetraiodofluorescein:

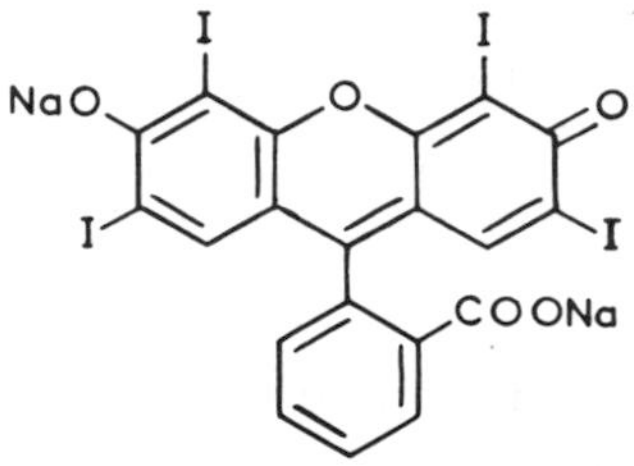

2. Colorimetric determinations by ion-pair extraction, see Section V.A, Notes 4-9.

B. Nitroethane and Phenyl Isocyanate

Principle (56,57)

Tertiary aliphatic amines catalyze the reaction between nitroethane and phenyl isocyanate to give dimethylfuroxane: yellow color.

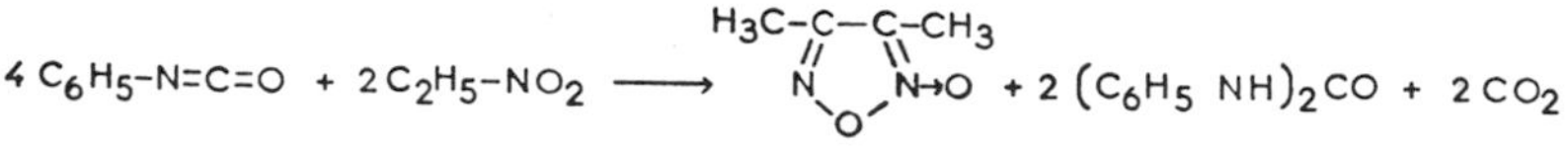

$$4\,C_6H_5{-}N{=}C{=}O + 2\,C_2H_5{-}NO_2 \longrightarrow \underset{\text{(3,4-dimethylfuroxane: } H_3C{-}C{=}N{-}O{-}N(\rightarrow O){=}C{-}CH_3\text{)}}{} + 2\,(C_6H_5NH)_2CO + 2\,CO_2$$

Reagent

A 1% solution of phenyl isocyanate in acetonitrile, freshly prepared, and kept protected from moisture.

Procedure

To 0.5 ml of sample solution in acetonitrile, add 0.5 ml of nitroethane and 2 ml of reagent. Heat at 70° for 15 min, and cool in a water bath. Read at 390 nm.

Results

As a rule, below a lower limit of concentration the amine does not initiate the reaction. The linear relationship holds above this limit. Beer's law is followed from the origin only with cinchonine and strychnine.

	Lower limit of determination, μg	A = 0.3 (1 cm cell) Sample, μg
Trimethylamine	4	8
Triethylamine	10	11.6
Tri-n-propylamine	20	41
Tri-n-butylamine	12	43
N,N-Dimethylbenzylamine	15	77

This reaction is also positive with alkaloids containing a saturated heterocyclic tertiary nitrogen atom.

	Lower limit of determination, μg	A = 0.3 (1 cm cell) Sample, μg
Cinchonine	0	100
Corynanthine	200	1040
Reserpine	160	630
Strychnine	0	137

Cinchonine, corynanthine, and strychnine, being insoluble in acetonitrile, can be dissolved in the minimum amount of ethanol and the solution diluted with acetonitrile.

No linear relationship is observed with hordenine (A = 0.13 for 16 μg) and with procaine (A = 0.3 for 19 μg).

Notes

1. In order to explain the catalytic effect of the tertiary amine, Mukaiyama and Oshino (57) proposed the following mechanism:

$$CH_3-CH_2-NO_2 + R_3N \rightleftharpoons (CH_3\ CH=NO_2)^- + R_3\overset{+}{N}H$$

$$C_6H_5-N=C=O + (CH_3-CH=NO_2)^- \rightleftharpoons (C_6H_5-N-\overset{O}{\overset{\|}{C}}-O-\overset{O}{\overset{\uparrow}{N}}=CH-CH_3)^-$$

$$\xrightleftharpoons{H^+} C_6H_5-NH-\overset{O}{\overset{\|}{C}}-O-\overset{O}{\overset{\uparrow}{N}}=CH-CH_3$$

(VIII)

By proton transfer, the adduct (VIII) decomposes according to

$$(VIII) \longrightarrow CH_3-C\equiv N\rightarrow O + C_6H_5-NH-COOH$$

$$2\,CH_3-C\equiv N\rightarrow O \longrightarrow H_3C-C=C-CH_3\ (\text{ring: } N-O-N\rightarrow O)$$

and the carbamic acid decomposes according to

$$C_6H_5-NH-COOH \longrightarrow C_6H_5-NH_2 + CO_2$$

$$C_6H_5-NH_2 \xrightarrow{C_6H_5-N=C=O} C_6H_5-NH-CO-NH-C_6H_5$$

2. The true structure of furoxanes does not seem as yet definitely established. They may be obtained through various bimolecular heterocyclizations (58), for instance, from two molecules of nitrile oxide or from oximes through an oxidative reaction.

C. 2,4,6-Trichlorobenzenediazonium Fluoborate

Principle (59,60)

Replacement of the halogen atom in the position para to the diazonium group, to give a quaternary ammonium salt: yellow color.

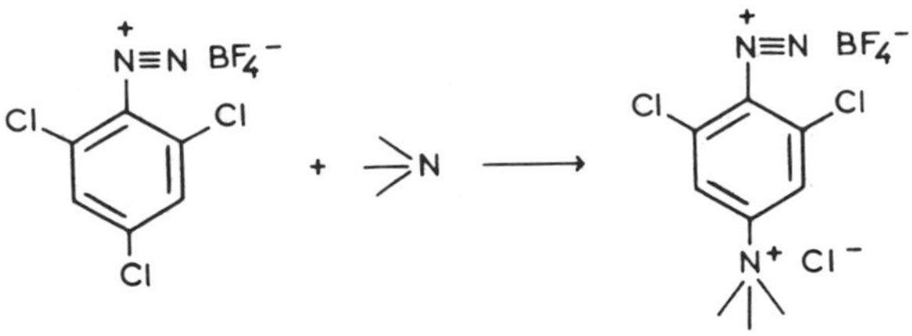

Reagent

Dissolve 0.100 g of 2,4,6-trichlorobenzenediazonium fluroborate (Chapter 17, Section I.H), and 0.020 g of cupric acetate monohydrate in 100 ml of acetone.

Procedure

To 1 ml of reagent, add 4 ml of sample solution in acetone. Let stand at room temperature for 2 hr, protected against bright light, and read at 400 nm.

Results

	A = 0.3 (1 cm cell) Sample, μg
Triethylamine	24
Tri-n-propylamine	53
Tri-n-butylamine	61
N,N-Dimethylbenzylamine	86
Amodiaquin	180

Tribenzylamine does not react.

Primary amines develop a very faint color; secondary amines do not react but interfere in the determination of tertiary amines.

Notes

1. The reaction can be applied to tertiary aromatic amines (A = 0.3 for 11 μg of N,N-dimethylaniline).

2. Quinoline, quinine, and strychnine also react, but Beer's law is not followed.

3. The compound obtained by reacting 2,4,6-trichlorobenzenediazonium fluoborate with trimethylamine allows the colorimetric (Chapter 2, Section IV.B) and fluorimetric

(Chapter 2, Section XIII.A) determination of primary alcohols.

4. Diazo compounds, see Chapter 16, Section I.

D. cis-Aconitic and Acetic Anhydrides

<u>Principle</u> (<u>61</u>)

Formation of an internal salt of aconitic anhydride: red-violet color.

$$\text{cis-aconitic anhydride } (CH_2\text{—}COOH) + R_3N \longrightarrow \text{internal salt } (CH_2\text{—}COOH,\ O^-,\ O\overset{+}{N}R_3)$$

<u>Reagent</u>

Dissolve 0.25 g of cis-aconitic anhydride in 40 ml of acetic anhydride, and dilute to 100 ml with toluene. (Stable for 24 hr, protected from moisture.)

<u>Procedure</u>

To 2 ml of sample solution in toluene, add 1 ml of reagent, heat at 100° for exactly 15 sec, and let stand for 15 min in a water bath at 20°. Add 5 ml of toluene, let stand for 15 min, and read at 500 nm or 525 nm.

The reaction is very sensitive to impurities. Toluene should be washed with concentrated sulfuric acid and distilled.

<u>Results</u>

	A = 0.3 (1 cm cell) Sample, µg	
	500 nm	525 nm
Triethylamine	17	14
Tri-n-propylamine	22	19
Tri-n-butylamine	27	23
2-Diethylaminoethanol	34	22

Quaternary ammonium salts also react (see next method).

Notes

1. The reaction was applied to various alkaloids (62,63).

2. Primary and secondary amines do not react, presumably because amide formation removes the amines.

3. Tertiary aliphatic amines and quaternary ammonium salts also develop a color with a mixture of citric acid and acetic anhydride (64), and the reaction may be applied to determinations (65).

4. Aconitic acid (IX) is normally prepared from citric acid by dehydration with sulfuric acid. It has then the trans configuration. On heating, it is readily decarboxylated to itaconic acid (X). On dehydration with acetic anhydride, it gives cis-aconitic anhydride. Distillation of this anhydride affords citraconic anhydride which, by hydration, gives citraconic or methylmaleic acid (XI). The trans isomer, mesaconic or methylfumaric acid (XII), is formed by heating citraconic acid with dilute nitric acid.

$(HOOC-CH_2)(HOOC)C=C(COOH)(H)$

(IX)

$HOOC-CH_2-C(COOH)=CH_2$

(X)

$(HOOC)(H)C=C(COOH)(CH_3)$

(XI)

$(HOOC)(H)C=C(CH_3)(COOH)$

(XII)

5. According to Groth and Dahlen (66), upon suitable action of acetic anhydride aconitic acid also affords α,γ-anhydroaconitic acid (XIII).

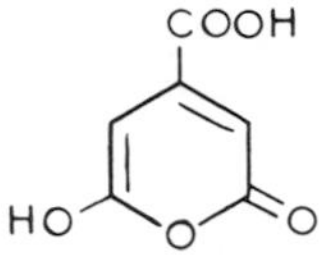

(XIII)

Compound (XIII), dissolved in acetic anhydride, develops a violet color in the presence of tertiary amines, as well as a fluorescence which allows the fluorimetric determinations of this class of compounds (see Section XIII.A).

IX. QUATERNARY AMMONIUM COMPOUNDS

A. cis-Aconitic and Acetic Anhydrides

Principle and Procedure

See the preceding method.

When the compound is only slightly soluble in toluene, dissolve it in a 1:1 mixture of acetic anhydride and toluene, or in the minimum amount of methanol, and dilute with the mixture. Read at 500 or 545 nm.

Results

	A = 0.3 (1 cm cell) Sample, μg	
	500 nm	545 nm
Acetylcholine chloride	48	34
Benzyltrimethylammonium chloride	22	17

Note

Quaternary benzyltrialkylammonium salts in aqueous solution can be characterized with a test paper impregnated with Methylene Blue tetraiodomercurate. Due to anion exchange, the dyestuff diffuses into the solution when such a salt is present (67).

X. ALIPHATIC AMINES (F)

A. 8-Hydroxyquinoline Sulfate and Zinc Sulfate

Principle (68)

Fluorescent zinc oxinate is formed when a neutral aqueous solution of 8-hydroxyquinoline and zinc sulfate is exposed to alkaline vapors: yellow-green fluorescence under ultraviolet light. This reaction is semiquantitative.

Reagent

Dissolve 0.70 g of neutral 8-hydroxyquinoline sulfate (Sunoxol) and 0.5 g of zinc sulfate heptahydrate in 100 ml of water. Then add dropwise 0.1 N sodium hydroxide until a turbidity persists, and filter.

Procedure

Microdiffusion device: Use a 100 ml glass-stoppered conical flask with three symmetrical indentations at half its height, so that they can hold by its rim a 2-3 ml thimble beaker in the middle of the flask. A small glass rod is soldered upright to the rim of the beaker, making easy its setting and removal.

Pipet 1 ml of reagent into the beaker, and set it into the flask. Pipet 4 ml of aqueous solution of the amine down to the bottom of the flask, add 3 ml of approximately 10 N sodium hydroxide without shaking, allowing the solution to run down the wall of the flask, stopper the flask, and mix with a circular motion. Let stand at room temperature for 1 hr. Then visually compare under ultraviolet light the fluorescence of the solution in the beaker to standards obtained from known amounts of the amine.

Results

	Lower limit of detection, μg
Ethylamine	210
Diethylamine	50
Triethylamine	45
Benzylamine	325
Cyclohexylamine	280

Notes

1. The reaction may also be performed with a paper strip impregnated with the reagent, as described for the detection of amines with silver nitrate and manganese sulfate (see Section III.A).

2. The zinc complex of 8-hydroxyquinoline is insoluble in water and develops a yellow-green fluorescence. Magnesium and aluminum complexes also develop a fluorescence, but the color is less handsome and of slighter intensity.

3. A saturated aqueous solution of nickel dimethylglyoximate was also proposed for the detection of alkaline compounds. Due to the shifting of the following equilibrium

$$Ni^{2+} + 2\ \text{dimethylglyoxime} \rightleftarrows Ni\ (\text{dimethylglyoxime})_2 + 2H^+$$

a red precipitate is formed (69).

XI. PRIMARY AND SECONDARY ALIPHATIC AMINES (F)

A. 4-Chloro-7-nitrobenzofurazan (7-Chloro-4-nitrobenzoxadiazole)

Principle (70,71)

Reaction with 4-chloro-7-nitrobenzofurazan to give a 4-dialkylamino-7-nitrobenzofurazan: yellow fluorescence.

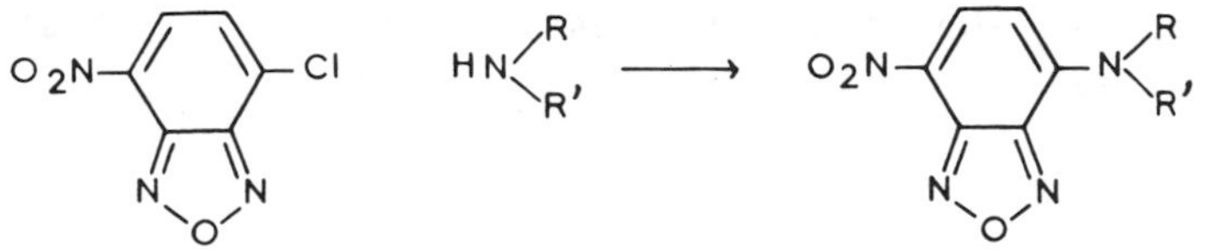

Reagent

An 0.05% solution of 4-chloro-7-nitrobenzofurazan in ethanol. Keep in the dark.

Procedure

To 1 ml of sample solution in ethanol, add 0.75 ml of reagent, and heat at 60° for 30 min, exposed to daylight. Let stand at room temperature for 3 min, and dilute to 4 ml with ethanol. Read at exc: 436 nm; em: 535 nm.

Standard: An aqueous solution of fluorescein sodium.

Results

	Determination limits, μg	Reading 50	
		Sample, μg	Standard, μg/ml
n-Propylamine	0.25 - 1.0	0.41	0.12
n-Butylamine	0.25 - 1.0	0.45	0.12
Benzylamine	0.25 - 1.0	0.41	0.083
Diethylamine	1 - 5	1.9	0.083
Di-n-propylamine	3 - 12	4.3	0.13
Di-n-butylamine	2 - 10	4.0	0.06

Notes

1. Amino acids also develop a fluorescence, but the linear relationship is not followed.

2. The reaction is negative with tertiary aliphatic amines; very weakly positive with aromatic amines, phenols, and mercaptans.

B. Sodium 1,2-Naphthoquinone-4-sulfonate and Potassium Borohydride

Principle (72)

Replacement of the sulfonate group of the naphthoquinonesulfonic acid by an amino group to give an N-alkylaminonaphthoquinone, which is then reduced to the corresponding aminodihydroxynaphthalene: blue fluorescence.

R R'NH + [1,2-naphthoquinone-4-sulfonate, SO_3Na] ⟶ [4-(NRR')-1,2-naphthoquinone] $\xrightarrow{KBH_4}$ [4-(NRR')-1,2-dihydroxynaphthalene]

Reagents

a. An 0.5% aqueous solution of sodium 1,2-naphthoquinone-4-sulfonate. Stable for a few hours at 0°, protected against light.

b. Dissolve 0.050 g of potassium borohydride in 5 ml of 0.02 N sodium hydroxide, and dilute to 100 ml with ethanol.

c. Dilute 0.2 ml of concentrated hydrochloric acid to 100 ml with ethanol.

Procedure

To 1 ml of aqueous solution of the amine hydrochloride or of the free amine neutralized with hydrochloric acid, add 0.2 ml of 0.02 N sodium hydroxide and 0.2 ml of reagent a. Heat at 50° for 5 min, and cool for 2 min in ice water. Perform these operations in the dark. Transfer the solution into a 10 ml separatory funnel containing 2 ml of methylene chloride, rinse the tube with 1 ml of water, and pour the rinsing into the funnel. Shake immediately for about 5 sec, let stand for 2 min, and collect the whole bulk of the organic layer. Add to it 0.1 ml of reagent b, mix, let stand at room temperature for 2 min, and dilute with 2 ml of reagent c. Read at exc: 366 nm; em: 410 nm.

Standard: A solution of quinine sulfate in 0.1 N sulfuric acid.

Results

	Determination limits, µg	Reading 50 Sample, µg	Reading 50 Standard, µg/ml
Methylamine (hydrochloride)	3 - 15	6.6	1.14
Ethylamine (hydrochloride)	1.5 - 7.5	3.3	1.06
n-Butylamine	1.4 - 7.0	3.0	1.25
Dimethylamine (hydrochloride)	0.6 - 3.0	1.32	1.0
Diethylamine (hydrochloride)	0.6 - 3.0	1.32	1.06
Di-n-propylamine	0.6 - 3.0	1.28	1.11
Diisopropylamine	83 - 415	174	0.81
Di-n-butylamine (hydrochloride)	1 - 5	2.25	1.13
N-Methylbenzylamine (hydrochloride)	4 - 20	8.5	0.88

Tertiary amines practically do not react, but interfere in the determination of primary and secondary amines.

Notes

1. Aniline develops a green fluorescence (exc: 366 nm; em: 470 nm) but the linear relationship is not followed. However, according to another procedure based on the same principle, primary aromatic amines can also be determined (Chapter 5, Section V.A).

2. Amino acids cannot be determined, since their naphthoquinone derivatives are not extractable into methylene chloride.

3. The o-diphenol obtained by reducing the quinone derivative is very easily back-oxidized when exposed to

air (73). However, it is fairly stable in the hydrogen-saturated reaction medium, thus allowing accurate determinations.

4. Reaction between amino compounds and 1,2-naphthoquinone-4-sulfonate, see Chapter 16, Section VII.

C. Dansyl Chloride

Principle (74,75)

Reaction with dansyl chloride (1-dimethylaminonaphthalene-5-sulfonyl chloride) to give a sulfonamide: yellow-green fluorescence.

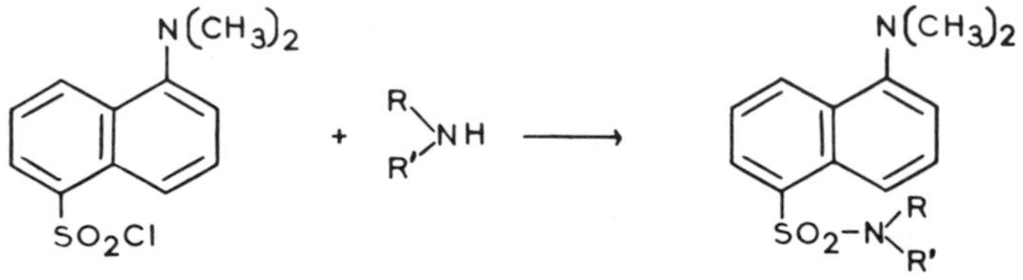

Reagents

a. Dilute 1 ml of 0.01 N ethanol solution of potassium hydroxide to 25 ml with acetonitrile.

b. An 0.05% solution of dansyl chloride in acetone.

Procedure

To 1 ml of sample solution in acetonitrile, add 0.2 ml of reagent *a* and 0.2 ml of reagent *b*, mix, and let stand in the dark at room temperature for 5 min. Add 3 ml of acetone, and read at exc: 366 nm; em: 510 nm.

Standard: A solution of 2,6-dimethyl-3,5-diacetyl-1,4-dihydropyridine in 50% ethanol.

Results

	Determination limits, μg	Reading 50 Sample, μg	Standard, μg/ml
Ethylamine	0.3 - 1.5	0.69	6.15
n-Propylamine	0.4 - 2.0	0.90	6.35
n-Butylamine	0.4 - 2.0	0.86	5.50
Benzylamine	1 - 5	2.05	6.05
Diethylamine	1 - 5	2.3	6.20
Di-n-propylamine	1.2 - 6.0	2.7	6.35
Di-n-butylamine	1.2 - 6.0	2.6	5.45
Dibenzylamine	50 -250	95	4.50

Dibenzylamine reacts but weakly, probably due to steric hindrance. The reaction is practically negative with primary and secondary aromatic amines up to 200 μg.

Notes

1. Dansyl chloride has been used as fluorescent tag in the thin-layer chromatography of proteins, amino acids, and amines. The spots can eventually be scraped from the plate, extracted, and the fluorescence measured (76-79).

2. The herein-described procedure cannot be applied to the determination of amino acids, on account of their poor solubility in the reaction medium. When the reaction is performed in the presence of water, the blank fluorescence becomes too intense to permit any accurate measurement.

XII. PRIMARY ALIPHATIC AMINES (F)

A. Acetylacetone and Formaldehyde

Principle (80,81)

Reaction with acetylacetone and formaldehyde to give an N-substituted 2,6-dimethyl-3,5-diacetyl-1,4-dihydropyridine: yellow-green fluorescence.

$H_3C-CO-CH_2-CO-CH_3$ + HCHO + H_2N-R + $H_2C(CO-CH_3)(OC-CH_3)$ $\longrightarrow$ H_3C-OC, $CO-CH_3$, H_3C, CH_3, N–R (1,4-dihydropyridine ring)

Reagent

Dissolve 0.68 g of sodium acetate trihydrate in about 5 ml of water, add 0.4 ml of acetylacetone and 1 ml of 30% aqueous solution of formaldehyde, and dilute to 15 ml with water. Prepare freshly.

Procedure

To 1 ml of nearly neutral aqueous solution of the amine, add 0.5 ml of reagent, plug the tubes with cotton wool, and heat at 100° for exactly 10 min, protected against bright light. To avoid discrepant results, the water of the bath must be stirred so that all the tubes be strictly at the same temperature. Chill for 3 min in ice water, and add 2.5 ml of water. Read at exc: 405 nm; em: 510 nm.

Standard: A solution of 2,6-dimethyl-3,5-diacetyl-1,4-dihydropyridine in ethanol.

Results

	Determination limits, μg	Reading 50 Sample, μg	Reading 50 Standard, μg/ml
Ethylamine (hydrochloride)	1 - 5	2.05	0.77
n-Propylamine	0.5 - 2.5	1.05	0.79
n-Butylamine (hydrochloride)	1 - 5	2.2	0.84
Benzylamine (hydrochloride)[a]	0.5 - 2.5	1.1	0.82

[a]Read at em: 500 nm.

Notes

1. This method is based on the Hantzsch reaction, which allows miscellaneous colorimetric and fluorimetric determinations (see Chapter 16, Section IX).

2. Application to the fluorimetric determination of α-amino acids, see Chapter 9, Section XI.A.

B. Ninhydrin and Phenylacetaldehyde

Principle (82)

Condensation with ninhydrin and phenylacetaldehyde in buffered medium: green fluorescence.

Reagents

a. A 3% aqueous solution of monosodium phosphate dihydrate adjusted to pH 8 with 2 N sodium hydroxide.

b. An 0.9% aqueous solution of ninhydrin.

c. An 0.12% solution of phenylacetaldehyde in ethanol.

Procedure

To 4 ml of buffer a, add 0.4 ml of reagent b, 0.2 ml of aqueous solution of the amine, and 0.2 ml of reagent c. Mix, heat at 60° for 15 min, cool in ice water for 2 min, let stand for 10 min at room temperature, and read at exc: 395 nm; em: 485 nm. (The same results are obtained when exciting at 405 nm.)

Standard: A solution of 2,6-dimethyl-3,5-diacetyl-1,4-dihydropyridine in ethanol.

Results

	Determination limits, μg	Reading 50	
		Sample, μg	Standard, μg/ml
n-Propylamine	0.48 - 2.4	0.9	0.67
n-Butylamine	0.6 - 3.0	1.08	1.18
Benzylamine	1.2 - 6.0	2.3	2.18
Ethanolamine	0.72 - 3.6	1.44	1.25

	Determination limits, μg	Sample, μg	Reading 50 Standard, μg/ml
Norepinephrine[a]	4 - 20	7.6	0.85
Histamine (hydrochloride)	1.6 - 8.0	2.9	1.68

[a]Dissolve in the minimum amount of 0.1 N sodium hydroxide, neutralize with 0.1 N hydrochloric acid, and dilute suitably.

Notes

1. Secondary and tertiary amines do not react. A slightly positive reaction is given by aniline.

2. The same reaction allows the fluorimetric determination of α-amino acids (Chapter 9, Section XI.C), and of 2-amino-2-deoxy hexoses (Chapter 14, Section XVI.A).

3. Recently, Weigele et al. (83) established that the major fluorescent component obtained from ethylamine has structure (XIV). A minor fluorophor possesses structure (XV). Moreover, they have shown that (XIV) is obtained from (XV) through an oxidative step, in which the excess of ninhydrin can act as the oxidant. From there, a new reagent, Fluorescamine (XVI) was proposed for the fluorimetric assay of primary amines (84), amino acids, peptides, and proteins (85). It reacts with amino compounds to yield fluorophors identical with those of the fluorogenic ninhydrin reaction.

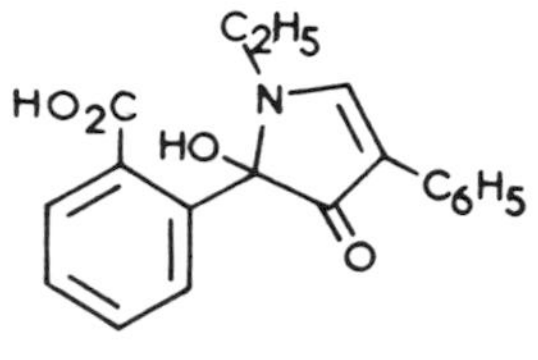

(XIV) 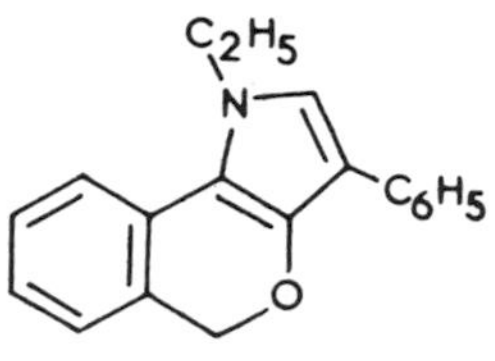 (XV) 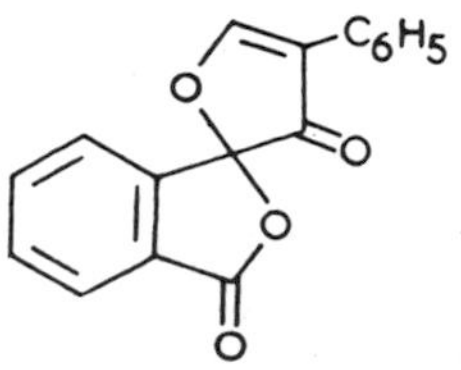 (XVI)

4. Reactions with ninhydrin, see Chapter 16, Section X.

C. o-Diacetylbenzene

Principle (71)

Reaction with o-diacetylbenzene in alkaline medium: blue fluorescence.

Reagents

a. Clark and Lubs buffer for pH 8.6, see Appendix I.C.

b. An 0.25% solution of o-diacetylbenzene in ethanol.

Procedure

To 1 ml of sample solution in ethanol (free amine or its hydrochloride), add 0.5 ml of buffer a and 0.2 ml of reagent b. Let stand at 30° for 1 hr, cool to room temperature in a water bath, and dilute with 2 ml of ethanol. Read at exc: 366 nm; em: 430 nm.

Standard: A solution of quinine sulfate in 0.1 N sulfuric acid.

Results

	Determination limits, μg	Reading 50 Sample, μg	Standard, μg/ml
Methylamine (hydrochloride)	0.2 - 1.0	0.43	0.256
Ethylamine	0.1 - 0.5	0.22	0.238
Ethylamine (hydrochloride)	0.2 - 1.0	0.44	0.227
n-Propylamine	0.1 - 0.5	0.21	0.164
n-Butylamine	0.2 - 1.0	0.44	0.222

Note

Application of o-diacetylbenzene to the colorimetric determination of primary aromatic amines, see Chapter 5, Section IV.I.

XIII. TERTIARY ALIPHATIC AMINES (F)

A. Aconitic Acid and Acetic Anhydride

Principle (75)

Reaction of the amine with a mixture of aconitic acid and acetic anhydride: green fluorescence.

This fluorescence is probably due to a species formed between the amine and the α,γ-anhydroaconitic acid (XVII) which arises from the dehydration of aconitic acid (66) (see Section VIII.D, Note 5).

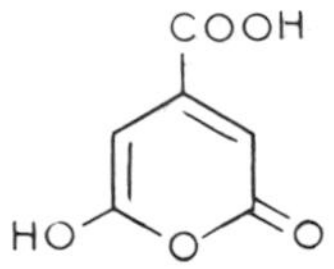

(XVII)

Reagent

Dissolve 0.25 g of aconitic acid in 2.5 ml of acetone, add 1 ml of acetic anhydride, and incubate at 40° for 20 min. Allow to cool, add 40 ml of methylene chloride, shake, chill for 5 min in ice water, filter, and dilute to 50 ml with methylene chloride. This reagent is stable for several hours at room temperature.

The aconitic acid should be perfectly colorless. If necessary, recrystallize from acetic acid.

Procedure

To 4 ml of sample solution in ethyl acetate, add in the dark 0.2 ml of reagent, mix, and read within 5 min at exc: 405 nm; em: 485 nm.

Standard: A solution of 2-methyl-5-carboxy-7-amino-quinoline in glacial acetic acid.

Results

	Determination limits, µg	Reading 50	
		Sample, µg	Standard, µg/ml
Triethylamine	1 - 5	2.3	0.029
Tri-n-propylamine	2 - 10	4.5	0.027
Tri-n-butylamine	2 - 10	4.4	0.024

Notes

1. The fluorescence is stable for about 5 min, provided that the tubes are kept in the dark. Only one reading

per tube is possible, since the fluorescence is rapidly destroyed by the exciting light.

2. The sensitivity is somewhat higher when operating at 0°, but secondary amines then afford a rather stable fluorescence, which is very transient at room temperature.

REFERENCES

1. J. Bartos, *Talanta*, 16, 551 (1969).

2. M. Pesez, J. Bartos, and J.C. Lampetaz, *Bull. Soc. Chim. Fr.*, 719 (1962).

3. F. Wöhler, *Ann. Phys. Chem.*, 41, 344 (1837).

4. E. Iliescu, *Pharm. Praxis*, No. 3, 47 (1961). Cf. *Farmacia (Bucarest)*, 7, 565 (1959).

5. N.A. Tananaeff and J. Tananaeff, *Z. Anorg. Allgem. Chem.*, 170, 113 (1928).

6. P. Mauer, *Mikrochim. Ichnoanal. Acta*, 17, 882 (1964).

7. J. Verdier, private communication.

8. J. Deltombe and G. Leboutte, *J. Pharm. Belg.*, 13, 38 (1958); J. Deltombe, G. Leboutte, and N. Rosier, *ibid.*, 17, 236 (1962); J. Bosly, *ibid.*, 18, 162 (1963); J. Bosly, F. Dutrieux, and R. Stainier, *ibid.*, 18, 168 (1963).

9. J. Bosly, *Farmaco, Ed. Pract.*, 19, 448 (1964).

10. J. Lagubeau and P. Mesnard, *Bull. Soc. Chim. Fr.*, 2815 (1965).

11. E.G. Brown and T.J. Hayes, *Analyst*, 80, 755 (1955).

12. M. Pesez and J. Bartos, *Ann. Pharm. Fr.*, 23, 781 (1965).

13. W. Brackman and E. Havinga, *Rec. Trav. Chim. Pays-Bas*, 74, 937, 1021, 1070, 1100, 1107 (1955).

14. L. Ekladius and H.K. King, *Biochem. J.*, 65, 128 (1957).

15. C. Mannich, *Arch. Pharm.*, 273, 97 (1935).

16. D.T. Dubin, *J. Biol. Chem.*, 235, 783 (1960).

17. F.C. McIntire, L.M. Clements, and M. Sproull, *Anal. Chem.*, 25, 1757 (1953).

18. From K. Blau and W. Robson, *Chem. Ind. (London)*, 424 (1957).

19. J. Bartos and M. Pesez, *Bull. Soc. Chim. Fr.*, 1627 (1970).

20. From J. Bartos, *Talanta*, 8, 619 (1961).

21. J. Bartos and M. Pesez, *Ann. Pharm. Fr.*, 28, 459 (1970).

22. M. Pesez and J. Bartos, *Ann. Pharm. Fr.*, 23, 65 (1965).

23. Cf. D. Buckley, H.B. Henbest, and P. Slade, *J. Chem. Soc.*, 4891 (1957).

24. Y. Tashima, H. Hasegawa, H. Yuki, and K. Takiura, *Japan Analyst*, 19, 43 (1970).

25. Cf. T.F. Chin and J.L. Lach, *J. Pharm. Sci.*, 54, 1550 (1965); K. Gustavii and G. Schill, *Acta Pharm. Suecica*, 3, 241 (1966); K. Howorka, *Pharm. Zentralhalle*, 108, 322 (1969).

26. O. Prudhomme, *J. Pharm. Chim.*, [9] 1, 8 (1940).

27. M.E. Auerbach, *Ind. Eng. Chem., Anal. Ed.*, 15, 492 (1943); 16, 739 (1944).

28. M. Tsubouchi, *J. Pharm. Sci.*, 60, 943 (1971).

29. Cf. F. Pellerin, J.A. Gautier, and O. Barrat, *Bull. Soc. Chim. Fr.*, 1027 (1960); F. Pellerin, *ibid.*, 1071 (1961).

30. F. Matsui and W.N. French, *J. Pharm. Sci.*, 60, 287 (1971).

31. D. Lojodice and B.M. Colombo, *Farmaco, Ed. Pract.*, 25, 763 (1970).

32. Z.I. Chalaya and L.S. Mikhailova, *Zh. Anal. Khim.*, 25, 1829 (1970).

33. V.E. Starostenko and N.T. Bubon, *Farmatsiya*, 20, 41 (1971).

34. K.O. Borg and D. Westerlund, *Z. Anal. Chem.*, 252, 275 (1970).

35. From A. Häussler, *Deut. Apotheker Ztg.*, 97, 729 (1957).

36. From E. Rimini, *Ann. Farmacoterap. Chim.*, 27-28, 193 (1898); *Z. Anal. Chem.*, 41, 438 (1902).

37. V.N. Bernshtein and V.G. Belikov, *Russ. Chem. Rev.*, 30, 227 (1961).

38. From M. Pesez and J. Bartos, *Talanta*, 5, 216 (1960).

39. M. Pesez and J. Bartos, *Talanta*, 5, 216 (1960).

40. F.E. Critchfield and J.B. Johnson, *Anal. Chem.*, 28, 436 (1956).

41. From M. Pesez and P. Poirier, *Bull. Soc. Chim. Fr.*, 754 (1953).

42. K.H. Beyer and J.T. Skinner, *J. Pharm. Exptl. Therap.*, 68, 419 (1940).

43. K.H. Beyer, *J. Amer. Chem. Soc.*, 64, 1318 (1942).

44. From E. Sawicki and H. Johnson, *Chemist-Analyst*, 55, 101 (1966).

45. F. Fontani and F. Morandini, *J. Pharm. Pharmacol.*, 22, 411 (1970).

46. Cf. E. Sawicki and H. Johnson, *Chemist-Analyst*, 55, 101 (1966).

47. Cf. J. Bartos, *Pharm. Weekblad*, 93, 594 (1958).

48. From M. Pesez and J. Bartos, *Ann. Pharm. Fr.*, 15, 467 (1957).

49. J. Bartos, *Ann. Pharm. Fr.*, 22, 383 (1964).

50. Cf. F. Feigl and V. Anger, *Mikrochim. Acta*, 1, 138 (1937).

51. From C.F. Cullis and D.J. Waddington, *Anal. Chim. Acta*, 15, 158 (1956).

52. G.R. Umbreit, *Anal. Chem.*, 33, 1572 (1961).

53. J. Bartos, *Ann. Pharm. Fr.*, 20, 478 (1962).

54. Cf. L. Horner and H. Lang, *Chem. Ber.*, 89, 2768 (1956).

55. J. Bartos, *Bull. Soc. Chim. Fr.*, 331 (1966).

56. From J. Bartos, *Ann. Pharm. Fr.*, 19, 610 (1961).

57. Cf. T. Mukaiyama and T. Oshino, *J. Amer. Chem. Soc.*, 82, 5339 (1960).

58. Cf. J.V.R. Kaufman and J. Picard, *Chem. Rev.*, 59, 429 (1959).

59. M. Pesez and J. Bartos, *Bull. Soc. Chim. Fr.*, 2333 (1963).

60. Cf. H. Meerwein, K. Wunderlich, and K.F. Zenner, *Angew. Chem.*, 74, 807 (1962).

61. S. Sass, J.J. Kaufman, A.A. Cardenas, and J.J. Martin, *Anal. Chem.*, 30, 529 (1958).

62. M. Palumbo and S. Sacca, *Farmaco, Ed. Pract.*, 17, 65 (1962).

63. Cf. *F.S.A.O.V.*, 273-275.

64. S. Ohkuma, *J. Pharm. Soc. Japan*, 75, 1124 (1955).

65. W.D. Langley, *Anal. Chem.*, 39, 199 (1967).

66. A.B. Groth and M.E. Dahlen, *Acta Chem. Scand.*, 21, 291 (1967).

67. M. Pesez, *Ann. Pharm. Fr.*, 16, 441 (1958).

68. From L. Velluz and M. Pesez, *Ann. Pharm. Fr.*, 4, 10 (1946).

69. F. Feigl, *Spot Tests in Organic Analysis*, Elsevier (Amsterdam), 6th Ed., 113 (1960).

70. Cf. P.B. Ghosh and M.W. Whitehouse, *Biochem. J.*, 108, 155 (1968).

71. J. Bartos and M. Pesez, *Talanta*, 19, 93 (1972).

72. M. Pesez and J. Bartos, *Ann. Pharm. Fr.*, 27, 161 (1969).

73. H. Goldstein and G. Genton, *Helv. Chim. Acta*, 20, 1413 (1937).

74. Cf. N. Seiler and M. Wiechmann, *Z. Anal. Chem.*, 220, 109 (1966).

75. M. Pesez and J. Bartos, *Talanta*, 16, 331 (1969).

76. G. Weber, *Biochem. J.*, 51, 155 (1952).

77. B.S. Hartley and V. Massey, *Biochim. Biophys. Acta*, 21, 58 (1956).

78. W.R. Gray and B.S. Hartley, *Biochem. J.*, 89, 59 P (1963).

79. See C.E. White and R.J. Argauer, *Fluorescence Analysis*, Marcel Dekker Inc., New York, 180-195 (1970).

80. Cf. S. Belman, *Anal. Chim. Acta*, 29, 120 (1963).

81. From M. Pesez and J. Bartos, *Talanta*, 14, 1097 (1967).

82. From K. Samejina, W. Dairman, J. Stone, and S. Udenfriend, *Anal. Biochem.*, 42, 237 (1971).

83. M. Weigele, J.F. Blount, J.P. Tengi, R.C. Czajkowski, and W. Leimgruber, *J. Amer. Chem. Soc.*, 94, 4052 (1972).

84. M. Weigele, S.L. DeBernardo, J.P. Tengi, and W. Leimgruber, *J. Amer. Chem. Soc.*, 94, 5927 (1972).

85. S. Udenfriend, S. Stein, P. Böhlen, W. Dairman, W. Leimgruber, and M. Weigele, *Science*, 178, 871 (1972).

Chapter 5

AROMATIC AMINES

I. INTRODUCTION

A. Colorimetry

All classes of aromatic amines exhibit more or less reducing properties and can therefore be determined with ferric ferricyanide. However, the reaction is not specific for arylamines, and the oxidative condensation with flavylium perchlorate, giving sensitivities which vary from 5 to 75 μg for A = 0.3, also suffers from a lack of specificity.

Diazo coupling is widely used. As a rule, the highest sensitivity is reached with N,N-dialkylarylamines, which can sometimes be determined at the level of 2 μg. Coupling can be achieved with p-nitrobenzenediazonium fluoborate, 4-azobenzenediazonium fluoborate, diazotized 2-aminobenzothiazole, with the diazo compound obtained by the dehydration of phenitrazole (3-phenyl-5-nitrosamino-1,2,4-thiadiazole) in acid medium, or with the diazo derivative formed by oxidizing 3-methylbenzo-thiazolin-2-one hydrazone with ferric chloride. With the exception of diazotized 2-aminobenzothiazole, all these diazonium salts or their precursors are stable in the solid state, thus making the determinations highly reproducible.

Because primary aromatic amines afford poorly sensitive results, it is worth converting them first into their corresponding diazo compounds with sodium nitrite in acid medium. The diazonium salt is then revealed with a phenol or an arylamine. As a rule, the excess of nitrous ion should be destroyed with ammonium sulfamate before coupling in order to avoid side reactions, as in the case of development with N-(1-naphthyl)ethylenediamine (Bratton-Marshall reagent), which allows determinations within the range 6-9 μg (for A = 0.3). However, when the amine is diazotized in the presence of p-aminosalicylic acid (PAS), the diazo derivative of PAS is very unstable and, by loss of nitrogen, is converted to 2,4-dihydroxybenzoic acid, which then couples with the diazotized amine; in this case no elimination of the excess nitrous acid is necessary. Determinations are possible at levels of 14-25 μg.

Besides diazotization, other reactions are specific for primary arylamines. They react with disodium pentacyanoaquoferrate (III). The mechanism of this reaction is unknown. It allows determinations of 13-58 μg of amine, with the exception of sulfanilamide, which reacts but weakly.

Only primary amines can afford Schiff bases by reaction with p-dimethylaminobenzaldehyde or p-dimethylaminocinnamaldehyde, or with sodium glutaconic aldehyde enolate, which is obtained from 4-pyridylpyridinium dichloride in alkaline medium. These reactions are highly sensitive making determinations possible at levels of 1.5-7.5 μg (for A = 0.3).

Condensation with succinic dialdehyde affords N-aryl-substituted pyrroles, which can be developed either with p-dimethylaminobenzaldehyde or with flavylium perchlorate. The reactions are highly sensitive, but primary aliphatic amines also react.

Through oxidative reaction with thymol, primary arylamines give benzoquinoneanils, whose blue color allows the determination of 20-105 μg of compound.

With o-diacetylbenzene in acid medium, a violet color is developed, allowing determinations at levels of 3.4-6.2 μg. The mechanism of the reaction is unknown.

The hydrogen bonded to the nitrogen atom in primary and secondary aromatic amines allows determinations through reactions with 1-fluoro-2,4-dinitrobenzene or p-nitrophenylazobenzoyl chloride. However, primary and secondary aliphatic amines react the same.

Two other reactions should theoretically be applicable to secondary as well as to primary arylamines, but only the primary ones give satisfactory results:

Reaction with 9-chloroacridine in acid medium gives 9-arylaminoacridines which permit determinations within the range 14.5-33.6 μg.

The amino group can be substituted for the sulfonate group of 1,2-naphthoquinone-4-sulfonic acid. The N-arylaminonaphthoquinones formed are extractable into methylene chloride and can be then developed with 2,4-dinitrophenylhydrazine. Determinations are made at levels of 3.0-14.5 μg.

B. Fluorimetry

Methods are available only for primary aromatic amines. The N-arylaminonaphthoquinones obtained from the reaction with naphthoquinonesulfonic acid can be reduced to the corresponding fluorescent aminodihydroxynaphthalenes after extraction into methylene chloride, allowing determinations within the range 0.48-0.72 μg for reading 50.

Diazotized primary arylamines can be coupled with 2,6-diaminopyridine. Upon oxidation, the azo compound gives a fluorescent 8-substituted 6-aminopyridinotriazole, permitting determinations at levels of 1.25-2.45 μg.

II. AROMATIC AMINES

A. Ferric Ferricyanide

Principle and Procedure

See Chapter 3, Section II.B, determination of phenols.

Results

	A = 0.3 (1 cm cell) Sample, μg
Aniline	3.1
N-Methylaniline	2.1
N,N-Dimethylaniline	2.3
p-Aminobenzoic acid	6.0

The reaction is also positive with sulfanilamide and sulfathiazole, but Beer's law is not followed.

Note

Alternatively, uranyl ferricyanide can be used (Chapter 3, Section II.B, Note 3). A = 0.3 is given by 5 μg of aniline.

B. Flavylium Perchlorate

Principle (1)

Oxidative coupling with flavylium perchlorate, the oxidant being atmospheric oxygen: blue to violet color.

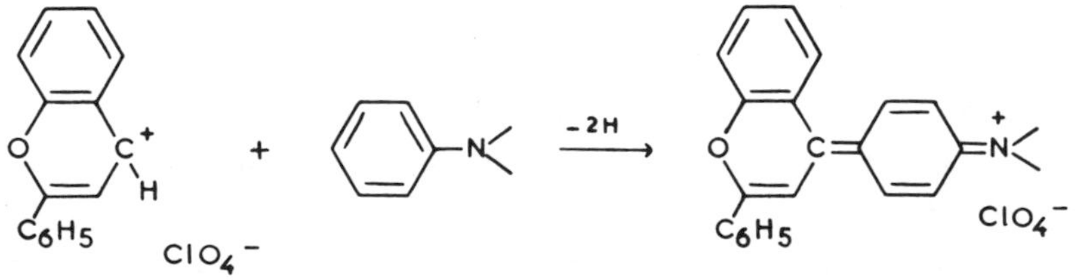

Reagent

An 0.25% solution of flavylium perchlorate (Chapter 17, Section I.D) in 2% solution of anhydrous sodium acetate in glacial acetic acid.

Procedure

To 5 ml of sample solution in glacial acetic acid, add 0.5 ml of reagent, heat at 100° for 30 min, cool to room temperature, and read.

Results

	Color	λ Max, nm	A = 0.3 (1 cm cell) Sample, μg
Aniline (hydrochloride)	Violet	550	13.9
N-Methylaniline	Blue-violet	570	3.8
N,N-Dimethylaniline	Blue	590	5.0
α-Napthylamine[a]	Blue	510	75

[a]1.5 ml of reagent.

Sulfanilamide and β-naphthylamine also react, but Beer's law is not followed.

Notes

1. The reaction is positive with a number of compounds possessing an active hydrogen atom, such as malonic acid, barbituric acid, antipyrine, indole, and pyrogallol and resorcinol through their keto tautomers (1,2).

2. In the absence of oxygen, no color is developed (3).

3. The mechanism of the reaction may sometimes be rather complicated. For instance, with malonic acid, the species responsible for the color is 4,4'-flavenylidenemethylflavylium perchlorate (2). This condensation thereby involves a complete decarboxylation, and accordingly mono- and diesters of malonic acid do not

react, thus allowing detection of trace amounts of the free acid in the presence of its esters (4).

4. Flavylium perchlorate was used as a spray reagent in paper chromatography (5).

5. Flavylium perchlorate also allows the colorimetric determination of primary aliphatic (Chapter 4, Section VI.G) and aromatic (Chapter 5, Section IV.F) amines and of amino acids (Chapter 9. Section VI.E) by development of the pyrrole ring formed with succinic dialdehyde, 2-deoxy sugars (Chapter 14, Section IX.B) and 2-amino-2-deoxy sugars condensed with acetylacetone (Chapter 14, Section X.C).

C. p-Nitrobenzenediazonium Fluoborate

Principle (6)

Coupling with p-nitrobenzenediazonium ion in alkaline medium: miscellaneous colors.

$$O_2N-C_6H_4-\overset{+}{N}\equiv N + C_6H_5-NH_2 \longrightarrow O_2N-C_6H_4-N=N-C_6H_4-NH_2$$

Reagent

An 0.5% aqueous solution of p-nitrobenzenediazonium fluoborate. Prepare fresh immediately before use, and filter.

Procedure

To 1 ml of aqueous solution of the amine, add 0.1 ml of reagent, 0.2 ml of 10% aqueous solution of tetramethylammonium hydroxide, and 10 ml of dimethylformamide. Mix, and read immediately.

Results

	Color	λ Max, nm	A = 0.3 (1 cm cell) Sample, μg
Aniline	Red	530	6.6
N-Methylaniline	Green	425	80
N,N-Dimethylaniline	Orange	500	29
p-Aminobenzoic acid	Pink-orange	520	20
Sulfanilamide[a]	Pink-orange	530	250

[a]Let stand for 15 min before reading.

Notes

1. Solvent effects in the spectrophotometric determination of weak organic acids in alkaline solution, with application to primary aromatic amines, see Ref. (6).

2. Other determinations with p-nitrobenzenediazonium fluoborate, see Chapter 3, Section II.H, Note 3.

3. Diazo coupling, see Chapter 16, Section I.

D. 4-Azobenzenediazonium Fluoborate

Principle (7)

Coupling with 4-azobenzenediazonium ion and acidification: blue, sometimes green or violet color.

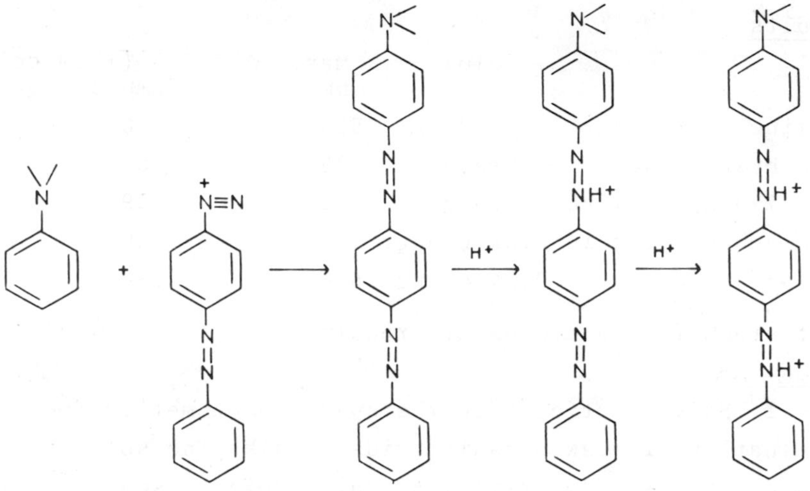

Yellow Red Blue

Reagent

An 0.1% solution of 4-azobenzenediazonium fluoborate (Chapter 17, Section I.A) in 2-methoxyethanol (Methyl Cellosolve).

Procedure

To 1 ml of sample solution in 2-methoxyethanol, add 2 ml of reagent, and keep at the given temperature for the given length of time (see Results). Chill in ice water, still at 0° add 10 ml of concentrated hydrochloric acid, and read.

Results

	Reaction time, min	Temperature, °C	λ Max, nm	A = 0.3 (1 cm cell) Sample, μg
Aniline	5	100	630	25.0
N-Methyl-aniline	10	70	620	11.0
N,N-Dimethyl-aniline	6	100	630	4.4
α-Naphthylamine	10	20-24	640	4.5

p-Aminobenzoic acid gives an almost negative reaction.

Note

Diazo coupling, see Chapter 16, Section I.

E. Diazotized 2-Aminobenzothiazole

Principle and Procedure

See Chapter 3, Section II.I, determination of phenols.

Results

	Color	λ Max, nm	A = 0.3 (1 cm cell) Sample, μg
Aniline	Yellow	395	12
M-Methylaniline[a]	Violaceous red	390	50
N,N-Dimethyl-aniline	Violaceous red	540	5.6

[a]Add at first sodium hydroxide, then the reagent.

Novocaine and sulfanilamide do not react.

F. Phenitrazole

Principle and Procedure

See Chapter 3, Section II.J, determination of phenols: pink or orange color.

Results

	Color	λ Max, nm	A = 0.3 (1 cm cell) Sample, μg
Aniline[a]	Orange	495	28
N-Methylaniline[a]	Pink	510	27
N,N-Dimethylaniline	Violaceous pink	535	6

[a]Let stand for 5 min at 0° instead of room temperature.

p-Aminobenzoic acid gives a very faint color; sulfanilamide does not react.

G. 3-Methylbenzothiazolin-2-one Hydrazone and Ferric Chloride

Principle (8)

Condensation with the diazo derivative obtained by oxidizing 3-methylbenzothiazolin-2-one hydrazone with ferric chloride in the reaction medium: green-yellow color.

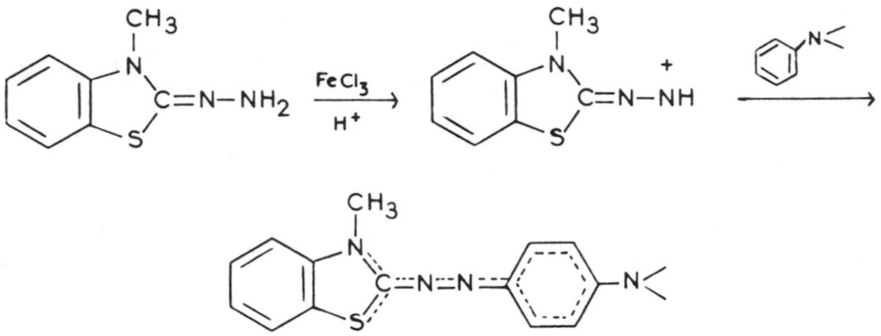

CH_3 N C=N—NH_2 S $\xrightarrow[H^+]{FeCl_3}$ CH_3 N C=N—NH S + $C_6H_5N<$ ⟶

CH_3 N C=N=N=C_6H_4—N< S

Reagents

a. An 0.35% aqueous solution of 3-methylbenzothiazolin-2-one hydrazone hydrochloride.

b. A 1.2% aqueous solution of ferric chloride hexahydrate.

Procedure

To 0.5 ml of sample solution in methanol, add 0.5 ml of reagent a and 1 ml of reagent b. Let stand for 30 min in the dark, and add 3 ml of methanol. Mix, and read at 570 nm.

Results

	A = 0.3 (1 cm cell) Sample, μg
Aniline	2.3
o-Toluidine	1.9
m-Toluidine	2.2
p-Toluidine	45
N-Methylaniline	1.9

	A = 0.3 (1 cm cell) Sample, μg
N,N-Dimethylaniline	1.9
p-Aminobenzoic acid	25
Sulfanilamide	24

Note

Properties of 3-methylbenzothiazolin-2-one hydrazone, see Chapter 16, Section III.

III. PRIMARY AND SECONDARY AROMATIC AMINES

A. 1-Fluoro-2,4-dinitrobenzene (in Ethanol)

Principle and Procedure

See Chapter 4, Section IV.C, determination of primary and secondary aliphatic amines: yellow-orange color.

Results

	λ Max, nm	A = 0.3 (1 cm cell) Sample, μg
Aniline	340	19
N-Methylaniline	360	210

B. p-Nitrophenylazobenzoyl Chloride and Benzyltrimethylammonium Hydroxide

Principle and Procedure

See Chapter 4, Section IV.F, determination of primary and secondary aliphatic amines: orange-red color.

Results

	λ Max, nm	A = 0.3 (1 cm cell) Sample, μg
Aniline	525	54
N-Methylaniline	535	19

The reaction is weakly sensitive with p-aminobenzoic acid, and Beer's law is not followed.

IV. PRIMARY AROMATIC AMINES

A. Trisodium Pentacyanoamminoferrate(III) and Bromine Water

Principle (9)

Reaction with disodium pentacyanoaquoferrate(III) obtained by adding bromine water to a solution of trisodium pentacyanoamminoferrate(III) (sodium amminoprusside): brown-green to green color.

Reagent

To 10 ml of 2% aqueous solution of trisodium pentacyanoamminoferrate(III), add bromine water dropwise until a violet color develops (about 1 ml). Add sodium nitroprusside by small amounts until disappearance of the color (about 0.1 g), then add 0.200 g of anhydrous sodium carbonate.

Procedure

To 5 ml of aqueous solution of the amine, add the given volume of reagent, let stand for the given length of time (see Results), and read.

Results

	Reagent, ml	Reaction time, min	λ Max, nm	A = 0.3 (1 cm cell) Sample, μg
Aniline	0.5	90	660	20
p-Toluidine	0.2	45	700	58
p-Anisidine	0.2	15	670	13.5
Sulfanilamide	1.0	30	520	780

Notes

1. Relations between pentacyanoferrates:

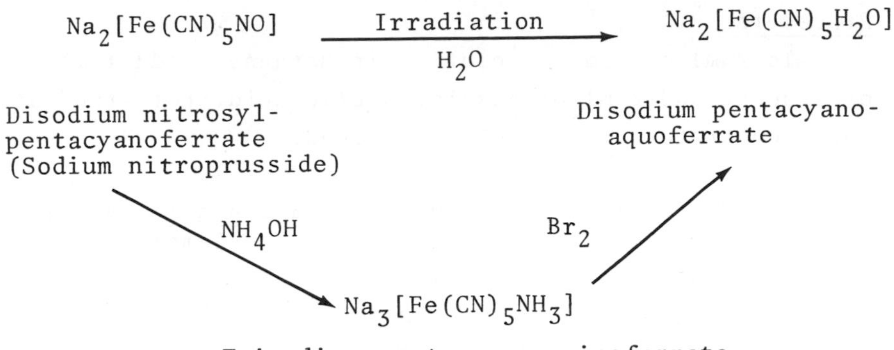

Trisodium pentacyanoamminoferrate
(Sodium amminoprusside)

2. Green color is developed when sodium amminoprusside is reacted with nitroso derivatives such as p-nitroso-N,N-dimethylaniline, p-nitrosophenol, 1-nitroso-2-naphthol, 2-nitroso-1-naphthol, isonitrosoacetylacetone, isonitrosoacetophenone (10,11). The reaction was attributed to the formation of derivatives of the type

$$Na_3[Fe(CN)_5(R - NO)]$$

3. Trisodium pentacyanoamminoferrate(III) also allows the colorimetric determination of pyridine derivatives (Chapter 13, Section IV.A).

B. p-Dimethylaminobenzaldehyde

Principle

Reaction with p-dimethylaminobenzaldehyde in acetic acid medium to give a Schiff base: yellow color.

$$Ar-NH_2 + OCH-C_6H_4-N(CH_3)_2 \longrightarrow Ar-N=CH-C_6H_4-N(CH_3)_2$$

Reagent

A 1% solution of p-dimethylaminobenzaldehyde in methanol.

Procedure

To 2 ml of sample solution in methanol, add 1 ml of reagent and 2 ml of glacial acetic acid. Let stand at room temperature for 10 min, and read.

Results

	λ Max, nm	A = 0.3 (1 cm cell) Sample, μg
Aniline	435	2.0
o-Toluidine	420	3.0
m-Toluidine	435	2.5
p-Toluidine	435	3.0
p-Aminobenzoic acid	445	3.5
Sulfanilamide	445	5.0

Notes

1. p-Dimethylaminobenzaldehyde has led to numerous applications as analytical reagent (12). It also allows the determination of trace amounts of hydrazine (13).

2. In this book, the reagent was applied to the colorimetric determination of primary alkylamines (Chapter 4, Section VI.F), primary arylamines (Chapter 5, Section IV.E), and amino acids (Chapter 9, Section VI.D) converted to pyrrole derivatives, primary arylamines as Schiff bases (Chapter 5, Section IV.B), nitroaromatic derivatives reduced beforehand to amines (Chapter 7, Section VI.A), indole derivatives (Chapter 13, Section III.C), and 2-amino-2-deoxy sugars condensed with acetylacetone (Chapter 14, Section X.B) or phenylacetone (Chapter 14, Section X.E).

C. p-Dimethylaminocinnamaldehyde

Principle (14)

Reaction with p-dimethylaminocinnamaldehyde in the presence of acetic acid to give a Schiff base: orange to red color.

$$Ar{-}NH_2 + OCH{-}CH{=}CH{-}C_6H_4{-}N(CH_3)_2 \longrightarrow Ar{-}N{=}CH{-}CH{=}CH{-}C_6H_4{-}N(CH_3)_2$$

Reagent

An 0.1% solution of p-dimethylaminocinnamaldehyde in methanol.

Procedure

To 1 ml of sample solution in methanol, add 0.5 ml of reagent and 0.5 ml of glacial acetic acid. Let stand for 10 min at room temperature, add 3 ml of methanol, mix, and read.

Results

	λ Max, nm	A = 0.3 (1 cm cell) Sample, μg
Aniline	520	1.5
p-Toluidine	520	2.0
p-Anisidine	520	2.5
p-Aminobenzoic acid	540	3.0
Sulfanilamide	540	6.0

Notes

1. p-Dimethylaminocinnamaldehyde was used for the colorimetric determination of 4-aminoantipyrine (15) and of methampyrone (16).

2. m-Diphenols do not react under the above conditions, whereas they develop a green color when the same reagent is used in hydrochloric acid medium (Chapter 3, Section V.B). Contrariwise, in this medium, aromatic amines afford only poorly sensitive colors.

3. p-Dimethylaminocinnamaldehyde also allows the colorimetric determination of primary aliphatic amines (Chapter 4, Section VI.C) and indole derivatives (Chapter 13, Section III.D).

D. 4-Pyridylpyridinium Dichloride

Principle (17)

4-Pyridylpyridinium dichloride is decomposed in alkaline medium, giving rise to sodium glutaconic aldehyde enolate which then reacts with the amine to give an imino derivative of Schiff base type: yellow color.

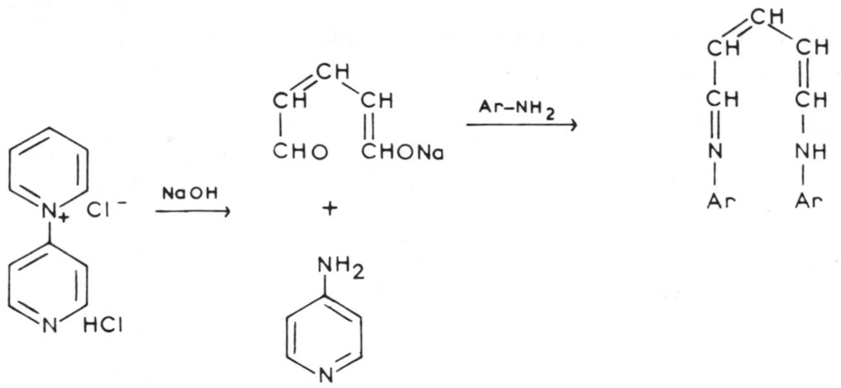

Reagent

A 1% aqueous solution of 4-pyridylpyridinium dichloride.

Procedure

To 1 ml of aqueous solution of the amine, add 1 ml of 2 N sodium hydroxide, 1 ml of reagent, and 2 ml of 2 N hydrochloric acid. Let stand for 5 min, and read.

Results

	λ Max, nm	A = 0.3 (1 cm cell) Sample, μg
Aniline	455	5.4
p-Toluidine	455	5.0
p-Anisidine	455	5.6
p-Aminobenzoic acid	470	4.2
Procaine (hydrochloride)	470	7.5
Sulfanilamide	470	6.5

Notes

1. 4-Aminopyridine formed together with glutaconic aldehyde enolate does not behave like an aromatic amine and therefore does not interfere.

2. Determinations through opening of the pyridine ring, see Chapter 16, Section XI.

E. 2,5-Diethoxytetrahydrofuran and p-Dimethylaminobenzaldehyde

Principle and Procedure

See Chapter 4, Section VI.F, determination of primary aliphatic amines.

Results

	T_1, min	T_2, min	A = 0.3 (1 cm cell) Sample, μg
Aniline	2	15	2.1
p-Toluidine	2	5	2.1
Sulfanilamide	2	45	5.0

F. 2,5-Diethoxytetrahydrofuran and Flavylium Perchlorate

Principle and Procedure

See Chapter 4, Section VI.G, determination of primary aliphatic amines. Read at 500 nm.

Results

	T_1, min	T_2, min	A = 0.3 (1 cm cell) Sample, μg
Aniline	2	45	5
p-Toluidine	2	20	6
Sulfanilamide	2	120	17

G. 9-Chloroacridine

Principle (18)

Reaction with 9-chloroacridine to give a 9-arylaminoacridine: yellow color.

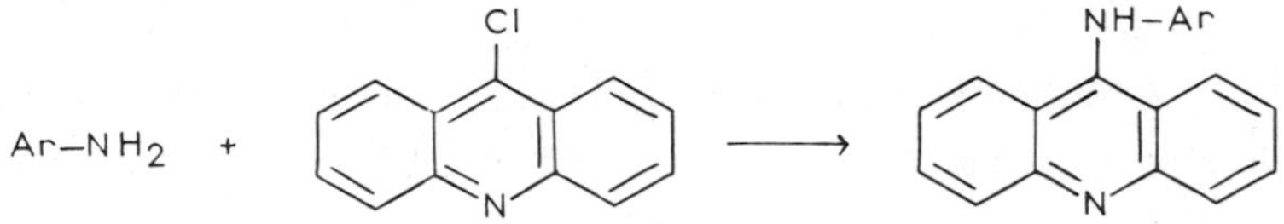

$$Ar{-}NH_2 + \text{9-chloroacridine} \longrightarrow \text{9-(NH-Ar)acridine}$$

<u>Reagent</u>

An 0.04% solution of 9-chloroacridine in ethanol. Keep in the dark.

<u>Procedure</u>

To 1 ml of neutral sample solution in ethanol, add 1 ml of reagent and 0.1 ml of 0.1 N hydrochloric acid. Let stand for 45 min at room temperature, and add 3 ml of ethanol. Mix, and read at 435 nm.

<u>Results</u>

	A = 0.3 (1 cm cell) Sample, μg
p-Toluidine	14.5
o-Anisidine	19.0
p-Anisidine	16.5
β-Naphthylamine	16.5
o-Aminophenol	19.5
p-Aminophenol	14.7
p-Aminobenzoic acid	15.0
Sulfanilamide	19.6
Sulfamethoxypyridazine	33.6
Sulfathiazole	28.5
Sulfamethizole	31.0

Aniline and o-toluidine do not react under these conditions.

H. Sodium 1,2-Naphthoquinone-4-sulfonate and 2,4-Dinitrophenylhydrazine

<u>Principle</u>

See Chapter 4, Section IV.E, determination of primary and secondary aliphatic amines.

Reagents

a. An 0.5% aqueous solution of sodium 1,2-naphthoquinone-4-sulfonate. Stable for a few hours at 0°, protected against light.

b. To 0.20 g of 2,4-dinitrophenylhydrazine, add 20 ml of ethanol and 0.5 ml of concentrated sulfuric acid. When the crystals are dissolved, dilute to 100 ml with ethanol.

Procedure

To 1 ml of nearly neutral aqueous solution of the amine, add 0.2 ml of reagent a, and let stand at room temperature for the given length of time (see Results), protected against light. Transfer the solution into a 10 ml separatory funnel containing 2 ml of methylene chloride, rinse the tube with 1 ml of water, and pour the rinsing into the funnel. Shake immediately for about 5 sec, let stand for 2 min, and collect the whole bulk of the organic layer. Add to it 3 ml of reagent b, fit the tubes with air condensers, and heat at 50° for 10 min. Chill for a few minutes in ice water, and still at 0° add slowly 0.5 ml of concentrated sulfuric acid. Mix, and read at 500 nm.

Results

	Reaction time, min	A = 0.3 (1 cm cell) Sample, μg
Aniline	0	3.0
o-Toluidine	2	3.2
p-Anisidine	0	4.2
Ethyl p-aminobenzoate	5	7.5
Procaine (hydrochloride)	5	14.5

Notes

1. Compounds substituted on the aromatic ring with groups which favor solubility in water, such as aminobenzoic acids or sulfonamides, cannot be determined,

since their naphthoquinone derivatives are not extractable into methylene chloride under the above conditions.

2. Secondary aromatic amines also react, but Beer's law is not followed.

3. Reaction between amino compounds and 1,2-naphthoquinone-4-sulfonic acid, see Chapter 16, Section VII.

I. o-Diacetylbenzene

Principle

Condensation with o-diacetylbenzene: violet color.

Reagents

a. A 2.5% solution of potassium acetate in glacial acetic acid.

b. An 0.5% solution of o-diacetylbenzene in ethanol.

Procedure

To 1 ml of sample solution in ethanol, add 0.5 ml of solution a and 0.2 ml of reagent b. Let stand at 30° for 20 min, cool to room temperature in a water bath, and add 2 ml of ethanol. Read at 550 nm.

Results

	A = 0.3 (1 cm cell) Sample, μg
Aniline	3.4
m-Toluidine	4.0
p-Toluidine	3.9
p-Anisidine	4.7
β-Naphthylamine	6.2
o-Aminophenol	6.2
p-Aminophenol	4.4

Notes

1. With primary arylamines substituted at the para position, the reaction rate depends on the nature of the substituent. A weak sensitivity and a slow reaction rate are observed with electron-attracting groups: This

is the case of p-aminobenzoic acid and sulfanilamide (A = 0.3 for 25 μg after 2 hr at 30°). Some ortho-substituted derivatives behave the same (A = 0.3 for 10 μg of o-toluidine after 90 min at 30°).

2. A very faint color is obtained with 1500 μg of N-methylaniline; N,N-dimethylaniline does not react.

3. Primary aliphatic amines react but very weakly. Nevertheless, under suitable conditions, the same reagent allows their fluorimetric determination (see Chapter 4, Section XII.C).

4. o-Diacetylbenzene was proposed as reagent for amino acids, proteins (19-25), and amines (26,27). With amino acids, the sensitivity is rather weak and markedly varies from one compound to the other (24).

J. Thymol and Sodium Hypochlorite

Principle (28)

Oxidative reaction with thymol, to give a benzoquinoneanil: blue color.

$Ar-NH_2$ + [thymol structure]–OH $\xrightarrow{ClO^-}$ $Ar-N=$[quinone ring]$=O$

Reagents

a. A dilute aqueous solution of sodium hypochlorite: 100 ml contains about 0.1 g of active chlorine.

b. A 5% solution of thymol in ethanol.

Procedure

Operate under bright light. To 5 ml of aqueous solution of the amine, add 0.2 ml of 50% sulfuric acid, 0.9 ml of reagent a, and 0.5 ml of reagent b. Let stand for 5 min with occasional shaking, then add 0.5 ml of 10 N sodium hydroxide, and let stand in a water bath at 20-25° for 5 min. Read at 600 nm.

Results

	A = 0.3 (1 cm cell) Sample, μg
Aniline	20
p-Aminobenzoic acid	46
Sulfanilamide	105

Notes

1. Hydrogen peroxide in the presence of copper was also used as oxidant (29).

2. Phenol was proposed instead of thymol for the detection and determination of sulfonamides (30).

3. According to the same principle, aniline inversely allows the colorimetric determination of phenols (see Chapter 3, Section II.K).

K. p-Aminosalicylic Acid and Sodium Nitrite

Principle (31)

Diazotization of a mixture of the amine and p-aminosalicylic acid (PAS). By loss of nitrogen, very unstable diazotized PAS is converted to 2,4-dihydroxybenzoic acid, which then couples with the diazotized amine. No elimination of the excess nitrous ion is necessary: yellow-orange color.

$$Ar{-}NH_2 \xrightarrow{HNO_2} Ar{-}\overset{+}{N}{\equiv}N$$

$$\text{COOH, OH, } NH_2 \text{ (PAS)} \xrightarrow{HNO_2} \text{COOH, OH, } N{\equiv}\overset{+}{N} \xrightarrow{-N_2} \text{COOH, OH, OH} \quad \longrightarrow \text{Azo dye}$$

Reagents

a. An 0.1% aqueous solution of sodium p-aminosalicylate dihydrate.

b. A 1% aqueous solution of sodium nitrite.

Procedure

To 2 ml of aqueous solution of the amine, add 1 ml of reagent *a*, chill to 0°, and add 0.5 ml of reagent *b* and 1 ml of 10% sulfuric acid. Let stand at 0° for 3 min, and add 2 ml of 20% aqueous solution of anhydrous sodium carbonate. Let stand for 15 min at room temperature, and dilute to 10 ml with water. Read at 425 nm.

Results

	A = 0.3 (1 cm cell) Sample, μg
Aniline	14.7
p-Aminobenzoic acid	14.5
p-Aminohippuric acid	15.0
Procaine (hydrochloride)	25.0
Sulfanilamide	16.0

Notes

1. The instability of diazotized PAS allows the colorimetric determination of m-aminophenol in impure PAS (32).

2. When heated in acid medium, PAS is readily decarboxylated to m-aminophenol, and is so determined in biological fluids (33).

3. Some side reactions intervene during the decomposition of diazotized PAS, but under suitable conditions they do not interfere in the above determinations (34).

4. Diazo coupling, see Chapter 16, Section I.

L. Sodium Nitrite and N-(1-Naphthyl)ethylenediamine Dihydrochloride

Principle (35)

Diazotization with nitrous acid, and development of the diazo compound with N-(1-naphthyl)ethylenediamine (Bratton-Marshall reagent) in acid medium: red-purple color.

$$Ar{-}NH_2 \xrightarrow{NO_2^-} Ar{-}\overset{+}{N}{\equiv}N \xrightarrow{\text{1-naphthyl-}NH{-}CH_2{-}CH_2{-}NH_2} \text{4-}(Ar{-}N{=}N)\text{-1-naphthyl-}NH{-}CH_2{-}CH_2{-}NH_2$$

Reagents

a. An 0.25% aqueous solution of sodium nitrite.

b. A 2.5% aqueous solution of ammonium sulfamate.

c. A 3 M aqueous solution of sodium acetate.

d. An 0.1% aqueous solution of N-(1-naphthyl)ethylenediamine dihydrochloride.

Procedure

To 1 ml of neutral aqueous solution of the amine, add 1 ml of 2 N hydrochloric acid and 1 ml of reagent a. Let stand for 15 min, add 1 ml of reagent b, shake vigorously, let stand for 1 min, add 2 ml of reagent c and 1 ml of reagent d. Let stand for 5 min, add 0.5 ml of concentrated hydrochloric acid, mix, and read.

Results

	λ Max, nm	A = 0.3 (1 cm cell) Sample, μg
Aniline	535	6.0
p-Aminobenzoic acid	525	8.0
Sulfanilamide	515	9.2

Notes

1. Bratton-Marshall reagent was particularly used for the colorimetric determination of sulfonamides (36).

2. Diazo coupling, see Chapter 16, Section I.

V. PRIMARY AROMATIC AMINES (F)

A. Sodium 1,2-Naphthoquinone-4-sulfonate and Potassium Borohydride

Principle (37)

See Chapter 4, Section XI.B, fluorimetric determination of primary and secondary aliphatic amines.

Reagents

a. An 0.5% aqueous solution of sodium 1,2-naphthoquinone-4-sulfonate. Prepare freshly, and keep at 0°, protected against light.

b. Dissolve 0.050 g of potassium borohydride in 1 ml of 0.1 N sodium hydroxide, and dilute to 100 ml with water.

Procedure

To 1 ml of neutral aqueous solution of the amine, add 0.2 ml of reagent a, mix, transfer the solution into a 10 ml separatory funnel containing 2 ml of methylene chloride, rinse the tube with 1 ml of water, and pour the rinsing into the funnel. Shake immediately for about 5 sec, let stand until complete decantation, and collect the whole bulk of the organic layer. The tubes being then protected against light, add 2 ml of ethanol, 0.1 ml of reagent b, mix, and let stand for 1 min. Add 0.5 ml of 0.01 N hydrochloric acid, mix, and read at exc: 366 nm; em: 470 nm.

Standard: A solution of 2,6-dimethyl-3,5-dicarbethoxy-1,4-dihydropyridine in ethanol-water 1:49.

Results

	Determination limits, μg	Reading 50 Sample, μg	Standard, μg/ml
Aniline	0.2 - 1.0	0.48	1.89
o-Toluidine	0.3 - 1.5	0.72	1.92
m-Toluidine	0.2 - 1.0	0.48	1.64
p-Toluidine	0.2 - 1.0	0.48	1.67
p-Anisidine	0.3 - 1.5	0.73	1.85
o-Aminophenol	0.2 - 1.0	0.48	1.82

Notes

1. Compounds substituted on the aromatic ring with groups which favor the solubility in water, such as aminobenzoic acids or sulfonamides, cannot be determined, since their naphthoquinone derivatives are not extractable into methylene chloride under the above conditions.

2. Secondary aromatic amines react but weakly (5-50 μg), and the emission maximum is located at 450 nm. Tertiary arylamines do not react.

3. According to this procedure, the reaction is negative with primary and secondary aliphatic amines.

4. Reaction between amino compounds and 1,2-naphthoquinone-4-sulfonic acid, see Chapter 16, Section VII.

B. Sodium Nitrite, 2,6-Diaminopyridine, and Ammoniacal Cupric Sulfate Solution

Principle (38)

Diazotization with nitrous acid, coupling with 2,6-diaminopyridine, and oxidation to give an 8-substituted 6-aminopyridinotriazole: blue to green fluorescence.

$$Ar{-}NH_2 \xrightarrow{NO_2^-} Ar{-}\overset{+}{N}{\equiv}N \xrightarrow{H_2N\text{-}C_5H_3N\text{-}NH_2} \text{(2,6-diamino-3-(Ar-N=N)-pyridine)} \xrightarrow{\text{Oxid.}} \text{(6-amino-2-Ar-pyridinotriazole)}$$

Reagents

a. An 0.1 M aqueous solution of sodium nitrite.

b. A 2% aqueous solution of ammonium sulfamate.

c. Prepare a buffer for pH 5: Adjust 400 ml of 1 M aqueous solution of sodium acetate to pH 5 with 1 N hydrochloric acid (about 80 ml), and dilute to 1 liter with water. The reagent is an 0.05% solution of 2,6-diaminopyridine in this buffer. The 2,6-diaminopyridine should be colorless. If necessary, recrystallize once from benzene.

d. Dissolve 1 g of anhydrous cupric sulfate in 10 ml of a 7:3 mixture of water and aqueous ammonium hydroxide solution (d = 0.92, 22° Bé).

Procedure

To 2 ml of sample solution in 0.1 N hydrochloric acid, add 0.05 ml of 6 N hydrochloric acid and 0.3 ml of reagent a. Mix, and maintain at 0° for 10 min in ice water. Add 1 ml of reagent b, mix, and let stand for 5 min. Add 2 ml of reagent c, 0.4 ml of 1 N sodium hydroxide, mix, and let stand for 20 min at room temperature. Transfer the solution into a 25 ml separatory funnel, rinse the tube with 2 ml of water, and pour the rinsing into the funnel. Add 5 ml of benzene or ethyl acetate (see Results), and shake for 1 min. Discard the aqueous layer, and wash the organic phase twice with 6 ml portions of the buffer for pH 5, then twice with 6 ml portions of water. Transfer the organic layer into a 10 ml volumetric flask, rinse the funnel with 3 ml of the corresponding solvent, and pour the rinsing into the flask. Evaporate the solvent to dryness on a steam bath, with the aid of a stream of nitrogen. Dissolve the residue in 2 ml of water, add 0.4 ml of reagent d, stopper the flask, and heat at 100° for 35 min. Cool rapidly in ice water to about room temperature, add 0.4 ml of 6 N hydrochloric acid, dilute to 10 ml with water, let stand for 10 min, and read.

Standard: According to the compound tested (see Results), two standards are used:

(a) A solution of esculin in ethanol.

(b) A solution of esculin in a 1:1 mixture of ethanol and 1 N sulfuric acid.

Results

	Solvent[a]	Exc, nm	Em, nm	Determination limits, μg	Reading 50 Sample, μg	Reading 50 Standard, μg/ml
Aniline	B	353	410	0.5 - 2.5	1.25	8.15 (a)
m-Toluidine	B	355	415	0.7 - 3.5	1.75	8.15 (a)
p-Anisidine	B	365	475	1 - 5	2.45	1.07 (b)
p-Aminobenzoic acid	EA	355	405	0.7 - 3.5	1.75	8.35 (a)
Sulfanilamide	EA	353	400	0.7 - 3.5	1.71	6.45 (a)
β-Naphthylamine	B	345	495	1 - 5	2.4	0.065 (b)

[a] B = benzene, EA = ethyl acetate.

Notes

1. The method allows indeed much more sensitive determinations (38), but at concentrations lower than those given under Results, the linear relationship does not hold.

2. Diazo coupling, see Chapter 16, Section I.

REFERENCES

1. J. Bartos, *Pharm. Weekblad,* 93, 594 (1958).

2. R. Wizinger and H.V. Tobel, *Helv. Chim. Acta,* 40, 1305 (1957).

3. M. Blackburn, G.B. Sankey, A. Robertson, and W.B. Whalley, *J. Chem. Soc.,* 1573 (1957).

4. M. Pesez and M. Brunet, unpublished results.

5. K. Formanek and H. Höller, *J. Chromatog.*, 2, 652 (1959).

6. Cf. E. Sawicki, T.R. Hauser, and T.W. Stanley, *Anal. Chem.*, 31, 2063 (1959).

7. E. Sawicki, J.L. Noe, and F.T. Fox, *Talanta*, 8, 257 (1961).

8. From E. Sawicki, T.W. Stanley, T.R. Hauser, W. Elbert, and J.L. Noe, *Anal. Chem.*, 33, 722 (1961).

9. Cf. V. Anger, *Mikrochim. Acta*, 2, 3 (1937).

10. O. Baudisch, *Ber.*, 54, 413 (1921).

11. F. Feigl, V. Anger, and O. Frehden, *Mikrochemie*, 15, 181 (1934).

12. *F.S.A.O.V.*, 179-193.

13. M. Pesez and A. Petit, *Bull. Soc. Chim. Fr.*, 122 (1947).

14. M. Pesez and J. Bartos, *Bull. Soc. Chim. Fr.*, 3802 (1966).

15. M. Strell and S. Reindl, *Artzneim.-Forsch.*, 11, 552 (1961).

16. K. Kato, M. Umeda, and S. Tsubota, *J. Pharm. Soc. Japan*, 24, 116 (1964).

17. Cf. F. Feigl, V. Anger, and R. Zappert, *Mikrochemie*, 16, 67 (1934).

18. From J.T. Stewart, T.D. Shaw, and A.B. Ray, *Anal. Chem.*, 41, 360 (1969); J.T. Stewart, A.B. Ray, and W.B. Fackler, *J. Pharm. Sci.*, 58, 1261 (1969).

19. G. Hillmann, *Z. Physiol. Chem.*, 277, 222 (1943).

20. W. Winkler, *Chem. Ber.*, 81, 256 (1948).

21. R. Riemschneider and C. Weygand, *Monatsh. Chem.*, 86, 201 (1955).

22. H. Wartenberg, *Acta Histochem.*, 3, 145 (1956).

23. R. Riemschneider and J. Wierer, *Z. Anal. Chem.*, 193, 186 (1963).

24. H. Jensen, M. Bourhis, and E. Neuzil, *Bull. Soc. Pharm. Bordeaux*, 108, 177 (1969).

25. M. Bourhis, H. Jensen, and E. Neuzil, *Ann. Pharm. Fr.*, 28, 561 (1970).

26. F. Weygand, H. Weber, E. Maekawa, and G. Eberhardt, *Chem. Ber.*, 89, 1994 (1956).

27. M. Kagawa, *Bunseki Kagaku*, 16, 669, 671 (1967).

28. Cf. M. Pesez, *Ann. Chim. Anal.*, 25, 37 (1943).

29. V. Arreguine, *Anales Asoc. Quim. Argentina*, 31, 38 (1943).

30. L. Vignoli, B. Cristau, and J.P. Defretin, *Ann. Pharm. Fr.*, 23, 715 (1965).

31. M. Pesez, *Bull. Soc. Chim. Biol.*, 33, 195 (1951).

32. M. Pesez, *Bull. Soc. Chim. Fr.*, 918 (1949).

33. M. Pesez, *Bull. Soc. Chim. Biol.*, 31, 1369 (1949).

34. Cf. A. Sezerat, *Ann. Pharm. Fr.*, 13, 350 (1955).

35. J.W. Daniel, *Analyst*, 86, 640 (1961).

36. A.C. Bratton and E.K. Marshall, Jr., *J. Biol. Chem.*, 128, 537 (1939).

37. J. Bartos and M. Pesez, *Talanta*, 19, 93 (1972).

38. From L.J. Dombrowski and E.L. Pratt, *Anal. Chem.*, 43, 1042 (1971).

Chapter 6

GUANIDINES AND UREAS

I. INTRODUCTION

A. Colorimetry

Monosubstituted guanidines can be determined through the Sakaguchi reaction with 8-hydroxyquinoline and sodium hypochlorite, in the presence of thymine as color-stabilizing agent. The sensitivity varies with the compound tested (from 9 to 95 μg for A = 0.3). The mechanism of the reaction is at least partly elucidated.

Monosubstituted ureas react with diacetyl monoxime and sodium N-phenylsulfanilate in the presence of an oxidant to give a violaceous color allowing determinations at levels of 7.5 - 17.5 μg (nitrourea reacts but weakly). The mechanism of this reaction is unknown.

B. Fluorimetry

Guanidine, mono-, and N,N-disubstituted guanidines develop a yellow-green fluorescence with ninhydrin in alkaline medium. The sample size varies with the compound tested (1.9 - 28.8 μg). The mechanism of the reaction is but partly known.

Beside guanidine itself, only monosubstituted guanidines react with phenanthrenequinone in alkaline medium, according to an unknown mechanism. The sensitivity varies with the compound tested.

Urea, mono-, and N,N'-disubstituted ureas are dehydrated to the corresponding carbodiimides, which afford fluorescent adducts of unknown structure with 2,4,6-trinitrobenzoic acid. The sensitivity varies from 0.24 to 2 μg (for reading 50).

N,N'-Disubstituted thioureas are determined according to the same principle: Carbodiimides are obtained by reacting the thioureas with mercuric oxide.

II. MONOSUBSTITUTED GUANIDINES

A. Thymine, 8-Hydroxyquinoline, and Sodium Hypochlorite

Principle (1,2,3)

Oxidative condensation with 8-hydroxyquinoline in the presence of thymine as color-stabilizing agent: orange to pink-orange color.

Reagents

a. A 1% solution of thymine in 1 N sodium hydroxide.

b. A 2% solution of 8-hydroxyquinoline in ethanol.

c. A dilute aqueous solution of sodium hypochlorite: 100 ml contains 0.316 g active chlorine.

Procedure

To 1 ml of aqueous solution of the guanidine derivative, add 0.5 ml of reagent *a* and 0.1 ml of reagent *b*. Mix, immerse the tubes in ice water, and after 2 min add 0.5 ml of prechilled reagent *c*. Let stand for 3 min at o°, add 3 ml of methanol, and read.

Results	λ Max, nm	A = 0.3 (1 cm cell) Sample, μg
3,4,5-Trimethoxybenzyl-guanidine (sulfate)[a]	495	20
Guanethidine (sulfate)[b]	490	95
Glycocyamine	490	9
Arginine	490	16
Streptomycin (sulfate)	510	49

[a]Operate at room temperature, and let stand for 1 min only.

[b]Beer's law is followed up to A = 0.5.

Compounds insoluble in water can be dissolved in the minimum amount of ethanol or pyridine, and the solution diluted with water so that the final concentration of the solvent be less than or equal to 1%.

Notes

1. In the original reaction developed by Sakaguchi for arginine in 1925 (1), the reagents were sodium hypochlorite and α-naphthol, but the color was very unstable. Modifications of this reaction have been many.

Hypobromite (4), N-bromosuccinimide (5,6), and 1,3-dibromo-5,5-dimethylhydantoin (6) were proposed instead of hypochlorite, 1-naphthol-8-sulfonic acid (7), and 8-hydroxyquinoline (2) instead of α-naphthol. It was established that the color can be stabilized, either by extracting it into butanol (8), or by destroying the excess hypohalic acid with urea (2), metabisulfite, thiosulfate (9), urethan (10), or sulfosalicylic acid (11), and it was shown that thymine further increases the stability and intensifies the color (3).

Many authors were mainly interested in the determination of arginine and of some of its closely related biological derivatives, and the procedures which have been devised for this purpose must often be somewhat modified when applied to other guanidine compounds.

The procedure herein described was developed in our laboratories, and the work has been aimed at the setting of a method allowing the colorimetric determination of a variety of monosubstituted guanidines. Oxine was found to give more satisfactory results than α-naphthol, and it was observed that the efficiency of urea (analytical grade) sometimes depends on the origin of the sample used. The most reproducible data were obtained with thymine.

Under the above conditions, the color developed by arginine is stable for about 2 hr.

2. The structure of the dye formed by reacting butylguanidine, α-naphthol, and hypobromite was elucidated (I), and the following mechanism was tentatively proposed (12):

$$R{-}N{=}C(NH_2){-}NH_2 \xrightarrow{BrO^-} R{-}N{=}C(NH_2){-}NHBr \xrightarrow{OH^-} R{-}N{=}C(NH_2){-}\bar{N}{-}Br \xrightarrow{-Br^-} R{-}N{=}C{=}N{-}NH_2$$

$$\text{α-naphthol (OH)} \xrightarrow{BrO^-} \text{2-bromo-1,4-naphthoquinone (Br)} + R{-}N{=}C{=}N{-}NH_2 \longrightarrow \text{(Br, O) naphthoquinone} {=}N{-}NH{-}CO{-}NH{-}R$$

(I)

3. Accordingly, the reaction is specific for monosubstituted guanidines. Creatine does not react, nor guanidine, probably on account of its complete oxidation by hypohalogenite.

4. The reaction is but weakly positive with p-chlorophenylguanidine and sulfaguanidine, and negative

with 3,4,5-trimethoxybenzoylguanidine. These compounds bear electron-attracting substituents on the benzene ring.

On the other hand, the reaction is positive with chlorguanide hydrochloride (II) (A = 0.3 for 76 μg), which is not a monosubstituted guanidine, but according to our experiments thymine might play a particular role in this case.

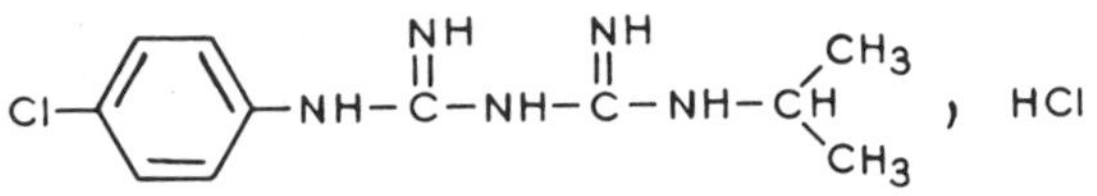

$$Cl-C_6H_4-NH-\overset{NH}{\overset{\|}{C}}-NH-\overset{NH}{\overset{\|}{C}}-NH-CH(CH_3)_2 \text{ , HCl}$$

(II)

III. MONOSUBSTITUTED UREAS

A. Diacetyl Monoxime, Sodium N-phenylsulfanilate, and Potassium Persulfate

Principle (13)

Reaction with diacetyl monoxime and sodium N-phenylsulfanilate (diphenylamine-p-sulfonate) in sulfuric acid medium, and oxidation by potassium persulfate: violaceous color.

Reagents

a. A 1% aqueous solution of sodium N-phenylsulfanilate. Keep in the dark.

b. A 3% aqueous solution of diacetyl monoxime.

c. A 1% aqueous solution of potassium persulfate.

Procedure

To 3 ml of aqueous solution of the urea derivative, add 6 ml of 50% sulfuric acid, 0.1 ml of reagent a, and 0.25 ml of reagent b. Shake vigorously, fit the tubes with air-condensers, and heat at 100° for 10 min. Chill for 2 min in ice water, add 0.25 ml of reagent c, mix, and heat at 100° for exactly 1 min, then cool for 10 min in a water bath at room temperature. Read at 550 nm.

Results

	A = 0.3 (1 cm cell) Sample, μg
Monomethylurea	7.5
Monoethylurea	9.6
Nitrourea	124
Citrulline	17.5

Notes

1. The heating and cooling times are critical. All the tubes must be treated exactly in the same fashion, otherwise Beer's law is not satisfactorily observed.

2. Beer's law is not followed with urea.

3. The absorption maximum is located at 560 nm, but more accurate results are obtained when reading at 550 nm.

4. Only a very faint color is given by disubstituted ureas, such as N,N'-dicyclohexyl- or diphenylurea.

IV. GUANIDINE, MONO-, AND N,N-DISUBSTITUTED GUANIDINES (F)

A. Ninhydrin

Principle (14,15)

Condensation with ninhydrin in alkaline medium: yellow-green fluorescence.

Reagent

An 0.25% aqueous solution of ninhydrin. Keep in the dark.

Procedure

To 1 ml of aqueous solution of the guanidine derivative, add 2 ml of reagent and 2 ml of 0.5 N sodium hydroxide, and let stand in the dark for 30 min at 19-21°. Read at exc: 405 nm.

Standard: A solution of 2,6-dimthyl-3,5-diacetyl-1,4-dihydropyridine in ethanol.

Results

	Em, nm	Determination limits, μg	Reading 50 Sample, μg	Standard, μg/ml
Guanidine (sulfate)	505	6 - 30	14.4	0.46
3,4,5-Trimethoxy-benzylguanidine (sulfate)	510	12 - 60	28.8	0.52
Guanethidine (sulfate)	510	8 - 40	19.2	0.63
Glycocyamine	505	1.6 - 8.0	3.76	0.53
Arginine	505	4 - 20	9.8	0.50
Creatine	505	0.8 - 4.0	1.92	0.44

Notes

1. The reaction is but weakly positive with p-chlorophenylguanidine, sulfaguanidine, and streptomycin, and negative with 3,4,5-trimethoxybenzoylguanidine. These compounds bear electron-attracting substituents on the benzene ring (with the exception of streptomycin).

2. Conn and Davis (14) postulated that the fluorescent species is a Schiff base-type adduct formed by reaction between the free $-NH_2$ group of the guanidine derivative and the aldehyde group of o-carboxyphenylglyoxal (IV) proceeding from the opening of the five-membered ring of ninhydrin (III) in alkaline medium (16). According to Samejina et al. (17), the analytical data would agree with a ternary compound corresponding to a combination of the guanidine, one equivalent of o-carboxyphenylglyoxal, and one equivalent of o-carboxyphenylglyoxylic aldehyde.

o-Carboxyphenylglyoxal strongly absorbs the exciting light (405 nm). However, in alkaline medium, it is slowly converted into o-carboxymandelic acid (V), which is colorless, nonfluorescent, and has a negligible absorbance at the excitation wavelength. The fluorescent species itself being unstable under these solvent conditions, the reaction time and the temperature are highly critical.

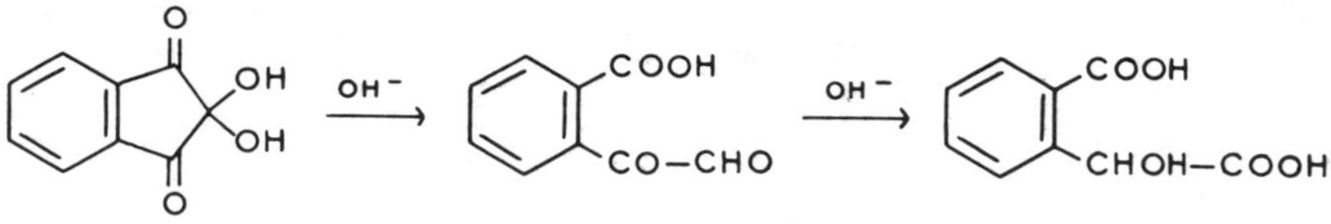

(III) (IV) (V)

3. Ninhydrin, see Chapter 16, Section X.

V. GUANIDINE AND MONOSUBSTITUTED GUANIDINES (F)

A. Phenanthrenequinone

Principle (18,19)

According to an unknown mechanism, condensation with phenanthrenequinone in alkaline medium to give 2-amino-1H-phenenthro[9,10-d]imidazole: blue-violet fluorescence.

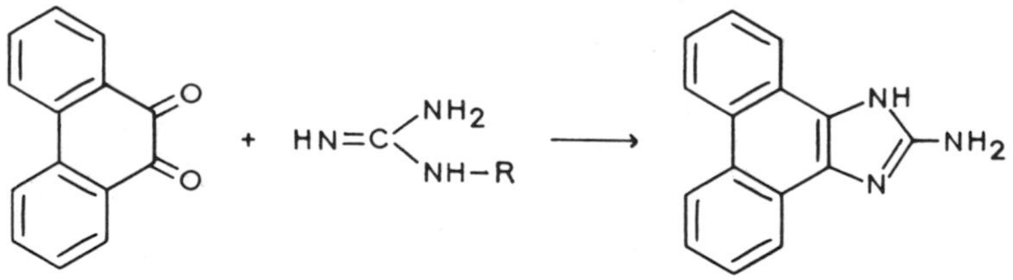

Reagent

An 0.05% solution of phenanthrenequinone in ethanol.

Procedure

To 0.5 ml of aqueous solution of the guanidine derivative, add 0.1 ml of approximately 10 N sodium hydroxide and 0.5 ml of reagent. Mix, and let stand for 30 min at room temperature. Add 0.25 ml of concentrated hydrochloric acid and 2.5 ml of water. Read at exc: 360 nm; em: 400 nm.

Standard: A solution of quinine sulfate in 0.1 N sulfuric acid.

Results

	Determination limits, μg	Reading 50 Sample, μg	Reading 50 Standard, μg/ml
Guanidine (sulfate)	1.5 - 6.0	2.5	2.7
p-Chlorophenyl-guanidine	37 - 150	63.7	2.28
3,4,5-Trimethoxy-benzylguanidine (sulfate)	2 - 10	4.2	2.2
Guanethidine (sulfate)	2 - 10	4.0	2.27
Glycocyamine	0.8 - 4.0	1.6	1.89
Arginine	1.5 - 7.5	3.0	1.66
Streptomycin (sulfate)	15 - 60	22.5	2.1
Dihydrostreptomycin (sulfate)	16 - 80	32.0	1.82

Notes

1. After acidification and dilution with water, the solution is often slightly turbid. Filtration is not necessary, since this turbidity does not interfere with measurements.

2. The reaction is negative with creatine and chlorguanide.

VI. UREA AND MONOSUBSTITUTED UREAS (F)

A. 2,4,6-Trinitrobenzoic Acid and p-Toluenesulfonyl Chloride

Principle (20)

Upon reaction with p-toluenesulfonyl chloride in the presence of pyridine, dehydration of the urea to the corresponding carbodiimide, and development with 2,4,6-trinitrobenzoic acid: yellow fluorescence.

$$R{-}NH{-}CO{-}NH_2 \xrightarrow[C_5H_5N]{H_3C{-}C_6H_4{-}SO_2Cl} R{-}N{=}C{=}NH \xrightarrow{\text{2,4,6-}(NO_2)_3C_6H_2COOH} \text{Fluorescent species}$$

Reagents

a. A 2% solution of 2,4,6-trinitrobenzoic acid in acetone. Prepare fresh immediately before use.

b. Dilute a mixture of 1.0 g of p-toluenesulfonyl chloride and 0.7 ml of pyridine to 10 ml with methylene chloride. Prepare fresh immediately before use.

Procedure

Evaporate to dryness at 60-70° the solution of the sample in methylene chloride or in a mixture of methanol and methylene chloride. From then on, protect the process against light, with the tubes immersed in a water bath at 20°. To the residue of evaporation, add 0.5 ml of reagent *a* and 0.1 ml of reagent *b*, and immediately shake for about 10 sec. Let stand for 30 min, add 4 ml of ethyl acetate, mix, and let stand for 2 min. The precipitate formed during the manipulation settling out very readily, no filtration is necessary. Read at exc: 405 nm; em: 570 nm.

The fluorescence is stable for about 10 min at 20°, protected against light.

Standard: A solution of erythrosine sodium in 99% ethanol.

Results

	Determination limits, μg	Reading 50 Sample, μg	Reading 50 Standard, μg/ml
Urea	0.2 - 1.0	0.44	5.6
Monomethylurea	0.4 - 2.0	0.84	5.5
Monoethylurea	0.4 - 2.0	0.80	6.0
Nitrourea	1 - 5	2.0	4.4

Notes

1. N,N'-Disubstituted ureas give only a very faint fluorescence. However, they can be determined with a somewhat different procedure based on the same principle (see following method).

2. Some samples of 2,4,6-trinitrobenzoic acid afford highly fluorescent blanks. This drawback can be overcome by washing the acid with hot benzene.

3. Although, as a rule, fluorescence is quenched by nitro groups attached to the benzene ring, some trinitro compounds develop a strong yellow or orange, sometimes green fluorescence. They often are picryl derivatives of biguanide, N-oxiamidine, or guanylisourea, and some of them may be related to Meisenheimer-like complexes (21). The structure of the species formed by reacting a carbodiimide with trinitrobenzoic acid has not been elucidated.

VII. N,N'-DISUBSTITUTED UREAS (F)

A. p-Toluenesulfonyl Chloride and 2,4,6-Trinitrobenzoic Acid

Principle

See preceding method, determination of urea and monosubstituted ureas.

Reagents

a. Dilute a mixture of 1.0 g of p-toluenesulfonyl chloride and 0.7 ml of pyridine to 10 ml with methylene chloride. Prepare fresh immediately before use.

b. A 2% solution of 2,4,6-trinitrobenzoic acid in acetone. Prepare fresh immediately before use.

Procedure

Evaporate the sample solution to dryness at 50° in methylene chloride. Let cool to room temperature, add 0.1 ml of reagent a, and let stand for the given length of time (see Results). Then, protected against light, add

4 ml of methylene chloride, 0.5 ml of reagent b, mix, and read at exc: 405 nm; em: 570 nm.

The fluorescence is stable for about 10 min at room temperature, protected against light.

Standard: A solution of erythrosine sodium in 99% ethanol.

Results

	Reaction time, min	Determination limits, μg	Reading 50 Sample, μg	Standard, μg/ml
N,N'-Dimethylurea	10	0.1 - 0.5	0.24	6.0
N,N'-Dicyclohexylurea	20	0.1 - 0.5	0.24	5.25
N,N'-Dibenzylurea	45	0.4 - 2.0	1.0	5.4

Notes

1. Dixanthylurea also reacts in the range 2-10 μg but affords irregular results.

2. The reaction is negative with N,N'-aromatic disubstituted ureas, when the benzene ring is bonded directly to the nitrogen atom.

3. Monosubstituted ureas give only a very faint fluorescence. They can be determined as described in the preceding method.

4. Comparison of results obtained from N,N'-dicyclohexylurea and from N,N'-dicyclohexylcarbodiimide established that the yield of the dehydration step is almost quantitative.

5. Dehydration of N,N'-disubstituted ureas with p-toluenesulfonyl chloride in the presence of pyridine makes N,N'-disubstituted carbodiimides accessible on the preparative and plant scales (22).

6. Determination of N,N'-disubstituted thioureas, see following method.

VIII. N,N'-DISUBSTITUTED THIOUREAS (F)

A. Mercuric Oxide and 2,4,6-Trinitrobenzoic Acid

Principle (20)

Reaction with mercuric oxide, and development of the carbodiimide formed with 2,4,6-trinitrobenzoic acid (see Section VI.A): yellow fluorescence.

$$R-NH-CS-NH-R \xrightarrow{HgO} R-N{=}C{=}N-R \xrightarrow{\text{2,4,6-}(NO_2)_3C_6H_2COOH} \text{Fluorescent species}$$

Reagent

An 0.2% solution of 2,4,6-trinitrobenzoic acid in acetone. Prepare fresh immediately before use.

Procedure

To 4 ml of sample solution in methylene chloride, add approximately 0.05 g of yellow mercuric oxide, shake for 10 sec, and let stand for 5 min. Filter without shaking, and add 0.5 ml of reagent to the filtrate protected against light. Mix, and read immediately at exc: 405 nm; em: 570 nm.

Standard: A solution of erythrosine sodium in 99% ethanol.

Results

	Determination limits, μg	Reading 50 Sample, μg	Standard, μg/ml
N,N'-Dicyclohexylthiourea	0.1 - 0.5	0.25	5.8
N,N'-Dibenzylthiourea	0.2 - 1.0	0.48	6.5

Notes

1. The reaction is practically negative with N,N'-aromatic disubstituted thioureas, when the benzene ring

is bonded directly to the nitrogen atom (di-p-tolylthiourea), and with thiourea itself.

2. Desulfurization of N,N'-disubstituted thioureas with mercuric oxide makes the corresponding carbodiimides accessible on the preparative scale (23).

3. Fluorimetric determination of urea, monosubstituted ureas, and N,N'-disubstituted ureas, see Sections VI.A and VII.A.

REFERENCES

1. Cf. S. Sakaguchi, *J. Biochem. (Tokyo)*, 5, 25, 133 (1925).

2. Cf. S. Sakaguchi, *Japan Med. J.*, 1, 278 (1948); *J. Biochem. (Japan)*, 37, 231 (1950); 38, 91 (1951).

3. Cf. J.F. Van Pilsum, R.P. Martin, E. Kito, and J. Hess, *J. Biol. Chem.*, 222, 225 (1956).

4. C.J. Weber, *J. Biol. Chem.*, 86, 217 (1930).

5. I. Szilagyi and I. Szabo, *Nature*, 181, 52 (1958).

6. K. Satake and J.M. Luck, *Bull. Soc. Chim. Biol.*, 40, 1743 (1958).

7. H. Kraut, E.N. Schrader-Beielstein, and M. Weber, *Z. Physiol. Chem.*, 286, 248 (1951).

8. G. Ceriotti and L. Spandrio, *Biochem. J.*, 66, 603 (1957).

9. J.F. Van Pilsum, E. Kito, J. Hess, and K. Eik-Nes, *Federation Proc.*, 12, 432 (1953).

10. J.P. Salta and Y. Khouvine, *Bull. Soc. Chim. Biol.*, 35, 697 (1953).

11. S. Akamatsu and T. Watanabe, *J. Biochem. (Tokyo)*, 49, 566 (1961).

12. A. Heesing and K. Hope, *Chem. Ber.*, 100, 3649 (1967).

13. S.B. Koritz and P.P. Cohen, *J. Biol. Chem.*, 209, 145 (1954).

14. Cf. R.B. Conn, Jr., and R.B. Davis, *Nature*, 183, 1053 (1959).

15. K. Beyermann and H. Wisser, *Z. Anal. Chem.*, 245, 311 (1969).

16. S. Ruhemann, *J. Chem. Soc.*, 97, 2025 (1910).

17. K. Samejina, W. Dairman, and S. Udenfriend, *Anal. Biochem.*, 42, 222 (1971).

18. Cf. S. Yamada and H.A. Itano, *Biochem. Biophys. Acta*, 130, 538 (1966).

19. J. Bartos and M. Pesez, *Talanta*, 19, 93 (1972).

20. J. Bartos, *Ann. Pharm. Fr.*, 28, 321 (1970).

21. Cf. J. Bartos, *Bull. Soc. Chim. Fr.*, 3694 (1965).

22. G. Amiard and R. Heymes, *Bull. Soc. Chim. Fr.*, 1360 (1956).

23. E. Schmidt, F. Hitzler and E. Lahde, *Ber.*, 71, 1933 (1938).

Chapter 7

MISCELLANEOUS NITROGEN DERIVATIVES

I. INTRODUCTION

A. Colorimetry

Aliphatic and aromatic nitro compounds, and nitriles, can be reduced to the corresponding primary amines, to which are then applied reactions already described for alkyl- and arylamines.

Amines so obtained from aliphatic nitro compounds can be estimated with the silver nitrate-manganese sulfate test paper (semiquantitative determination); those obtained from nitriles can be estimated with the same reagent, or determined by coupling with diazotized p-nitroaniline. Amines obtained from aromatic nitro compounds are determined as Schiff bases with 3,3-diphenylacrolein or p-dimethylaminobenzaldehyde, or by diazotization and subsequent coupling with β-naphthol. Sensitivities depend on the method selected and the compound tested.

Fluorene acts as a mild reductant and reduces aromatic dinitro compounds in alkaline medium to give quinoidal aci-nitro-nitroso dianions.

Nitrite ion is liberated when primary and secondary aliphatic nitro compounds are reacted with hydrogen peroxide in alkaline medium, thus allowing determination of these compounds through diazotization of p-sulfanilic acid and coupling with 3-hydroxy-2-naphthoyic acid. The reaction is of good sensitivity (4.4-5.5 μg for A = 0.3).

One hydrogen atom of the methylene group adjacent to the nitro group of primary aliphatic nitro compounds being highly active, these compounds can be converted into their aci form in alkaline medium, then developed with ferric ion (A = 0.3 is given by about 100 μg of the compound tested). The active hydrogen atom also allows their determination with diazotized 2-aminobenzothiazole, but the reaction is far from being specific for this class of compounds.

Aromatic nitro compounds also are converted into their aci form in alkaline medium, affording miscellaneous colors whose sensitivity varies widely with the compound tested.

Aliphatic 2-nitro-1-hydroxy compounds are decomposed in alkaline medium, and the liberated formaldehyde can be developed which chromotropic acid (A = 0.3 for 16.5-29.5 μg).

B. Fluorimetry

Fluorimetric methods are available only for mono- and N,N-disubstituted hydrazines, and aliphatic 2-nitro-1-hydroxy compounds and nitriles, and allow determinations at levels of a few micrograms.

Mono- and N,N-disubstituted hydrazines reduce 1,2-naphthoquinone-4-sulfonic acid to the highly fluorescent 1,2-dihydroxynaphthalene-4-sulfonic acid.

Formaldehyde liberated by aliphatic 2-nitro-1-hydroxy compounds (see above) is determined through the Hantzsch reaction with ethyl acetoacetate and ammonia.

A similar reaction allows the determination of nitriles after their reduction to primary amines. The reagents are acetylacetone and formaldehyde.

II. ALIPHATIC NITRO COMPOUNDS

A. Titanium Trichloride, Silver Nitrate, and Manganese Sulfate

Principle (1)

Reduction of the nitro compound with titanium trichloride, and estimation of the amine formed with a test paper impregnated with a mixture of silver nitrate and manganese sulfate (see Reagent b).

Reagents

a. A 1% aqueous solution of titanium trichloride.

b. Silver-manganese test paper, see Chapter 4, Section III.A, under Reagent.

Procedure

Operate with the device described in b. Pipet 1 ml of sample solution in water or in a mixture of water and ethanol into the vial, add 0.5 ml of reagent a, and let stand at room temperature for 5 min. Add 1 ml of approximately 10 N sodium hydroxide without shaking, allowing the solution to run down the wall of the vial, stopper with the cap in which the test paper has been inserted, and mix with a circular motion, avoiding projections of droplets on the paper. Heat at 50° for 1 hr. The gray color of the paper can be compared with standards obtained from known amounts of the nitro compound, either visually or spectrophotometrically with the aid of a reflectance accessory.

Results

	Lower limit of detection, μg
Nitromethane	40
Nitroethane	50
1-Nitropropane	65
2-Nitropropane	65

With higher molecular weight compounds, it may be necessary to prolong the heating time.

Note

Nitriles can also be estimated (Section VIII.A).

III. PRIMARY AND SECONDARY ALIPHATIC NITRO COMPOUNDS

A. Hydrogen Peroxide, p-Sulfanilic and 3-Hydroxy-2-naphthoic Acids

Principle (2,3)

The nitrite ion liberated by reaction with hydrogen peroxide in the presence of sodium hydroxide is developed by diazotization of p-sulfanilic acid and coupling with 3-hydroxy-2-naphthoic acid: pink color.

$$>CH-NO_2 \xrightarrow{OH^-} >C=NO_2^- \xrightarrow{H_2O_2} >C=O + NO_2^-$$

The excess hydrogen peroxide is destroyed by nickel chloride.

$$NO_2^- + H_2N-C_6H_4-SO_3H \rightarrow \overset{+}{N}\equiv N-C_6H_4-SO_3H \rightarrow$$ 3-hydroxy-2-naphthoic acid (COOH, OH) with $N=N-C_6H_4-SO_3H$

Reagents

a. A 6% aqueous solution of hydrogen peroxide (20 volumes oxygen).

b. An 0.1% aqueous solution of nickel chloride hexahydrate.

c. A 1% aqueous solution of p-sulfanilic acid.

d. A 1% solution of 3-hydroxy-2-naphthoic acid in 2 N sodium hydroxide.

Procedure

Operate protected against light. To 0.5 ml of aqueous solution of the nitro compound, add 0.2 ml of

4 N sodium hydroxide and 0.2 ml of reagent a. Heat at 100° for 15 min, add 0.1 ml of reagent b, and maintain heating at 100° for 30 min more. Add 0.5 ml of water and cool for 5 min in ice water. Add 0.25 ml of reagent c, 0.25 ml of 1:1 hydrochloric acid, mix, and let stand for 1 min at room temperature. Add 0.3 ml of reagent d, 2 ml of ethanol, mix, and read at 525 nm.

Results

	A = 0.3 (1 cm cell) Sample, μg
Nitromethane	4.4
Nitroethane	4.6
1-Nitropropane	5.5
2-Nitropropane	5.5

Note

Diazo coupling, see Chapter 16, Section I.

IV. PRIMARY ALIPHATIC NITRO COMPOUNDS

A. Sodium Hydroxide and Ferric Chloride

Principle (4)

Conversion of the nitro compound into its aci form in alkaline medium, and development with ferric chloride in acid medium: brown-red color.

Reagent

A 20% solution of ferric chloride hexahydrate in ethanol.

Procedure

To 1 ml of sample solution in dimethylformamide, add 0.4 ml of 4 N sodium hydroxide. Let stand at 20-25° for 15 min, cool to 0 + 5° in ice water, add 1.1 ml of 2 N hydrochloric acid, check the pH, and adjust it if necessary between 1 and 2 with hydrochloric acid or sodium hydroxide. Within 2 min, add 0.5 ml of reagent, and let stand for 15 min. Read at 500 nm.

Results

	A = 0.3 (1 cm cell) Sample, μg
Nitroethane	105
1-Nitropropane	120

Nitromethane practically does not react. The reaction is sensitive with 2-nitropropane, but the color is too transient to allow a colorimetric determination.

Note

In solution, primary and secondary aliphatic nitro compounds can be represented by an equilibrium mixture of the nitro form (I) and the aci form (II). In neutral or acidic medium, the aci form is the less stable and the equilibrium is shifted to the left. In alkaline medium, the mesomeric anion (III) is formed; it corresponds to the alkaline salt, and on acidification it gives the aci form (II); the equilibrium with (I) is then reached within a few minutes.

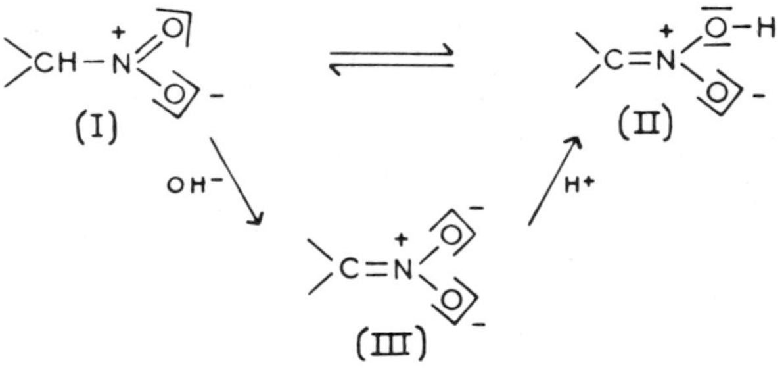

(I) ⇌ (II)

OH⁻ (I → III); H+ (III → II)

(III)

B. Diazotized 2-Aminobenzothiazole

Principle and Procedure

See Chapter 3, Section II.I, determination of phenols: violaceous red color.

Results

	λ Max, nm	A = 0.3 (1 cm cell) Sample, μg
Nitromethane	500	19
Nitroethane	500	31
1-Nitropropane	510	65

2-Nitropropane develops a faint color (A = 0.3 for 2000 μg).

V. ALIPHATIC 2-NITRO-1-HYDROXY COMPOUNDS

A. Sodium Hyroxide and Sodium Salt of Chromotropic Acid

Principle (5)

Decomposition in alkaline medium, and development of the formaldehyde formed with chromotropic acid: violet color.

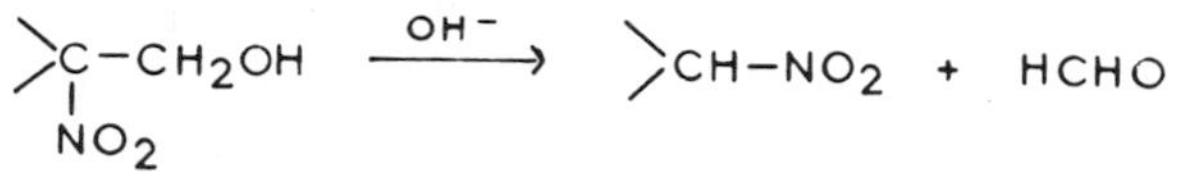

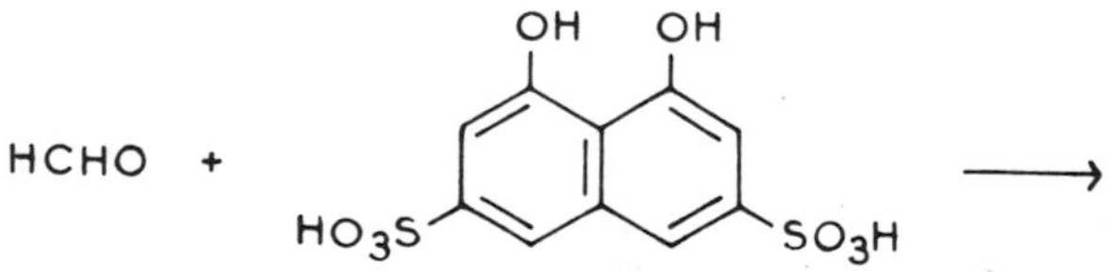

$$>C(NO_2)-CH_2OH \xrightarrow{OH^-} >CH-NO_2 + HCHO$$

$$HCHO + \text{(4,5-dihydroxynaphthalene-2,7-disulfonic acid, } HO_3S, SO_3H, OH, OH) \longrightarrow$$

Dibenzoxanthylium-type dyestuff

Reagent

A 2% aqueous solution of sodium salt of chromotropic acid.

Procedure

To 1 ml of aqueous solution of the hydroxynitro compound, add 1 ml of 0.5 N sodium hydroxide. Let stand at 25° for 5 min, add 1 ml of reagent, and 10 ml of concentrated sulfuric acid. Heat at 100° for 10 min, cool in a water bath, and read at 580 nm.

Results

	A = 0.3 (1 cm cell) Sample, μg
2-Nitro-1-butanol	29.5
2-Nitro-2-methyl-1,3-propanediol	24.0
2-Nitro-2-ethyl-1,3-propanediol	16.5

Note

Colorimetric determination of formaldehyde, see Chapter 16, Section V.

VI. AROMATIC NITRO COMPOUNDS

A. Potassium Borohydride and 3,3-Diphenylacrolein or p-Dimethylaminobenzaldehyde

Principle (6)

Reduction to primary aromatic amine by potassium borohydride in alkaline medium containing palladium, and development of the amine with 3,3-diphenylacrolein or p-dimethylaminobenzaldehyde to give a Schiff base: orange color.

$$Ar{-}NO_2 \xrightarrow[OH^-\ \ Pd]{KBH_4} Ar{-}NH_2 \begin{cases} \xrightarrow{(CH_3)_2N{-}C_6H_4{-}CHO} Ar{-}N{=}CH{-}C_6H_4{-}N(CH_3)_2 \\ \xrightarrow{(C_6H_5)_2C{=}CH{-}CHO} Ar{-}N{=}CH{-}CH{=}C(C_6H_5)_2 \end{cases}$$

Reagents

a. An 0.5% aqueous solution of palladium chloride.

b. An 0.2% solution of potassium borohydride in 1 N sodium hydroxide.

c_1. An 0.5% solution of 3,3-diphenylacrolein (Chapter 17, Section I.C) in ethanol.

c_2. An 0.5% solution of p-dimethylaminobenzaldehyde in ethanol.

Procedure

To 1 ml of sample solution in ethanol, add 0.05 ml of reagent a and 1 ml of reagent b. Let stand for 10 min, add 2 ml of 10% sulfuric acid, allow the hydrogen to evolve, filter, and dilute to 5 ml with water. Add 2 ml of reagent c_1 or c_2, and let stand for 15 min. Read at 400 nm with reagent c_1, 440 nm with reagent c_2.

Results

	A = 0.3 (1 cm cell) Sample, μg	
	Reagent c_1	Reagent c_2
p-Nitrotoluene	67	62
m-Dinitrobenzene	90	40
p-Nitrobenzoic acid	130	37[a]
Chloramphenicol	126	66

[a]Read at 450 nm.

Notes

1. A 1% aqueous solution of cupric sulfate pentahydrate (0.05 ml) can be used instead of palladous chloride, but the results are less sensitive.

2. Aqueous solutions of potassium borohydride are rather unstable, but generation of hydrogen may be largely prevented by dissolving the compound in a slightly basic medium. Some metal salts catalyze the decomposition of the stable solutions so obtained and are effective independently of pH. The reaction is accompanied by the formation of a dark suspension of the metal or of its boride (7). In the above procedure, the precipitated palladium catalyzes both the decomposition of borohydride and the reduction of the nitro group.

3. Nitroso groups (see Chapter 3, Section II.O, reduction of p-nitroso-N,N-dimethylaniline), and nitriles (Section VIII.B) are also reduced.

4. Analytical uses of p-dimethylaminobenzaldehyde, see Chapter 5, Section IV.B, Notes 1 and 2.

5. 3,3-Diphenylacrolein also allows the colorimetric determination of m-diphenols (Chapter 3, Section V.A).

B. Potassium Borohydride, Sodium Nitrite, and β-Naphthol

Principle (6)

Reduction to primary aromatic amine by potassium borohydride in alkaline medium containing palladium, and development of the amine by diazotization with nitrous acid and coupling with β-naphthol: orange color.

$$Ar{-}NO_2 \xrightarrow[OH^-\ Pd]{KBH_4} Ar{-}NH_2 \xrightarrow{HNO_2} Ar{-}\overset{+}{N}{\equiv}N \xrightarrow{\text{β-naphthol (OH)}} \text{1-(N=N–Ar)-2-naphthol (OH)}$$

Reagents

a. An 0.5% aqueous solution of palladium chloride.

b. An 0.2% solution of potassium borohydride in 1 N sodium hydroxide.

c. A 1% aqueous solution of sodium nitrite.

d. An 0.25% solution of β-naphthol in aqueous ammonium hydroxide (d = 0.92, 22° Bé). Prepare fresh just before use.

Procedure

To 1 ml of sample solution in ethanol, add 0.05 ml of reagent *a* and 1 ml of reagent *b*. Let stand for 10 min, add 1 ml of 10% sulfuric acid, allow the hydrogen to evolve, filter, and dilute to 5 ml with water. Chill to 0°, add 1 ml of reagent *c*, and let stand at 0° for 15 min. Add 1 ml of reagent *d*, mix, and read at 485 nm.

Results

	A = 0.3 (1 cm cell) Sample, μg
p-Nitrotoluene	31
m-Dinitrobenzene	66
p-Nitrobenzoic acid	18
Chloramphenicol	40

Notes

1. The amine formed can also be developed with p-aminosalicylic acid (Chapter 5, Section IV.K). The orange color is read at 435 nm, and A = 0.3 is given by 100 μg of chloramphenicol.

2. Reduction of borohydride in the presence of a metal salt, see Section VI.A, Note 2.

3. Diazo coupling, see Chapter 16, Section I.

C. Tetraethylammonium Hydroxide

Principle (8)

Reaction with tetraethylammonium hydroxide in dimethylformamide: miscellaneous colors.

Reagent

A 25% aqueous solution of tetraethylammonium hydroxide.

Procedure

To 5 ml of sample solution in dimethylformamide, add 0.1 ml of reagent. Let stand for the given length of time (see Results), and read.

Results

	Reaction time, min	Color	λ Max, nm	A = 0.3 (1 cm cell) Sample, μg
p-Nitrotoluene	2	Pink	510	11.5
o-Dinitrobenzene	2	Yellow-orange	425	12.5
m-Dinitrobenzene	2	Red	520	18.0

	Reaction time, min	Color	λ Max, nm	A = 0.3 (1 cm cell) Sample, μg
p-Dinitrobenzene	2	Yellow	430	11.5
2,4-Dinitrotoluene	10	Green	640	40.0
p-Nitroaniline	2	Yellow-orange	465	6.0
o-Nitrobenzaldehyde	2	Yellow	385	450
p-Nitrobenzaldehyde	2	Pink	515	12.5
p-Nitrobenzoic acid	5	Pink	510	3000
2,4-Dinitrophenol	10	Yellow	430	16.0
3.5-Dinitrobenzoic acid	2	Red-violet	540	30

The reaction is practically negative with nitrobenzene.

Note

Development of aromatic nitro compounds, see Chapter 16, Section II.

VII. AROMATIC DINITRO COMPOUNDS

A. Fluorene and Tetraethylammonium Hydroxide

Principle (9)

Reaction with fluorene in alkaline dimethylformamide: miscellaneous colors.

Reagents

a. An 0.1% solution of fluorene in dimethylformamide.

b. A 10% aqueous solution of tetraethylammonium hydroxide.

Procedure

To 10 ml of sample solution in dimethylformamide, add 1.2 ml of reagent *a* and 0.25 ml of reagent *b*. Read immediately.

Results

	Color	λ Max, nm	A = 0.3 (1 cm cell) Sample, μg
o-Dinitrobenzene	Green-blue	690	60
m-Dinitrobenzene	Red	525	60
p-Dinitrobenzene	Green	700	63
2,4-Dinitrotoluene	Green	640	66
3,5-Dinitrobenzoic acid	Pink	540	50
1,3-Dinitronaphthalene	Red	520	116

Note

The mechanism of the color reaction conceivably could involve the reduction of the dinitro derivative in alkaline solution to a colored dianion (10).

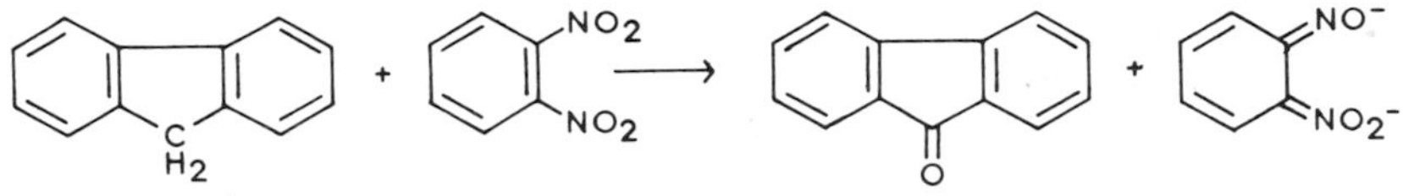

C H_2 + NO_2 NO_2 ⟶ O + NO^- NO_2^-

VIII. NITRILES

A. Titanium Trichloride, Silver Nitrate, and Manganese Sulfate

Principle (1)

Reduction of the nitrile with titanium trichloride, and estimation of the amine formed with a test paper impregnated with a mixture of silver nitrate and manganese sulfate (see Chapter 4, Section III.A).

Reagents and Procedure

See Section II.A, estimation of aliphatic nitro compounds. Heat at 70°.

Results

	Lower limit of detection, μg
Acetonitrile	60
Propionitrile	80
Isobutyronitrile	100
Benzonitrile	75
2,2'-Azobis(isobutyronitrile)	60
Ethyl cyanoacetate	80

B. Potassium Borohydride and Diazotized p-Nitroaniline

Principle (6)

Reduction to primary amine by potassium borohydride in alkaline medium containing palladium, and development of the amine with p-nitrobenzenediazonium ion: red-orange color.

$$R-CN \xrightarrow[OH^-\ Pd]{KBH_4} R-CH_2-NH_2 \xrightarrow[OH^-]{O_2N-C_6H_4-\overset{+}{N}\equiv N}$$

See Chapter 4, Section VI.E

Reagents

a. An 0.2% solution of potassium borohydride in 1 N sodium hydroxide.

b. An 0.5% aqueous solution of palladium chloride.

c. Buffer solution: 100 ml of aqueous solution contains 4.08 g of monopotassium phosphate, 1.6 g of sodium tetraborate decahydrate, and 12.7 ml of 2.5 N sodium hydroxide.

d. Dissolve 0.5 g of p-nitroaniline in 100 ml of 1 N hydrochloric acid, chill to 0°, and add 10 ml of 2.7% aqueous solution of sodium nitrite. Use between 30 min and 3 hr after preparation. Alternatively, the reagent may be an 0.2% solution of p-nitrobenzene-diazonium fluoborate in 0.5 N hydrochloric acid.

Procedure

To 1 ml of sample solution in a mixture of water and ethanol (the ratio of the solvents varying suitably with the compound to be determined), add 1 ml of reagent a and 0.05 ml of reagent b. Let stand for 10 min at room temperature, add 1 ml of 10% sulfuric acid, allow the hydrogen to evolve, neutralize with dilute sodium hydroxide to pH 4-5, filter, and dilute to 5 ml with water. Add 6 ml of buffer c, immerse the tubes in ice water for 5 min, add 1 ml of reagent d, and let stand at 0° for 20 min. Add dropwise 2 ml of prechilled 2.5 N sodium hydroxide, let stand for 10 min at room temperature, and read at 500 nm.

Results

	A = 0.3 (1 cm cell) Sample, μg
Benzonitrile	42
Benzyl cyanide	81

Unsatisfactory results are obtained with nitriles which, upon reduction, lead to volatile amines, such as acetonitrile.

Notes

1. Reduction with borohydride in the presence of a metal salt, see Section VI.A, Note 2.

2. Diazo coupling, see Chapter 16, Section I.

IX. MONO- AND N,N-DISUBSTITUTED HYDRAZINES (F)

A. Sodium 1,2-Naphthoquinone-4-sulfonate

Principle (11)

Reduction of 1,2-naphthoquinone-4-sulfonate by the hydrazine to 1,2-dihydroxynaphthalene-4-sulfonate: blue-green fluorescence.

O, O, SO_3Na → ($\rangle N-NH_2$) → OH, OH, SO_3Na

Reagents

a. Buffer: Mix equal volumes of 2.72% aqueous solution of monopotassium phosphate and of 5.36% aqueous solution of disodium phosphate heptahydrate.

b. An 0.02% solution of sodium 1,2-naphthoquinone-4-sulfonate in 0.1 N hydrochloric acid. Prepare fresh immediately prior to use.

Procedure

To 4 ml of sample solution in buffer a, add 0.4 ml of reagent b. Let stand for 5 min, and read at exc: 340 nm; em: 470 nm.

Standard: A solution of 2,6-dimethyl-3,5-dicarbethoxy-1,4-dihydropyridine in a 97:3 mixture of water and ethanol.

Results

	Determination limits, μg	Sample, μg	Reading 50 Standard, μg/ml
Methylhydrazine	0.04 - 0.2	0.074	2.12
N,N-Dimethylhydrazine	0.08 - 0.4	0.16	1.78
Benzylhydrazine (dihydrochloride)	0.2 - 1.0	0.36	2.27
Phenylhydrazine	0.4 - 2.0	0.76	3.9
N,N-Diphenylhydrazine (hydrochloride)	1.6 - 8.0	3.12	1.05

Note

Properties of 1,2-naphthoquinone-4-sulfonic acid, see Chapter 16, Section VII.

X. ALIPHATIC 2-NITRO-1-HYDROXY COMPOUNDS (F)

A. Sodium Hydroxide, Ethyl Acetoacetate, and Ammonium Acetate

Principle (12)

Decomposition in alkaline medium, and development of the formaldehyde formed with ethyl acetoacetate and ammonia: blue fluorescence.

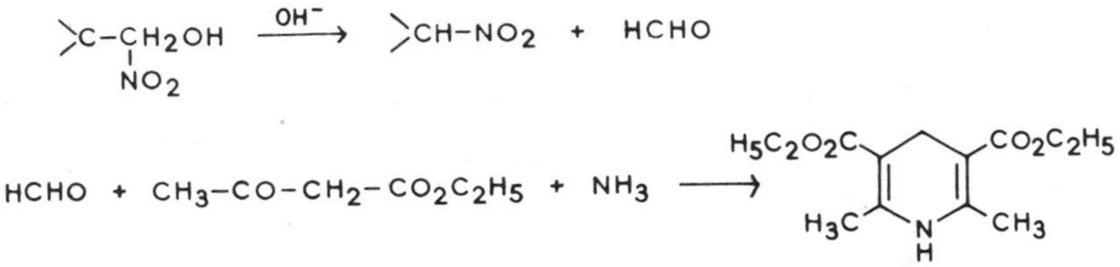

$$>C(NO_2)-CH_2OH \xrightarrow{OH^-} >CH-NO_2 + HCHO$$

$$HCHO + CH_3-CO-CH_2-CO_2C_2H_5 + NH_3 \longrightarrow \text{2,6-dimethyl-3,5-di}(CO_2C_2H_5)\text{-1,4-dihydropyridine}$$

Reagent

A 2% solution of ethyl acetoacetate in 20% aqueous solution of ammonium acetate.

Procedure

To 1 ml of aqueous solution of the nitrohydroxy compound, add 1 ml of 0.5 N sodium hydroxide. Let stand at room temperature for 5 min, add 1 ml of 1 N hydrochloric acid and 1 ml of reagent. Heat at 60° for 20 min, and cool to room temperature in ice water. Read at exc: 366 nm; em: 470 nm.

Standard: A solution of 2,6-dimethyl-3,5-dicarbethoxy-1,4-dihydropyridine in a 99:1 mixture of water and ethanol.

Results

	Determination limits, μg	Reading 50 Sample, μg	Standard, μg/ml
2-Nitro-1-butanol	2 - 10	4.0	0.94
2-Nitro-2-methyl-1,3-propanediol	1 - 5	2.1	1.09
2-Nitro-2-ethyl-1,3-propanediol	1 - 5	2.2	1.07

Note

Colorimetric and fluorimetric determination of formaldehyde, see Chapter 16, Section V.

XI. NITRILES (F)

A. Potassium Borohydride, Acetylacetone, and Formaldehyde

Principle (12)

Reduction to primary amine by potassium borohydride in alkaline medium containing palladium, and development of the amine with acetylacetone and formaldehyde: yellow-green fluorescence.

$$R{-}CN \xrightarrow[OH^-\ Pd]{KBH_4} R{-}CH_2{-}NH_2 \xrightarrow[HCHO]{CH_3{-}CO{-}CH_2{-}CO{-}CH_3}$$ 3,5-diacetyl-2,6-dimethyl-1,4-dihydropyridine bearing $N{-}CH_2{-}R$ (ring substituents: $H_3C{-}CO$, $CO{-}CH_3$, H_3C, CH_3)

Reagents

a. An 0.2% solution of potassium borohydride in 1 N sodium hydroxide.

b. An 0.5% aqueous solution of palladium chloride.

c. To 10 ml of 1 M aqueous solution of sodium acetate, add 0.8 ml of acetylacetone and 2 ml of 30% aqueous solution of formaldehyde, and dilute to 30 ml with 1 M sodium acetate. Prepare immediately before use.

Procedure

To 1 ml of sample solution in a mixture of water and ethanol (the ratio of the solvents varying suitably with the compound to be determined), add 1 ml of reagent a and 0.05 ml of reagent b. Let stand for 10 min at room temperature, add 1 ml of 2 N hydrochloric acid, mix, allow the hydrogen to evolve, and add 2 ml of 1 M aqueous solution of sodium acetate and 2 ml of reagent c. Heat at 100° for 10 min, allow to cool, and filter. Read at exc: 405 nm; em: 490 nm.

Standard: A solution of 2,6-dimethyl-3,5-diacetyl-1,4-dihydropyridine in ethanol.

Results

	Determination limits, μg	Reading 50 Sample, μg	Standard, μg/ml
Benzonitrile	2 - 10	4.0	3.12
Benzyl cyanide	3 - 15	6.0	3.03

Unsatisfactory results are obtained with nitriles which upon reduction lead to volatile amines, such as acetonitrile.

Notes

1. Reduction with borohydride in the presence of a metal salt, see Section VI.A, Note 2.

2. Fluorimetric determination of primary aliphatic amines with acetylacetone and formaldehyde, see Chapter 4, Section XII.A.

REFERENCES

1. M. Pesez, J. Bartos, and J.C. Lampetaz, *Bull. Soc. Chim. Fr.*, 719 (1962).

2. J. Bartos, *Ann. Pharm. Fr.*, 27, 159 (1969).

3. Cf. T. Meisel and L. Erdey, *Mikrochim. Acta*, 1148 (1966).

4. From E.W. Scott and J.F. Treon, *Ind. Eng. Chem., Anal. Ed.*, 12, 189 (1940).

5. L.R. Jones and J.A. Riddick, *Anal. Chem.*, 28, 254 (1956).

6. M. Pesez and J.F. Burtin, *Bull. Soc. Chim. Fr.*, 1996 (1959).

7. H.I. Schlesinger, H.C. Brown, A.E. Finholt, J.R. Gilbreath, H.R. Hoekstra, and E.K. Hyde, *J. Amer. Chem. Soc.*, 75, 215 (1953).

8. C.C. Porter, *Anal. Chem.*, 27, 805 (1955).

9. E. Sawicki and T.W. Stanley, *Anal. Chim. Acta*, 23, 551 (1960).

10. E. Sawicki, T.W. Stanley, and J. Noe, *Anal. Chem.*, 32, 816 (1960).

11. From M. Roth and J. Rieder, *Anal. Chim. Acta*, 27, 20 (1962).

12. M. Pesez and J. Bartos, *Talanta*, 14, 1097 (1967).

Chapter 8

CARBONYL COMPOUNDS

I. INTRODUCTION

A. Colorimetry

With some exceptions depending mainly on the compounds tested, all the methods proposed in this chapter afford sensitivities which allow determinations in the range 3-150 μg for A = 0.3.

Colorimetric determination of carbonyl compounds often takes advantage of the formation of hydrazones. The derivatives obtained with oxalhydrazide are developed as blue cupric complexes. Differential measurements at selected pH values may allow some specific determinations. In acetic-hydrochloric acid medium, 2,4-dinitrophenylhydrazine form yellow to yellow-orange hydrazones. With the latter reagent, condensation and development of a red to violet color in nonaqueous medium result in sensitive determinations without prior extraction of the hydrazone or elimination of the excess reagent. As a rule, the sensitivity may be reduced by steric hindrance, mainly in the case of ketones.

The 2,4-dinitrophenylhydrazones of ketones are developed through the Janovsky reaction with nitromethane in alkaline medium. The excess reagent does not interfere, and the sample weight giving A = 0.3 is fairly inversely proportional to the molecular weight of the ketone.

2,4-Dinitrophenylhydrazine also allows the determination of p-quinones, the development being achieved with ammonium hydroxide or with diethanolamine. In this latter case, the excess reagent is eliminated by reacting it with acetylacetone to give a pyrazole derivative.

The hydrazones of aldehydes can be converted to strongly colored formazans. Phenylhydrazones are so revealed with the diazonium salt generated by the dehydration of 3-phenyl-5-nitrosamino-1,2,4-thiadiazole (phenitrazole). 2-Hydrazinobenzothiazole and 3-methylbenzothiazolin-2-one-hydrazone afford hydrazones which are converted to formazans by the diazonium salt generated in the reaction medium by oxidation of the excess reagent.

The formation of Schiff bases allows the determination of aliphatic and aromatic aldehydes (with dimethyl-p-phenylenediamine), 2,3-unsaturated and aromatic aldehydes (with p-aminophenol), and 2,3-unsaturated aldehydes (with m-phenylenediamine).

Aliphatic aldehydes condense with anthrone in sulfuric acid medium to give colored species whose structures as yet are not well known. They can also be condensed with diethyl acetonedicarboxylate and ammonia, presumably to give 4-substituted-3,5-dicarbethoxy-1,4-dihydro-2,6-pyridinediacetic acid diethyl esters. Although a slightly yellow color is developed, the absorption maximum is located in the ultraviolet range.

α-Methylene ketones can be determined through the Zimmermann reaction with m-dinitrobenzene in alkaline medium. According to an unknown mechanism, they also react with sodium nitroprusside and ammonium hydroxide, but the color is poorly sensitive.

In acid medium, β-diketones are determined as red benzodiazepines upon reaction with o-phenylenediamine.

B. Fluorimetry

Aliphatic aldehydes are developed through the Hantzsch reaction with cyclohexane-1,3-dione and ammonia, to give 9-substituted 1,8-dioxodecahydroacridines which afford a blue fluorescence in acid medium and a yellow-green one in alkaline medium. Determinations are possible at levels lower than 1 μg.

Through the Doebner-Miller reaction, α-methylene aldehydes react with 3,5-diaminobenzoic acid in acid medium to give yellow-green fluorescent 2,3-dialkyl-5-carboxy-7-aminoquinolines, allowing determinations at the level of 5 μg.

Through their active methylene group, dialkyl α-methylene ketones can be substituted for the sulfonate group of 1,2 naphthoquinone-4-sulfonic acid. The corresponding dihydroxynaphthalene derivative obtained by reduction develops a blue-green fluorescence. The sensitivity of the method depends on the compound tested.

II. ALDEHYDES AND KETONES

A. Oxalhydrazide and Cupric Acetate

Principle (1)

The hydrazone formed by reacting oxalhydrazide with a carbonyl compound is developed as cupric complex: blue color.

$$>C{=}O + H_2N{-}NH{-}CO{-}CO{-}NH{-}NH_2 \longrightarrow\ >C{=}N{-}NH{-}CO{-}CO{-}NH{-}NH_2$$

$$\xrightarrow{Cu^{2+}}\ >C{=}N{-}N{=}\underset{\underset{\frac{Cu}{2}}{O\quad O}}{C{-}C}{-}NH{-}NH_2$$

Reagents

a. Teorell and Stenhagen buffer for suitable pH: see Appendix I.G.

b. A 1:1 mixture of 0.25% aqueous solution of oxalhydrazide and 0.0156% aqueous solution of cupric acetate monohydrate.

Procedure

To 1 ml of aqueous solution of the carbonyl compound, add 2 ml of buffer a adjusted to the suitable pH (see Results) and 2 ml of reagent b. Let stand at room temperature for 1 hr (2 hr with cyclohexanone, 4 hr with acetone), and read.

Results

	pH	λ Max, nm	A = 0.3 (1 cm cell) Sample, μg
Formaldehyde	7.0	610	7.0
Acetaldehyde	8.0	610	11.5
Propionaldehyde	9.0	610	17
Isobutyraldehyde	9.0	600	30
Glyoxylic acid (sodium salt)	9.0	610	130
Streptomycin (sulfate)	8.0	600	515
Acetone	10.0	590	91
Cyclohexanone	10.0	590	37.5

Cinnamic and salicylic aldehydes, furfural, glyoxal, acetophenone, glucose, and fructose practically do not react. A = 0.07 is read with 100 μg of cyclopentanone at pH 10.0 (590 nm).

Notes

Differential measurements at selected pH values may allow some specific determinations. For example, at pH 7.0 (optimum value for formaldehyde) the reaction is almost negative with acetone, cyclohexanone, and glyoxylic acid. At pH 10.0 (optimum value for acetone and cyclohexanone), 8 μg of formaldehyde afford an optical density of only 0.07.

2. This reaction was at first applied to the colorimetric determination of cupric ion with oxalhydrazide in the presence of formaldehyde, acetaldehyde, or cyclohexanone (2).

3. It was established that in the presence of an excess of oxalhydrazide, the complex (I) was formed (blue color). With an excess of carbonyl compound, the violet color developed corresponds to structure (II) (3).

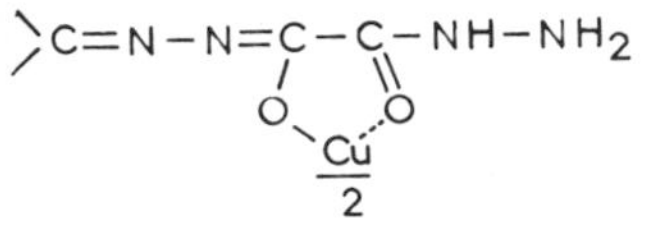

$$>C{=}N{-}N{=}C(O\cdots Cu/2)-C(=O\cdots)-NH{-}NH_2$$

(I)

$$>C{=}N{-}N{=}C(O\cdots Cu/2)-C(OH){=}N{-}N{=}C<$$

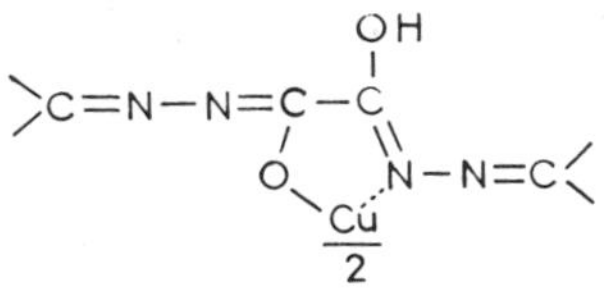

(II)

4. Determination of amino acids, see Chapter 9, Section VI.A.

B. 2,4-Dinitrophenylhydrazine in Acetic-Hydrochloric Acid

Principle (4)

Formation of the 2,4-dinitrophenylhydrazone: yellow-orange color.

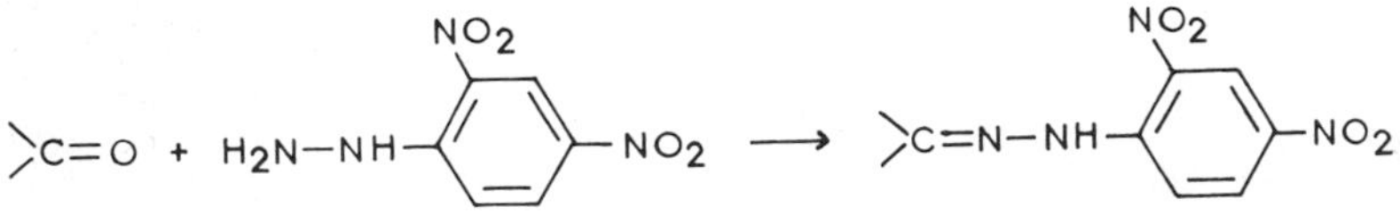

$$>C{=}O + H_2N{-}NH{-}C_6H_3(NO_2)_2 \longrightarrow >C{=}N{-}NH{-}C_6H_3(NO_2)_2$$

Reagent

An 0.1% solution of 2,4-dinitrophenylhydrazine in an 0.5% dilution of concentrated hydrochloric acid in glacial acetic acid.

Procedure

To 1 ml of sample solution in glacial acetic acid,

add 5 ml of reagent. Let stand for 1 hr at room temperature in the dark, and read at 412 nm.

Results

	A = 0.3 (1 cm cell) Sample, μg
A. Aldehydes	
Formaldehyde	20
Acetaldehyde	20
Propionaldehyde	21
Citral	30
Cinnamaldehyde	11
3,3-Diphenylacrolein	14
Benzaldehyde	21
Salicylaldehyde	15
p-Hydroxybenzaldehyde	12
Veratraldehyde	15
Piperonal	15
Vanillin	14
Furfural	14
B. Ketones	
Acetone	45
Methyl ethyl ketone	65
Cyclopentanone	56
Cyclohexanone	100
Acetophenone	35
Benzophenone[a]	430
Camphor[a]	3500
Ethyl acetoacetate	140

[a]Heat at 100° for 15 min, and read at 432 nm.

Notes

1. Ketosteroids can be determined upon the same principle (Chapter 15, Section IV.A).

2. Contrariwise to some other determinations of carbonyl derivatives with 2,4-dinitrophenylhydrazine,

elimination of the excess reagent or extraction of the formed species is not necessary, the blank being almost colorless.

C. p-Nitrophenylhydrazine in Acetic-Hydrochloric Acid

Principle

Formation of the p-nitrophenylhydrazone: yellow color.

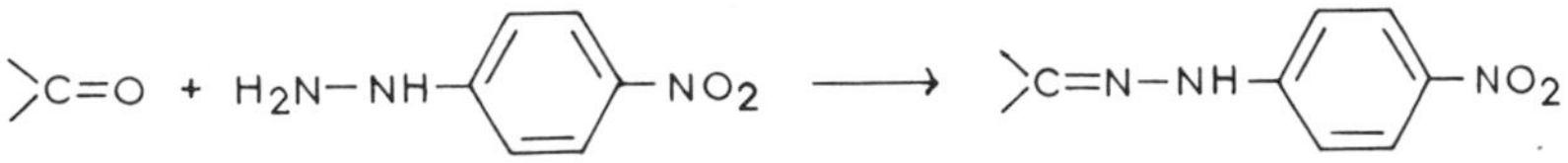

$$>C{=}O + H_2N{-}NH{-}C_6H_4{-}NO_2 \longrightarrow >C{=}N{-}NH{-}C_6H_4{-}NO_2$$

Reagent

An 0.05% solution of p-nitrophenylhydrazine in an 0.5% dilution of concentrated hydrochloric acid in glacial acetic acid, freshly prepared.

Procedure

To 1 ml of sample solution in glacial acetic acid, add 5 ml of reagent. Let stand for 3 hr in the dark (aldehydes), or heat at 100° for 15 min (ketones), and read.

Results

	λ Max, nm	A = 0.3 (1 cm cell) Sample, μg
A. Aldehydes		
Propionaldehyde	400	23
Cinnamaldehyde	410	11
Benzaldehyde	400	18
Salicylaldehyde	400	48
Veratraldehyde	410	20
Piperonal	405	19
Vanillin	405	21
B. Ketones		
α-Ionone	410	130
Acetophenone	405	140

The same procedure allows the colorimetric determination of ketosteroids (Chapter 15, Section IV.B).

D. p-Nitrophenylhydrazine and Benzyltrimethylammonium Hydroxide

Principle (5)

Formation of p-nitrophenylhydrazone, and without prior elimination of the excess reagent, development in alkaline nonaqueous medium: red to violet color.

Reagents

a. An 0.04% solution of p-nitrophenylhydrazine in an 0.1% dilution of concentrated hydrochloric acid in ethanol.

b. Dilute 1 ml of 40% solution of benzyltrimethylammonium hydroxide (see Chapter 17, Section III) in methanol to 100 ml with dimethylformamide. Prepare fresh just before use.

Procedure

To 0.5 ml of sample solution in ethanol, add 0.5 ml of reagent *a*. Heat at 70° for the given length of time (see Results), cool in a water bath, add 9 ml of reagent *b*, and read.

Results

	Reaction time, min	Color	λ Max, nm	A = 0.3 (1 cm cell) Sample, μg
1. Aldehydes				
Formaldehyde	20	Pink-red	510	20
Acetaldehyde	30	Pink-red	510	25
Propionaldehyde	30	Violet	520	28
Glyoxal (monohydrate)	30	Violet	710	6.4
Salicylaldehyde	20	Violet	580	8.8
Piperonal	20	Red	550	9.7
Vanillin	20	Violaceous pink	590	9.3

	Reaction time, min	Color	λ Max, nm	A = 0.3 (1 cm cell) Sample, μg
B. Ketones				
Acetone	30	Red	520	6.0
Methyl ethyl ketone	30	Red	520	13.0
Diacetyl	30	Violet	560	15.2
Cyclopentanone	30	Red	520	13.0
Cyclohexanone	20	Red	520	10.0
α-Ionone	30	Violet	550	20
Acetophenone	30	Violet	560	31

Acetylacetone, ethyl acetoacetate, benzophenone, camphor, glucose, and ribose give only a very faint color.

The same procedure allows the colorimetric determination of ketosteroids (Chapter 15, Section IV.C).

Note

Development of aromatic nitro compounds, see Chapter 16, Section II.

III. ALDEHYDES

A. Phenylhydrazine and Phenitrazole

Principle (6)

Formation of the phenylhydrazone and development as formazan with the diazonium salt generated by dehydration of 3-phenyl-5-nitrosamino-1,2,4-thiadiazole (phenitrazole) in the reaction medium: orange or violet color.

$$R{-}CHO \xrightarrow{C_6H_5{-}NH{-}NH_2} C_6H_5{-}NH{-}N{=}CH{-}R$$

$$\text{(3-phenyl-1,2,4-thiadiazol-5-yl)}{-}NH{-}NO \xrightarrow{H^+} \text{(3-phenyl-1,2,4-thiadiazol-5-yl)}{-}\overset{+}{N}{\equiv}N$$

$$\longrightarrow \text{(3-phenyl-1,2,4-thiadiazol-5-yl)}{-}N{=}N{-}C(R){=}N{-}NH{-}C_6H_5$$

Reagents

a. An 0.5% aqueous solution of phenylhydrazine hydrochloride.

b. A 1:1 mixture of 65-70% perchloric acid (d = about 1.62) and 0.4% solution of phenitrazole (Chapter 17, Section I.F) in ethanol. Mix, and keep at 0° for 5 min before use.

Procedure

To 1 ml of sample solution in ethanol, add 0.5 ml of reagent a, and let stand for 20 min at room temperature. Chill to 0°, add 1 ml of reagent b, let stand at 0° for 5 min, and add 3 ml of 5 N sodium hydroxide. Read after 5 min at 0°.

Results

	Color	λ Max, nm	A = 0.3 (1 cm cell) Sample, μg
Formaldehyde	Orange	490	3
Acetaldehyde	Orange	490	44
Propionaldehyde	Orange	490	105
Glyoxal (monohydrate)	Orange	490	14
Glyoxylic acid (sodium salt)	Orange	490	40
Benzaldehyde	Violet	550	23
Vanillin	Violet	565	92
Furfural	Orange	510	22

Notes

1. Phenitrazole was also applied to the colorimetric determination of phenols (Chapter 3, Section II.J) and aromatic amines (Chapter 5, Section II.F).

2. Tetrazolium salts and formazans, see Chapter 16, Section VIII.

IV. ALIPHATIC ALDEHYDES

A. 2-Hydrazinobenzothiazole and Potassium Ferricyanide

Principle (7)

Formation of the benzothiazole hydrazone and development as formazan with the diazonium salt generated in the reaction medium by oxidation of the excess of 2-hydrazinobenzothiazole: blue color.

N, S, C–NH–NH$_2$ + R–CHO ⟶ N, S, C–NH–N=CH–R

$Fe(CN)_6^{3-}$ ⟶ N, S, C–$\overset{+}{N}$≡N

N, S, C–N=N–C(R)=N–NH–C, N, S

OH^-

[N, S, C⋯N⋯N⋯C(R)⋯N⋯N⋯C, N, S]$^-$

Reagents

a. An 0.5% solution of 2-hydrazinobenzothiazole in a 1:9 mixture of concentrated hydrochloric acid and water.

b. A 1% aqueous solution of potassium ferricyanide.

Procedure

To 1 ml of aqueous solution of the aldehyde, add 1 ml of reagent *a*, let stand for 5 min, add 1 ml of reagent *b*, and let stand for 18 min. Add 2 ml of dimethylformamide, 2 ml of 1.8 N potassium hydroxide, and 13 ml of water. Read after 15 min.

Results

	λ Max, nm	A = 0.3 (1 cm cell) Sample, μg
Formaldehyde	582	4.0
Acetaldehyde	576	150
Propionaldehyde	577	250

Notes

1. 2-Hydrazinobenzothiazole also allows the colorimetric determination of aldoses and ketoses (Chapter 14, Section II.F) and 17-hydroxy-17-ketolsteroids (Chapter 15, Section XIV.B).

2. Tetrazolium salts and formazans, see Chapter XVI, Section VIII.

B. 3-Methylbenzothiazolin-2-one Hydrazone and Ferric Chloride

Principle (8)

Condensation with 3-methylbenzothiazolin-2-one hydrazone and development as formazan with the diazonium salt generated in the reaction medium by oxidation of the excess reagent: blue-green to green color.

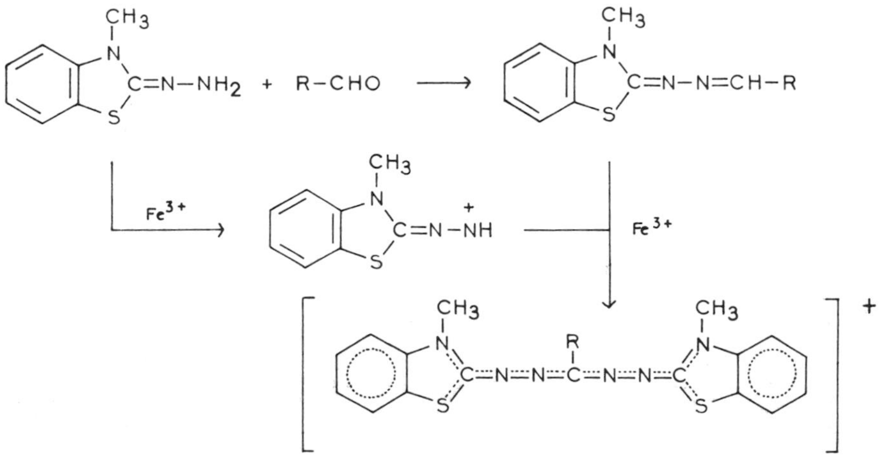

Reagents

a. An 0.4% aqueous solution of 3-methylbenzothiazolin-2-one hydrazone hydrochloride.

b. Aqueous solution containing 1% of ferric chloride hexahydrate and 1.6% of sulfamic acid.

Procedure

To 1 ml of sample solution in water or isopropyl alcohol, add 1 ml of reagent *a*, and let stand for 20 min at room temperature. Add 1 ml of reagent *b*, mix, and let stand for 10 min. Add 2 ml of water, mix, and read.

Results

	Solvent	λ Max, nm	A = 0.3 (1 cm cell) Sample, μg
Formaldehyde	Water	635	0.87
Acetaldehyde	Water	610	1.25
Propionaldehyde	Water	620	1.68
Butyraldehyde	Water	615	2.0
Caprylic aldehyde	Isopropanol	635	4.6
Lauric aldehyde	Isopropanol	635	33.6
Citronellal	Isopropanol	640	11.6
Citral	Isopropanol	640	13.5

Notes

1. Glyoxal and acrolein do not give satisfactory results. However, a modification of the above procedure allows their determination: To 1 ml of aqueous solution of the sample, add 1 ml of reagent *a* and 1 ml of reagent *b*. Let stand for 30 min at room temperature, add 2 ml of water, and read at 625 nm.

Beer's law is then followed; A = 0.3 is given by 2.73 μg of glyoxal or 1.32 μg of acrolein.

2. Aromatic aldehydes also react, but with rather variable results, and Beer's law is not always followed.

3. This method is an improvement over a previously described one (*9*), which was less sensitive because it necessitated a dilution with acetone to dissolve a

turbidity. In above procedure, this turbidity is avoided by adding sulfamic acid to the ferric chloride solution.

4. Properties of 3-methylbenzothiazolin-2-one hydrazone, see Chapter 16, Section III.

C. Anthrone

Principle (10)

Reaction with anthrone in sulfuric acid medium: yellow to brown color.

Reagent

An 0.2% solution of anthrone in concentrated sulfuric acid.

Procedure

To 2 ml of aqueous solution of the aldehyde, add 3 ml of reagent, pouring it directly into the center of the tube to ensure good mixing. Let stand for 10 min, and read.

Results

	λ Max, nm	A = 0.3 (1 cm cell) Sample, μg
Formaldehyde	488	61
Acetaldehyde	476	137
Propionaldehyde	455	160
Acrolein	510	12

Notes

1. The mechanism of the reaction is as yet not well known. The first step of the condensation certainly involves the reactive methylene group of anthrone (III), but afterwards the reaction probably proceeds in different ways, depending on the compound tested. It was established that acrolein affords benzanthrone (11).

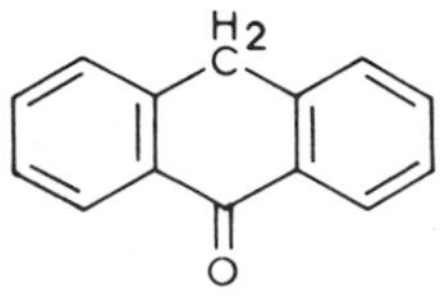

(III)

2. Anthrone in sulfuric acid medium was proposed as reagent for the characterization of 17-ketolsteroids (12). Among the steroids tested, only those having the $\Delta^4$3-keto configuration and lacking a ketonic group on carbon 11 yielded colored products with an absorption maximum at 600 nm. On the other hand, dehydroepiandrosterone reacts under similar conditions (13).

3. Anthrone in sulfuric acid also allows the colorimetric determination of pentoses (Chapter 14, Section VI.G) and hexoses (Chapter 14, Section VIII.B).

D. Dimethyl-p-phenylenediamine Oxalate

Principle (14)

Reaction with dimethyl-p-phenylenediamine in acetic acid medium to give a Schiff base: yellow or orange color.

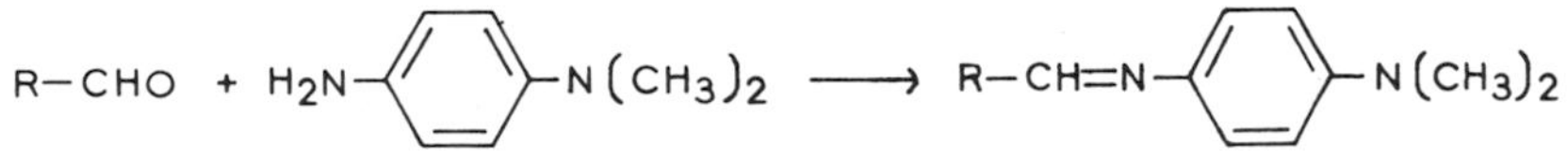

$$R{-}CHO + H_2N{-}C_6H_4{-}N(CH_3)_2 \longrightarrow R{-}CH{=}N{-}C_6H_4{-}N(CH_3)_2$$

Reagent

A 2% solution of dimethyl-p-phenylenediamine oxalate in glacial acetic acid.

Procedure

To 1 ml of sample solution in glacial acetic acid, cooled to just above the freezing point, add 2 ml of reagent precooled to the same temperature. Mix, and read immediately.

Results

	λ Max, nm	A = 0.3 (1 cm cell) Sample, μg
Acetaldehyde	390	11.5
Propionaldehyde	430	81.5
Butyraldehyde	390	25.5
Isobutyraldehyde[a]	370	109
Isovaleraldehyde	390	25
Enanthaldehyde	400	73
Caprylic aldehyde	435	129
2-Ethylcaproic aldehyde[a]	390	40
Lauric aldehyde	400	56
Crotonaldehyde[a]	450	10
Citronellal	390	276
Citral	445	9
2,2-Dimethyl-3-hydroxy-propionaldehyde	390	120
ω-Hydroxyvaleraldehyde	385	1044
Glyceraldehyde	410	10
Cinnamaldehyde	490	5.2

[a]Beer's law is not followed.

Notes

1. Some simple ketones also react, but the color is less sensitive. For example, A = 0.21 is given by 700 μg of acetone or 235 μg of methyl ethyl ketone.

2. Dimethyl-p-phenylenediamine also allows the colorimetric determination of aromatic aldehydes (Section VI.A), peroxides (Chapter 10, Section III.A), and keto-steroids (Chapter 15, Sections VIII.A and XI.B).

3. Stability of dimethyl-p-phenylenediamine, see Chapter 3, Section II.O, Note 2.

E. Diethyl Acetonedicarboxylate and Ammonium Acetate

Principle

Reaction with diethyl acetonedicarboxylate and ammonia. The formation of a 4-substituted 3,5-dicarbethoxy-1,4-dihydro-2,6-pyridinediacetic acid diethyl ester may be assumed: slightly yellow color, but the absorption maximum is located in the ultraviolet range.

R
CHO
$H_5C_2O_2C$–CH_2–CO–H_2C–$CO_2C_2H_5$ + $H_2C(CO_2C_2H_5)$–CO–CH_2–$CO_2C_2H_5$ + NH_3 ⟶ 4-R-3,5-($CO_2C_2H_5$)$_2$-2,6-(CH_2–$CO_2C_2H_5$)$_2$-1,4-dihydropyridine (N–H)

Reagent

Dissolve 8 g of ammonium acetate in 10 ml of water, add 8 ml of diethyl acetonedicarboxylate, and dilute to 50 ml with ethanol.

Procedure

To 1 ml of sample solution in glacial acetic acid, add 2 ml of reagent. Heat at 60° for 90 min, add 2 ml of 2 N sodium hydroxide, and read at 344 nm.

Results

	A = 0.3 (1 cm cell) Sample, μg
Formaldehyde	16
Acetaldehyde	4.6
Propionaldehyde[a]	6.2
Cinnamaldehyde	14

[a]Read at 340 nm.

Notes

1. Furfural also reacts: A = 0.3 for 6.2 μg at 344 nm.

2. Beer's law is not followed with citral and aromatic aldehydes.

3. Ketones do not react.

4. Inversely, a mixture of diethyl acetonedicarboxylate and formaldehyde allows the colorimetric and ultraviolet determination of primary aliphatic amines (Chapter 4, Section VI.I).

5. This method is based on the Hantzsch reaction, which allows miscellaneous colorimetric and fluorimetric determinations (see Chapter 16, Section IX).

V. 2,3-UNSATURATED ALDEHYDES

A. p-Aminophenol

Principle (15)

Reaction with p-aminophenol in trichloroacetic acid medium to give a Schiff base: yellow-green color.

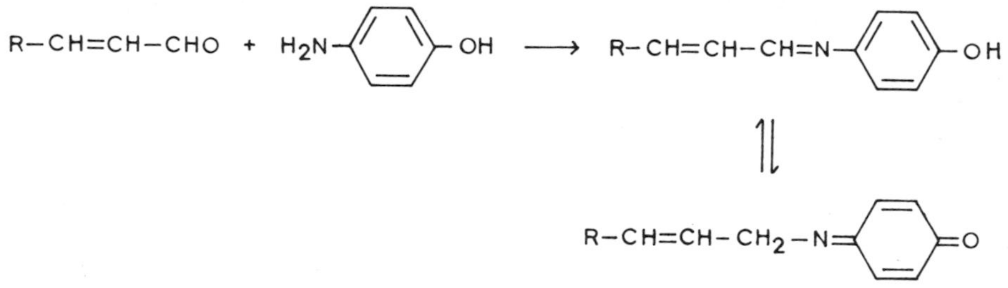

$$R-CH=CH-CHO + H_2N-C_6H_4-OH \longrightarrow R-CH=CH-CH=N-C_6H_4-OH$$

$$\rightleftharpoons$$

$$R-CH=CH-CH_2-N=C_6H_4=O$$

Reagent

A 3% solution of p-aminophenol in 6% aqueous trichloroacetic acid. If the solution is colored, treat with activated charcoal and filter.

Procedure

To 5 ml of sample solution in water or methanol, add 3 ml of 40% aqueous trichloroacetic acid and 2 ml of reagent. Read after 5 min.

Results

	λ Max, nm	A = 0.3 (1 cm cell) Sample, μg
Acrolein	350	2250
Crotonaldehyde	360	39
Cinnamaldehyde	400	9

Furfural (A = 0.3 for 26 μg at 385 nm) and vanillin (A = 0.3 for 52 μg at 395 nm) also react.

Note

p-Aminophenol also allows the colorimetric determination of aromatic aldehydes (see Section VI.B).

B. m-Phenylenediamine Oxalate

Principle (16)

Reaction with m-phenylenediamine to give a Schiff base: yellow-brown color.

Reagent

A 1% solution of m-phenylenediamine oxalate (Chapter 17, Section I.E) in a 4:1 mixture of alcohol and water.

Procedure

To 5 ml of sample solution in ethanol, add 5 ml of reagent. Let stand for 5 min at room temperature, and read.

Results

	λ Max, nm	A = 0.3 (1 cm cell) Sample, μg
Acrolein	350	76
Crotonaldehyde	325	220
Cinnamaldehyde	355	18

Furfural and vanillin also react.

Notes

1. m-Phenylenediamine was proposed as reagent for the characterization of α-methylene aldehydes. In

strongly acid medium, a substituted aminoquinoline is formed, which develops a yellow color and a yellow-green fluorescence (17). See also the fluorimetric determination of α-methylene aldehydes, Section XII.A.

2. As in the case of dimethyl-p-phenylenediamine (see Chapter 3, Section II.O, Note 2), m-phenylenediamine oxalate is more stable than the other common salts of this base.

VI. AROMATIC ALDEHYDES

A. Dimethyl-p-phenylenediamine Oxalate

Principle (14)

Reaction with dimethyl-p-phenylenediamine in acetic acid medium to give a Schiff base: yellow or orange color.

$$Ar{-}CHO + H_2N{-}C_6H_4{-}N(CH_3)_2 \longrightarrow Ar{-}CH{=}N{-}C_6H_4{-}N(CH_3)_2$$

Reagent

A 2% solution of dimethyl-p-phenylenediamine oxalate in glacial acetic acid.

Procedure

To 0.5 ml of sample solution in glacial acetic acid, add 2 ml of reagent. Let stand for 5 min in the dark, add 1.5 ml of acetic acid, mix, and read.

Results

	λ Max, nm	A = 0.3 (1 cm cell) Sample, μg
Benzaldehyde	465	19
p-Tolualdehyde	460	12
p-Nitrobenzaldehyde	450	33
p-Anisaldehyde	465	9
Salicylaldehyde	465	27
Protocatechualdehyde	465	7.5

	λ Max, nm	A = 0.3 (1 cm cell) Sample, μg
Veratraldehyde	465	10
Piperonal	465	10
Vanillin	460	9
o-Aminobenzaldehyde	480	16
2-Methoxy-6-naphthaldehyde	475	11.5

Furfural also reacts (A = 0.3 for 9μg at 480 nm).

Notes

1. Dimethyl-p-phenylenediamine also allows the colorimetric determination of aliphatic aldehydes (Section IV.D), peroxides (Chapter 10, Section III.A) and ketosteroids (Chapter 15, Section VIII.A and XI.B).

2. Stability of dimethyl-p-phenylenediamine, see Chapter 3, Section II.O, Note 2.

B. p-Aminophenol

Principle

Reaction with p-aminophenol in acetic acid medium to give a Schiff base: yellow color.

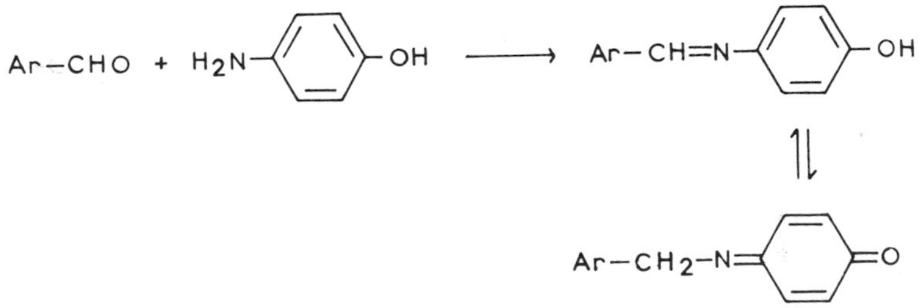

$$Ar{-}CHO + H_2N{-}C_6H_4{-}OH \longrightarrow Ar{-}CH{=}N{-}C_6H_4{-}OH$$

$$\rightleftharpoons$$

$$Ar{-}CH_2{-}N{=}C_6H_4{=}O$$

Reagent

A 2% solution of p-aminophenol in glacial acetic acid.

Procedure

To 3 ml of sample solution in glacial acetic acid, add 2 ml of reagent. Let stand for 15 min, and read.

Results

	λ Max, nm	A = 0.3 (1 cm cell) Sample, μg
Benzaldehyde	350	18
Salicylaldehyde	350	12
Veratraldehyde	390	9
Vanillin	390	7.6
p-Dimethylaminobenzaldehyde	440	3.8
2-Methoxy-6-naphthaldehyde	410	10.8

Furfural also reacts (A = 0.3 for 9 μg at 375 nm).

Note

p-Aminophenol also allows the colorimetric determination of 2,3-unsaturated aldehydes (Section V.A).

VII. KETONES

A. 2,4-Dinitrophenylhydrazine, Nitromethane, and Benzyltrimethylammonium Hydroxide

Principle (18)

Formation of the 2,4-dinitrophenylhydrazone, and without prior elimination of the excess reagent, development with nitromethane in alkaline medium (Janovsky reaction): violaceous pink color

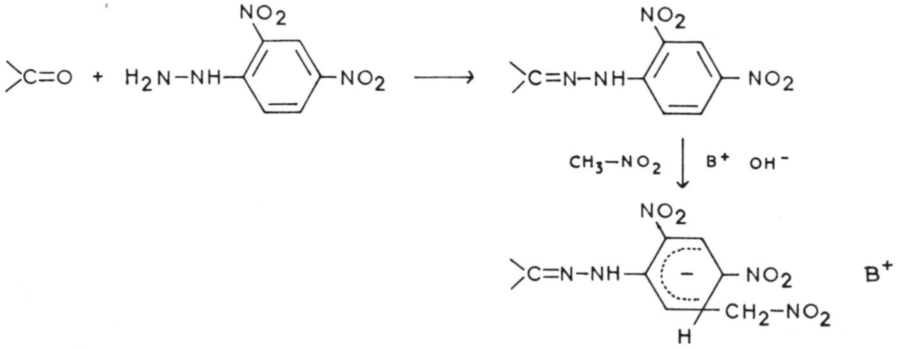

>C=O + H_2N–NH–(2,4-dinitrophenyl: NO_2, NO_2) ⟶ >C=N–NH–(2,4-dinitrophenyl: NO_2, NO_2)

CH_3–NO_2 ↓ B^+ OH^-

>C=N–NH–(ring: NO_2, NO_2, −, H, CH_2–NO_2) B^+

Reagents

a. Dissolve 0.025 g of 2,4-dinitrophenylhydrazine in a mixture of 10 ml of ethanol and 0.5 ml of concentrated hydrochloric acid, and dilute to 50 ml with ethanol.

b. A 40% solution of benzyltrimethylammonium hydroxide in methanol (see Chapter 17, Section III).

Procedure

Operate protected against light. To 0.5 ml of sample solution in ethanol, add 0.5 ml of reagent *a*, and let stand at the given temperature for the given length of time (see Results). Cool to room temperature if necessary, add 1.5 ml of nitromethane, 1.5 ml of dimethylformamide, and 0.3 ml of reagent *b*. Shake for 30 sec, and read immediately at 565 nm.

Results

	Temperature, °C	Reaction time, min	A = 0.3 (1 cm cell) Sample, μg
Acetone	18 - 24	20	4.3
Methyl ethyl ketone	18 - 24	20	5.3
Cyclopentanone	50	20	6.0
Cyclohexanone	18 - 24	10	7.4

Ketosteroids can also be determined (Chapter 15, Section IV.D).

Notes

1. The sample weight giving A = 0.3 is fairly inversely proportional to the molecular weight of the ketone.

2. β-Diketones afford a very transient color, making measurements impossible.

3. Aldehydes develop but a faint color, and the absorption maximum at 565 nm is observed only in pure nitromethane.

4. Janovsky reaction, see Chapter 16, Section VI.

VIII. α-METHYLENE KETONES

A. Sodium Nitroprusside and Ammonium Hydroxide

Principle (19)

Reaction with nitroprusside ion in the presence of ammonium hydroxide: violaceous red or blue color.

Reagents

a. A saturated aqueous solution of ammonium sulfate.

b. A 5% aqueous solution of sodium nitroprusside.

Procedure

To 1 ml of aqueous solution of the ketone, add 4 ml of reagent *a*, 0.2 ml of reagent *b*, and 2 ml of aqueous ammonium hydroxide (d = 0.92, 22° Bé). Let stand for the given length of time (see Results), and read.

Results

	Reaction time, min	λ Max, nm	A = 0.3 (1 cm cell) Sample, μg
Acetone	25	530	325
Methyl ethyl ketone	10	530	2300
Acetophenone	35	600	550

The color given by acetylacetone is very transient. Cyclohexanone does not react.

Notes

1. In 1883, Legal (20) observed that a solution of acetone developed a red color when treated successively with sodium nitroprusside, sodium hydroxide, and acetic acid. The reaction was then extended to other carbonyl compounds by von Bitto (21) and Deniges (22), who showed that it was bound up with the presence of a $-CH_2-CO-R$ group, the color being red when R = H or alkyl group, blue when R = aryl group. It may be concluded therefrom that the reaction probably depends on the ability of the ketone to enolize.

Sodium hydroxide can be replaced by ammonia, thus avoiding the further use of acetic acid, and it was established that more sensitive results were reached when operating in the presence of an excess of ammonium salt, such as ammonium sulfate (19). Such a method was used for the characterization of citric acid oxidized to acetonedicarboxylic acid, of pyruvic acid, and of lactic acid oxidized to pyruvic acid with bromine water (23). Under suitable conditions, an intensely blue color is displayed by these compounds.

Organic bases were also proposed: ethylenediamine (24) and piperidine, which leads to somewhat specific reactions (25). Acetaldehyde, propionaldehyde, and acrolein develop a blue color; cinnamaldehyde gives a rather faint reaction; enanthaldehyde, isobutyraldehyde, and phenylacetaldehyde do not react; neither do formaldehyde, chloral, benzaldehyde, salicylaldehyde, and furfural, but these compounds cannot enolize.

With acetaldehyde, it is worth using piperazine, which permits colorimetric determinations (26). The method also allows the characterization or determination of alanine and valine (27) oxidized to the corresponding aldehydes with hypochlorite, and of propylene glycol (28), ephedrine (29), and threonine (30) after periodate oxidation.

2. Nitroprusside also allows the colorimetric determination of primary (Chapter 4, Section VI.A) and secondary (Chapter 4, Section VII.A) aliphatic amines, imidazoline derivatives (Chapter 13, Section VI.A), and $\Delta^4$3-ketosteroids (Chapter 15, Section X.A).

3. Relations between nitroprusside and pentacyanoferrates, see Chapter 5, Section IV.A, Note 1.

B. m-Dinitrobenzene

Principle

Reaction with m-dinitrobenzene in alkaline medium to give a Meisenheimer-like σ complex, which is then oxidized by the excess reagent (Zimmermann reaction): violet color.

NO_2 ; $-CH_2-CO-$; B^+ OH^- ; H ; CH−CO− ; NO_2 ; B^+ ; Oxid. ; C−CO− ; NO_2 ; B^+

Reagents

a. A 3% solution of m-dinitrobenzene in pyridine.

b. Dilute 2.5 ml of 40% solution of benzyltrimethylammonium hydroxide in methanol (see Chapter 17, Section III) to 10 ml with pyridine.

c. Dilute 5 ml of 10% aqueous solution of ammonium acetate to 100 ml with pyridine.

Procedure

Operate protected against light. To 1 ml of sample solution in pyridine, add 0.5 ml of reagent a, chill to 0° in ice water, add 0.2 ml of reagent b, and let stand at 0° for 1 min. Add 3 ml of reagent c, mix, and read immediately at 570 nm.

Results

	A = 0.3 (1 cm cell) Sample, μg
Acetone	2.8
Methyl ethyl ketone	3.2
Cyclopentanone	3.0
Cyclohexanone	3.2
Acetophenone[a]	7.8

[a]Operate at room temperature.

Notes

1. Aldehydes react but weakly. A = 0.3 is given by 50 μg of propionaldehyde. Ketones with no methylene group in the α-position to the carbonyl group, such as benzil and benzophenone, do not react.

2. Janovsky and Zimmermann reactions, see Chapter 16, Section VI.

IX. β-DIKETONES

A. o-Phenylenediamine

Principle (31)

Condensation with o-phenylenediamine in acid medium to give a substituted benzodiazepine: red color.

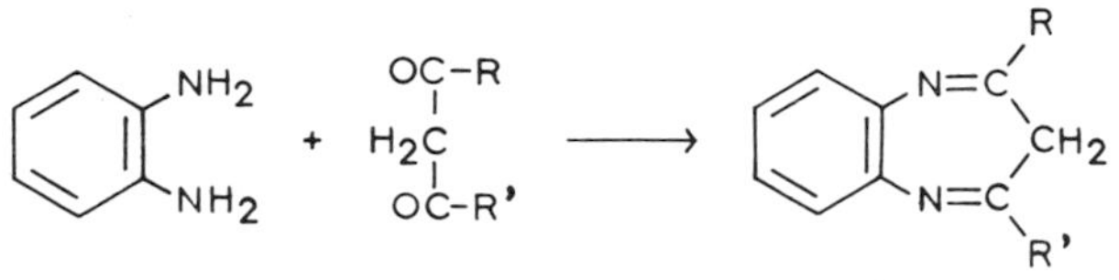

$$C_6H_4(NH_2)_2 + OC\text{-}R,\ H_2C,\ OC\text{-}R' \longrightarrow N{=}C(R),\ CH_2,\ N{=}C(R')$$

Reagent

Dissolve 0.40 g of o-phenylenediamine in a mixture of 4.35 g of dipotassium phosphate, 2 ml of 5 N phosphoric acid, and 50 ml of 2.4 N sulfuric acid, and dilute to 100 ml with water. Prepare fresh just before use.

Procedure

To 5 ml of aqueous solution of the β-diketone, add 1 ml of reagent. Let stand for 30 min, and read at 500 nm.

Results

	A = 0.3 (1 cm cell) Sample, μg
Acetylacetone	216
Hexane-2,4-dione	243
Heptane-3,5-dione	268

Note

o-Phenylenediamine also allows the fluorimetric determination of α-keto acids (Chapter 9, Section XII.A).

X. p-QUINONES

A. 2,4-Dinitrophenylhydrazine and Ammonium Hydroxide

Principle (32)

Formation of a 2,4-dinitrophenylhydrazone, and development with ammonium hydroxide: brown-red to green color.

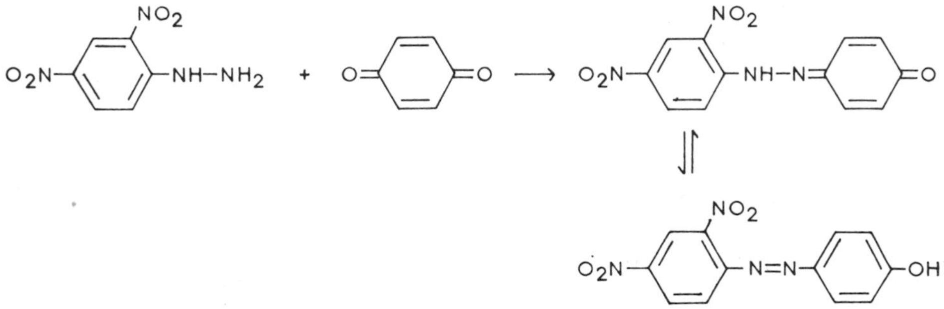

NO_2, O_2N–C₆H₃–NH–NH_2 + O=C₆H₄=O ⟶ O_2N–C₆H₃(NO_2)–NH–N=C₆H₄=O ⇌ O_2N–C₆H₃(NO_2)–N=N–C₆H₄–OH

Reagent

A saturated solution of 2,4-dinitrophenylhydrazine in ethanol. Prepare fresh just before use.

Procedure

To 1 ml of sample solution in 90% ethanol, add 0.1 ml of concentrated hydrochloric acid and 2 ml of reagent. Incubate at 37° for 2 hr, add 1 ml of aqueous ammonium hydroxide (d = 0.92, 22° Bé), dilute to 10 ml with ethanol, mix, and read.

Results

	Color	λ Max, nm	A = 0.3 (1 cm cell) Sample, μg
p-Benzoquinone	Brown-red	550	12
Menadione	Brown-green	630	9

With 2,3-dichloro-1,4 naphthoquinone, a green color is developed (λ Max, 660 nm) but the reaction is poorly sensitive (A = 0.2 for 100 μg).

B. 2,4-Dinitrophenylhydrazine and Diethanolamine

Principle (33)

Formation of a 2,4-dinitrophenylhydrazone, elimination of the excess reagent, and development with diethanolamine (scheme a): blue-green to blue-violet color.

The excess reagent is eliminated by reacting it with acetylacetone to give a pyrazole derivative according to the Knorr reaction (34) (scheme b).

(a)

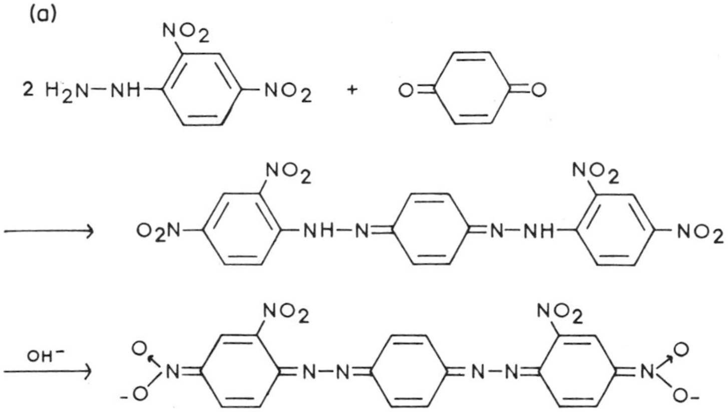

$$2\ H_2N{-}NH{-}C_6H_3(NO_2)_2 + O{=}C_6H_4{=}O \longrightarrow O_2N{-}C_6H_3(NO_2){-}NH{-}N{=}C_6H_4{=}N{-}NH{-}C_6H_3(NO_2){-}NO_2$$

$$\xrightarrow{OH^-} (O)({}^-O)N{=}C_6H_3(NO_2){=}N{-}N{=}C_6H_4{=}N{-}N{=}C_6H_3(NO_2){=}N(O)(O^-)$$

(b)

$$H_2N{-}NH{-}C_6H_3(NO_2)_2 + CH_3{-}CO{-}CH_2{-}CO{-}CH_3 \longrightarrow \text{3,5-dimethyl-1-(2,4-dinitrophenyl)pyrazole } (H_3C, CH_3, NO_2, NO_2)$$

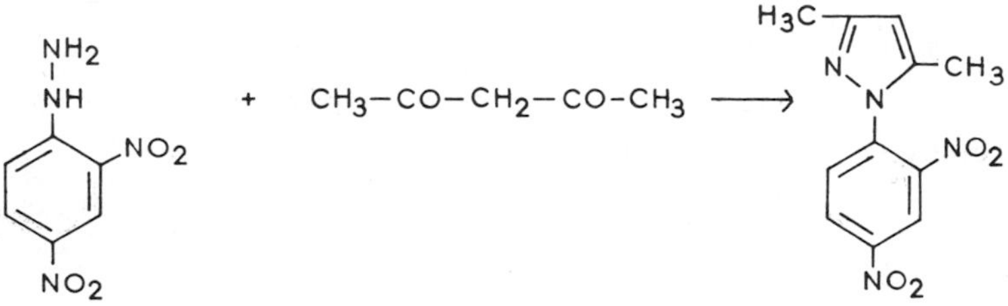

Reagents

a. Dissolve 0.1 g of 2,4-dinitrophenylhydrazine in 50 ml of methanol, add 4 ml of concentrated hydrochloric acid, and dilute to 100 ml with water.

b. A 2% solution of diethanolamine in pyridine.

Procedure

To 3 ml of sample solution in water or methylene chloride, add 1 ml of 10% sulfuric acid and 1 ml of reagent a. Let stand for the given length of time at the given temperature (see Results), add 2 drops of acetylacetone, mix, let stand for 2 min, cool to room temperature, and dilute to 15 ml with water. Add 5 ml of methylene chloride, shake vigorously, let the phases separate, pipet 3 ml of the lower layer, evaporate to dryness under vacuum, and add to the residue 10 ml of reagent b. Mix, and read.

Results

	Solvent	Temperature, °C	Reaction time, min	λ Max, nm	A = 0.3 (1 cm cell) Sample, μg
p-Benzoquinone	H_2O	25	30	615	7.6
Menadione	CH_2Cl_2	70	60	640	13.2

2,3-Dichloro-1,4-naphthoquinone (dissolved in methylene chloride) gives only a yellow color, and Beer's law is not followed.

Notes

1. α-Diketones also react. For example, A = 0.3 is given at 570 nm by 13 μg of diacetyl (dissolved in water, 30 min heating at 98°).

2. Development of aromatic nitrocompounds, see Chapter 16, Section II.

XI. ALIPHATIC ALDEHYDES (F)

A. Cyclohexane-1,3-dione and Ammonium Acetate

Principle (35)

Reaction with cyclohexane-1,3-dione and ammonia to give a 9-substituted-1,8-dioxodecahydroacridine: blue fluorescence in acidic medium, yellow-green fluorescence in alkaline medium.

R
CHO
O
O
NH_3
O
O
O R O
N
H

Reagent

Dissolve 0.25 g of cyclohexane-1,3-dione, 10 g of ammonium acetate, and 5 ml of glacial acetic acid in 80 ml of water and dilute to 100 ml with water.

Procedure

To 3 ml of aqueous solution of the aldehyde, add 1 ml of reagent, heat at 60° for 1 hr, chill for 2 min in ice water, and add either 1 ml of water (blue fluorescence) or 1 ml of 10 N sodium hydroxide (yellow-green fluorescence).

Read at exc: 366 nm; em: 470 nm (acidic medium) or exc: 436 nm; em, 520 nm (alkaline medium).

Standards: For the blue fluorescence, a solution of 1,8-dioxodecahydroacridine in a 1:49 mixture of ethanol and water. For the yellow-green fluorescence, a solution of 1,8-dioxodecahydroacridine in a 49:1 mixture of 2 N sodium hydroxide and ethanol.

Results

Reaction in acidic medium.

	Determination limits, μg	Reading 50	
		Sample, μg	Standard, μg/ml
Formaldehyde	0.5 - 2.5	1.0	0.51
Acetaldehyde	0.2 - 1.0	0.44	0.56
Propionaldehyde	0.3 - 1.5	0.62	0.44
Butyraldehyde	0.6 - 3.0	1.2	0.47

Reaction in alkaline medium.

	Determination limits, μg	Reading 50	
		Sample, μg	Standard, μg/ml
Formaldehyde	0.5 - 2.5	1.05	0.64
Acetaldehyde	0.2 - 1.0	0.45	0.80
Propionaldehyde	0.2 - 1.0	0.38	0.51
Butyraldehyde	0.5 - 2.5	1.0	0.60

Notes

1. Isobutyraldehyde, citral, glyoxal, and glyoxylic acid react much more weakly. Benzaldehyde can be determined in the range 2-10 μg, in acidic medium, em: 455 nm. No fluorescence is displayed by vanillin.

2. This method is based on the Hantzsch reaction, which allows miscellaneous colorimetric and fluorimetric determinations (see Chapter 16, Section IX).

XII. α-METHYLENE ALDEHYDES (F)

A. 3,5-Diaminobenzoic Acid Dihydrochloride

Principle (35,36)

Reaction with 3,5-diaminobenzoic acid in phosphoric acid medium to give a 2,3-dialkyl-5-carboxy-7-aminoquinoline: yellow-green fluorescence.

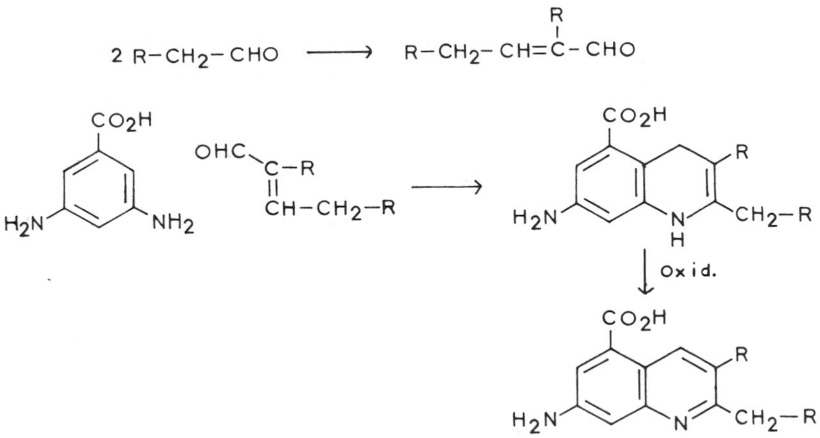

$$2\ R{-}CH_2{-}CHO \longrightarrow R{-}CH_2{-}CH{=}C(R){-}CHO$$

Reagent

To 2 g of 3,5-diaminobenzoic acid dihydrochloride, add 10 ml of phosphoric acid (d = 1.62), and dilute to 20 ml with water.

Procedure

To 2 ml of aqueous solution of the aldehyde, add 2 ml of reagent, and heat at 50° for 30 min, protected against bright light. Allow to cool to room temperature, and add 2 ml of water. Read at exc: 405 nm; em: 495 nm.

Standard: A solution of 2-methyl-5-carboxy-7-aminoquinoline in 10% acetic acid.

Results

	Determination limits, μg	Reading 50 Sample, μg	Standard, μg/ml
Acetaldehyde	2 - 10	5.7	0.31
Propionaldehyde	2 - 10	5.7	0.23
Butyraldehyde	2 - 10	5.7	0.19

Notes

1. The reaction allowed the fluorimetric determination of acetaldehyde formed during the enzymatic decarboxylation of pyruvic acid by washed bakers' yeast

in the presence of cocarboxylase or thiamine triphosphoric acid (TTP) (37), and of succinic semialdehyde (38).

2. The same reagent permits the fluorimetric determination of 2-deoxy sugars (Chapter 14, Section XV.A).

3. This reaction is based on the Doebner-Miller synthesis of quinoline derivatives (39). Its mechanism has been the subject of a number of papers (40). It has been currently accepted that it involves an aldol condensation and a dehydration of the aldol to an α,β-unsaturated aldehyde, followed by Michael addition of the aromatic amine. The conversion of the substituted dihydroquinoline so formed to the corresponding quinoline might be due to an intermolecular reaction giving rise to 1 mole of the quinoline and 1 mole of the tetrahydro derivative.

Recently Forrest et al. (41), working in deuterated media with aniline and acetaldehyde, demonstrated that crotonaldehyde is not necessarily formed as an intermediate. However, the whole mechanism of the Doebner-Miller synthesis does not seem as yet completely elucidated.

XIII. DIALKYL α-METHYLENE KETONES (F)

A. Sodium 1,2-Naphthoquinone-4-sulfonate and Potassium Borohydride

Principle (42)

Replacement of the sulfonate group of the naphthoquinone-sulfonic acid by the active methylene compound, and reduction to the corresponding dihydroxynaphthalene: blue-green fluorescence.

(1,2-naphthoquinone-4-SO_3Na) + $R-CO-CH_2-R'$ ⟶ (1,2-naphthoquinone-4-$CH(R')-CO-R$) $\xrightarrow{KBH_4}$ (1,2-dihydroxynaphthalene-4-$CH(R')-CO-R$)

Reagents

a. An 0.05% solution of sodium 1,2-naphthoquinone-4-sulfonate in 0.2 N sodium hydroxide. Prepare fresh immediately before use.

b. An 0.5% solution of potassium borohydride in 0.2 N sodium hydroxide. Prepare fresh immediately before use.

Procedure

Operate protected against light. To 0.5 ml of aqueous solution of the ketone, add 0.5 ml of reagent a, plug the tubes with cotton-wool, and heat at 50° for 20 min. Chill for 2 min in ice water, add 0.1 ml of reagent b, mix, and let stand for 2 min. Add 0.1 ml of 2 N hydrochloric acid, let stand for 2 min, and add 2 ml of pre-chilled 10 N sodium hydroxide. Read immediately.

Standard: A solution of 1,8-dioxodecahydroacridine in a 49:1 mixture of 2 N sodium hydroxide and ethanol.

Results

	Exc, nm	Em, nm	Determination limits, μg	Reading 50	
				Sample, μg	Standard, μg/ml
Acetone	405	520	5 - 25	8.5	0.32
Methyl ethyl ketone	436	530	0.5 - 2.0	0.85	0.18
Methyl isobutyl ketone	436	520	6 - 30	12.0	0.082
Ethyl acetoacetate	436	520	1.6 - 8.0	2.9	0.088

No linear relationship is observed with acetylacetone.

Notes

1. The reaction is weakly sensitive with aldehydes, but the linear relationship is followed.

2. Acetophenone, cyclopentanone, cyclohexanone, and ketosteroids do not lead to fluorescent species.

3. The quinone adduct obtained with cyclohexanone fluoresces without prior reduction, thus allowing a determination: Proceed as described under Procedure, with a concentration of 0.02% for reagent a. After heating and chilling, add immediately 3 ml of prechilled 10 N sodium hydroxide, and read at exc: 366 nm; em: 455 nm. A linear relationship is observed in the range 10-50 μg, with reading 50 corresponding to 20 μg of the ketone and to a 0.33 μg/ml solution of dioxodecahydroacridine in ethanol. The linearity is not held with cyclopentanone.

4. 1,2-Naphthoquinone-4-sulfonic acid, see Chapter 16, Section VII.

REFERENCES

1. J. Bartos and J.F. Burtin, *Ann. Pharm. Fr.*, 19, 769 (1961).

2. Cf. E. Jacobsen, F.J. Langmyhr, and A.R. Selmer-Olsen, *Anal. Chim. Acta*, 24, 579 (1961); see also R. Capelle, *Chim. Anal.*, 42, 69, 127 (1960).

3. A. Badinant and J.J. Vallon, *Chim. Anal.*, 48, 313, 396 (1966).

4. M. Pesez, *J. Pharm. Pharmacol.*, 11, 475 (1959).

5. From M. Pesez and J. Bartos, *Talanta*, 5, 216 (1960).

6. M. Pesez, J. Bartos, and J.F. Burtin, *Talanta*, 5, 213 (1960).

7. E. Sawicki and T.R. Hauser, *Anal. Chem.*, 32, 1434 (1960).

8. From T.R. Hauser and R.L. Cummins, *Anal. Chem.*, 36, 679 (1964).

9. E. Sawicki, T.R. Hauser, T.W. Stanley, and W. Elbert, *Anal. Chem.*, 33, 93 (1961).

10. T.W. Kwon and B.M. Watts, *Anal. Chem.*, 35, 733 (1963).

11. O. Bally and R. Scholl, *Ber.*, 44, 1656 (1911).

12. M.M. Graff, J.T. McElroy, and A.L. Mooney, *J. Biol. Chem.*, 195, 351 (1952).

13. M.M. Graff, *J. Biol. Chem.*, 197, 741 (1952).

14. M. Pesez and J. Bartos, *Talanta*, 10, 69 (1963).

15. F.L. Breusch and E. Ulusoy, *Z. Physiol. Chem.*, 291, 64 (1952).

16. From R.B. Wearn, W.M. Murray, Jr., M.P. Ramsey, and N. Chandler, *Anal. Chem.*, 20, 922 (1948).

17. L. Velluz, M. Pesez, and G. Amiard, *Bull. Soc. Chim. Fr.*, [5] 15, 680 (1948).

18. J. Bartos, *Chim. Anal.*, 53, 18 (1971).

19. Cf. A.C.H. Rothera, *J. Physiol.*, 37, 491 (1908).

20. E. Legal, *Breslauer Aerzliche Z.*, 5, 25, 38 (1883) [From E. Cattelain, *Ann. Pharm. Fr.*, 7, 395 (1949)].

21. B. von Bitto, *Justus Liebigs Ann. Chem.*, 267, 372 (1891-92).

22. G. Deniges, *Bull. Soc. Chim. Paris*, [3] 15, 1058 (1896).

23. H. Caron and D. Raquet, *J. Pharm. Chim.*, [9] 2, 332, 333 (1942).

24. H.J. Schaeffer, *Amer. J. Pharm.*, 98, 643 (1926).

25. L. Lewin, *Ber.*, 32, 3388 (1899).

26. C. Fromageot and P. Heitz, *Mikrochim. Acta*, 3, 52 (1938).

27. E. Aubel and J. Asselineau, *Bull. Soc. Chim. Fr.*, [5] 14, 114, 689 (1947).

28. P. Desnuelle and M. Naudet, *Bull. Soc. Chim. Fr.*, [5] 12, 871 (1945).

29. A. Wickström, *Ann. Pharm. Fr.*, 8, 86 (1950).

30. P. Desnuelle, S. Antonin, and M. Naudet, *Trav. Memb. Soc. Chim. Biol.*, 26, 1175 (1944).

31. R.F. Witter, J. Snyder, and E. Stotz, *J. Biol. Chem.*, 176, 493 (1948).

32. W.J. Canady and J.H. Roe, *J. Biol. Chem.*, 220, 563 (1956).

33. D.P. Johnson, F.E. Critchfield, and J.E. Ruch, *Anal. Chem.*, 34, 1389 (1962).

34. L. Knorr, *Ber.*, 16, 2587 (1883).

35. M. Pesez and J. Bartos, *Talanta*, 14, 1097 (1967).

36. Cf. L. Velluz, M. Pesez, and G. Amiard, *Bull. Soc. Chim. Fr.*, [5] 15, 680 (1948); L. Velluz, M. Pesez, and M. Herbain, *Bull. Soc. Chim. Fr.*, [5] 15, 681 (1948).

37. L. Velluz, G. Amiard, and J. Bartos, *J. Biol. Chem.*, 180, 1137 (1949).

38. R.A. Salvador and R.W. Albers, *J. Biol. Chem.*, 234, 922 (1959).

39. O. Doebner and W. von Miller, *Ber.*, 16, 2464 (1883).

40. Cf. F.W. Bergstrom, *Chem. Rev.*, 35, 153 (1944); G.M. Badger, H.P. Crocker, B.C. Ennis, J.A. Gayler, W.E. Matthews, W.G.C. Raper, E.L. Samuel, and T.M. Spotswood, *Australian J. Chem.*, 16, 814 (1963).

41. T.P. Forrest, G.A. Dauphinee, and W.F. Miles, *Can. J. Chem.*, 47, 2121 (1969).

42. J. Bartos, *Ann. Pharm. Fr.*, 27, 691 (1969).

Chapter 9

CARBOXYLIC ACIDS AND DERIVATIVES

I. INTRODUCTION

A. Colorimetry

Most determinations of carboxylic acids are based on the formation of esters.

The neutralized acid can be esterified with p-nitrophenacyl bromide, and the ester is then developed with diethylamine in nonaqueous medium. Upon the action of N,N'-dicyclohexylcarbodiimide, carboxylic acids in nitromethane solution are esterified with 2,4-dinitrophenol, and the ester is revealed with potassium borohydride. Both reactions allow determinations at levels of 7-90 μg for A = 0.3. Operation in nonaqueous media avoids prior elimination of the excess reagent or extraction of the ester.

In aqueous media, carboxylic acids are esterified with methanol in the presence of N,N'-dicyclohexylcarbodiimide, or with ethylene glycol in the presence of sulfuric acid. The esters are then allowed to react with hydroxylamine, and the hydroxamates so formed are developed with ferric ion. The determinations are not very sensitive (several hundred micrograms for A = 0.3).

The hydroxamate reaction can be extended to the determination of various carboxylic acid derivatives which are directly converted to hydroxamic acids upon the action of hydroxylamine: anhydrides, lactones, esters, and amides. Very often, samples of 0.2 mg or more are necessary for A = 0.3.

Free carboxylic acids can also be determined through diazo coupling. The nitrous acid which they liberate from sodium nitrite diazotizes sulfanilamide, and the diazonium salt is developed with α-naphthylamine. However, the linear relationship holds only above 0.2 microequivalent of acid.

Fatty acids react with the carbinol base of rosaniline, to give the red-orange rosaniline salt, making determinations possible at the level of about 70 μg.

Particular reactions allow determinations of carboxylic acids bearing other functional groups.

2-Hydroxy-1-carboxylic acids develop a yellow color with ferric ion. A = 0.3 is given by 125-350 μg.

Upon oxidation with periodate ion, 2,3-dihydroxy-1-carboxylic acids liberate glyoxylic acid, which is developed with resorcinol. The reaction is not very sensitive: 350-420 μg for A = 0.3.

Various reactions are available for α-amino acids. In what follows, the numbers in parentheses are the sample weights giving A = 0.3.

Upon reaction with insoluble cupric phosphate, α-amino acids give soluble cupric complexes. The copper so dissolved is then determined with oxalhydrazide and acetaldehyde (96-260 μg).

A reaction with p-nitrobenzoyl chloride involves both the carboxyl and the amino group to give a p-nitrophenyloxazolone (7.5-720 μg, with the exception of asparagine). It is therefore specific for α-amino acids.

Both groups are also involved in the reaction with ninhydrin in the presence of potassium cyanide (5-18.5 μg) or of ascorbic acid (4.8-20 μg). With a few exceptions, the recoveries vary from 92 to 105% (potassium cyanide), and from 89.6 to 97.7% (ascorbic acid). A 100% recovery was postulated for leucine.

It may be assumed that a somewhat similar mechanism allows determinations with ascorbic acid, which is at first oxidized to dehydroascorbic acid (8.5-35 μg at 390 nm, 30-106 μg at 525 nm).

α-Amino acids can also be determined through their amino group with p-nitrobenzenediazonium fluoborate (5-20 μg) and with succinic dialdehyde, the pyrrole ring so formed being developed either with p-dimethylaminobenzaldehyde (1.8-7.4 μg), or with flavylium perchlorate (6-12 μg). It must be emphasized that primary aliphatic amines behave the same.

B. Fluorimetry

Available methods are limited to 2,3-dihydroxy-1-carboxylic acids, α-amino acids, and α-keto acids.

The glyoxylic acid obtained by oxidizing 2,3-dihydroxy-1-carboxylic acids with periodate is developed with resorcinol through an oxidative reaction: 2.1-2.3 μg are necessary for reading 50.

The amino group of α-amino acids allows their determination through a Hantzsch reaction with acetylacetone and formaldehyde (4.5-43 μg for reading 50). However, the same reaction is given by aliphatic amines.

Reactions with o-phthaldialdehyde and 2-mercaptoethanol (0.42-0.88 μg), and with ninhydrin and phenylacetaldehyde (0.57-22 μg) seem to be more specific.

Upon condensation with o-phenylenediamine in sulfuric acid medium, α-keto acids afford hydroxyquinoxalines, which allow determinations at levels of 0.85-2.2 μg.

II. CARBOXYLIC ACIDS

A. Sodium Nitrite, Sulfanilamide, and α-Naphthylamine

Principle (1,2)

Upon the action of a free carboxylic acid, sodium nitrite releases nitrous acid which diazotizes sulfanilamide, and the diazonium salt is developed with α-naphthylamine: red color.

$$R-CO_2H \xrightarrow{NaNO_2} HNO_2 \xrightarrow{H_2N-C_6H_4-SO_2NH_2} N\equiv\overset{+}{N}-C_6H_4-SO_2NH_2$$

$$\xrightarrow{\alpha\text{-}C_{10}H_7NH_2} H_2N-C_{10}H_6-N=N-C_6H_4-SO_2NH_2$$

Reagents

a. A 2% solution of sulfanilamide in acetone.

b. A 2% aqueous solution of sodium nitrite.

c. A 2% solution of α-naphthylamine in acetone.

Procedure

To 1 ml of aqueous solution of the acid, add 0.1 ml of reagent a, 0.1 ml of reagent b, and 0.1 ml of reagent c. Heat at 70° for 5 min, immerse the tubes in a water bath at 20° for 2 min, add 4 ml of water and 2 ml of ethanol, and read at 475 nm.

Results

Below a lower limit of concentration which is of about 0.2 microequivalent of acid, the reaction is almost negative. The linear relationship holds above this limit.

Results

	A = 0.3 (1 cm cell) Sample, μg
Formic acid	15.0
Acetic acid	19.5
Propionic acid	21.0
Benzoic acid	30.0

Notes

1. It is obvious that inorganic acids also react (with the exception of boric acid).

2. Diazo coupling, see Chapter 16, Section I.

B. p-Nitrophenacyl Bromide and Diethylamine

Principle (3)

Esterification of the neutralized acid with p-nitrophenacyl bromide, and without prior elimination of the excess reagent development with diethylamine in non-aqueous medium: violet color.

$$R{-}COONa + Br{-}CH_2{-}CO{-}C_6H_4{-}NO_2 \longrightarrow R{-}COO{-}CH_2{-}CO{-}C_6H_4{-}NO_2$$

Reagents

a. An 0.3% solution of p-nitrophenacyl bromide in acetonitrile(neutral grade).

b. Pure diethylamine.

Procedure

To 0.25 ml of aqueous solution of the acid neutralized with sodium carbonate, add 0.1 ml of reagent a and let stand for 10 min. Add 5 ml of dimethyl sulfoxide, and after 5 min 0.5 ml of reagent b. Read at 550 nm.

Results

	A = 0.3 (1 cm cell) Sample, μg
Formic acid	10
Acetic acid	7

	A = 0.3 (1 cm cell) Sample, μg
Propionic acid	11
Butyric acid	12
Phenylacetic acid	19
Oxalic acid[a]	26
Succinic acid[a]	40
Citric acid[a]	90
Benzoic acid	14

[a]The concentration of reagent a is 0.5%.

Notes

1. Purification of dimethyl sulfoxide, see Chapter 17, Section IV.

2. Development of aromatic nitro compounds, see Chapter 16, Section II.

C. 2,4-Dinitrophenol, N,N'-Dicyclohexylcarbodiimide, and Potassium Borohydride

Principle (4)

Upon the action of N,N'-dicyclohexylcarbodiimide, esterification of the free acid with 2,4-dinitrophenol, and without prior elimination of the excess reagent development with potassium borohydride in nonaqueous medium: red-violet color.

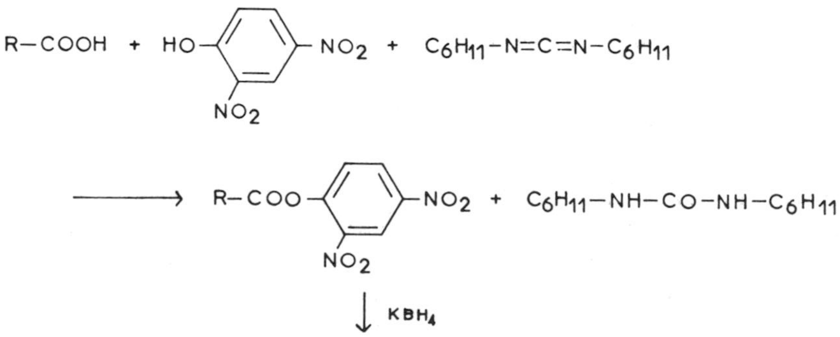

$$R{-}COOH + HO{-}C_6H_3(NO_2)_2 + C_6H_{11}{-}N{=}C{=}N{-}C_6H_{11}$$

$$\longrightarrow R{-}COO{-}C_6H_3(NO_2)_2 + C_6H_{11}{-}NH{-}CO{-}NH{-}C_6H_{11}$$

$$\downarrow KBH_4$$

Colored species

Reagents

a. 1:1 mixture of 0.5% solution of 2,4-dinitrophenol in nitromethane and 5% solution of N,N'-dicyclohexyl-carbodiimide in the same solvent. Prepare immediately before use.

b. An 0.1% solution of potassium borohydride in dimethylformamide.

Procedure

To 0.5 ml of sample solution in nitromethane, add 0.3 ml of reagent *a*, mix, and let stand for 15 min at room temperature in the dark. Add 3.5 ml of reagent *b*, mix, and read immediately at 560 nm (the blank develops a green color).

Results

	A = 0.3 (1 cm cell) Sample, μg
Acetic acid	19
Propionic acid	24
Palmitic acid	85
Benzoic acid	36

The reaction is negative with formic acid.

Notes

1. The sample weight giving A = 0.3 is inversely proportional to the molecular weight of the acid.

2. This method was derived from a paper published by Elmore and Smyth (5), who prepared p-nitrophenyl esters of acylamino acids by condensing the carboxylic group with p-nitrophenol in the presence of dicyclo-hexylcarbodiimide, and a work of Ramachandran (6), who established that 2,4-dinitrophenylamino compounds develop a red color which allows spectrophotometric determinations upon the action of sodium borohydride. He assumed that the reduction leads to an azo derivative.

Although it had been at first admitted that aromatic nitro compounds are not attacked by sodium borohydride (7), Severin et al. demonstrated in a series of papers (8) that a number of m-dinitro and trinitro aromatic compounds are reduced, but that the reduction concerns the benzene ring, the nitro group being unaffected. Therefore, without further experimental studies, no provisional formula can be attributed to the colored species formed in the herein-described method.

3. N,N'-Dicyclohexylcarbodiimide, proposed at first for the synthesis of ortho- and pyrophosphoric esters (9), has been widely used in peptide synthesis (10).

D. Ethylene Glycol, Hydroxylamine Hydrochloride, and Ferric Chloride

Principle (11)

Esterification of the free acid with ethylene glycol, conversion into the corresponding hydroxamate, and development with ferric ion: violaceous red color.

$$R{-}COOH \xrightarrow{HOCH_2{-}CH_2OH} \begin{array}{l} CH_2O{-}CO{-}R \\ | \\ CH_2O{-}CO{-}R \end{array} \xrightarrow{NH_2OH} R{-}CO{-}NHOH \xrightarrow{Fe^{3+}}$$

See Section VIII.A, Note 1

Reagents

a. Pure ethylene glycol.

b. A 10% aqueous solution of hydroxylamine hydrochloride.

c. Dissolve 20 g of ferric chloride hexahydrate in 500 ml of water, add 20 ml of concentrated sulfuric acid, and dilute to 1000 ml with water.

Procedure

To 0.5 ml of aqueous solution of the acid, add 1.5 ml of reagent a and 0.2 ml of 50% sulfuric acid. Heat at 100° for 3 min, cool in a water bath, add 0.5 ml of reagent b, 2 ml of 4.5 N sodium hydroxide, and after 1 min add 10 ml of reagent c. Let stand for 5 min, and read at 500 nm.

Results

	A = 0.3 (1 cm cell) Sample, μg
Formic acid	680
Acetic acid	500
Propionic acid	690
Oxalic acid	1165
Succinic acid	775
Lactic acid	1340
Malic acid	1210
Tartaric acid	2130
Citric acid	4260

Note

Hydroxamates, see Section VIII.A.

E. Methanol, N,N'-Dicyclohexylcarbodiimide, Hydroxylamine Hydrochloride, and Ferric Perchlorate

Principle (12)

Esterification with methanol in the presence of dicyclohexylcarbodiimide at pH 3, conversion into the corresponding hydroxamate, and development with ferric ion: red-violet color.

$$R{-}COOH + CH_3OH + C_6H_{11}{-}N{=}C{=}N{-}C_6H_{11}$$

$$\longrightarrow R{-}COOCH_3 + C_6H_{11}{-}NH{-}CO{-}NH{-}C_6H_{11}$$

$$\downarrow NH_2OH$$

$$R{-}CO{-}NHOH \xrightarrow{Fe^{3+}}$$

See Section VIII.A, Note 1

Reagents

a. A 1% solution of N,N'-dicyclohexylcarbodiimide in methanol.

b. A 1:1 mixture of 10% solution of hydroxylamine hydrochloride in methanol and 2.5 N solution of sodium hydroxide in methanol. Filter, and use freshly prepared.

c. Dissolve with heating 0.8 g of iron filings in 10 ml of 70% perchloric acid, allow to cool, add 10 ml of water, and dilute to 100 ml with ethanol. To 16 ml of this stock solution, add 4.8 ml of 70% perchloric acid, and dilute to 100 ml with ethanol.

Procedure

To 0.5 ml of aqueous solution of the acid, adjusted if necessary to pH 3 with hydrochloric acid, add 1 ml of reagent a. Let stand for 15 min, add 1 ml of reagent b, and after 15 min 2.5 ml of reagent c. Read immediately at 525 nm.

Results

	A = 0.3 (1 cm cell) Sample, μg
Acetic acid	220
Propionic acid	375
Butyric acid	450

Notes

1. At a pH of about 0, formic acid reacts with methanol and hydroxylamine even in the absence of dicyclohexylcarbodiimide.

2. Dicyclohexylcarbodiimide, see Section II.C, Note 3.

3. Hydroxamates, see Section VIII.A, Notes.

III. FATTY ACIDS

A. Rosaniline Carbinol Base

Principle (13)

Formation of a rosaniline salt: red-orange color.

Reagent

Reflux 2 g of finely powdered rosaniline carbinol base with 75 ml of benzene for 1 hr. Allow to cool, and filter or centrifuge.

Procedure

To 1 ml of sample solution in isopropanol, add 1 ml of reagent and heat at 75° for 30 min. Allow to cool, dilute to 10 ml with benzene, and read.

Results

	λ Max, nm	A = 0.3 (1 cm cell) Sample, μg
Palmitic acid	450	68
Stearic acid	470	73

Notes

1. A confusion may occur when using the word rosaniline. Many authors designate by this name the dyestuff C.I. Basic Violet 14, Magenta or Fuchsine (I), whereas Krainick and Müller (13) and other authors have used it for the corresponding carbinol base (II), or triaminodiphenyltolylcarbinol.

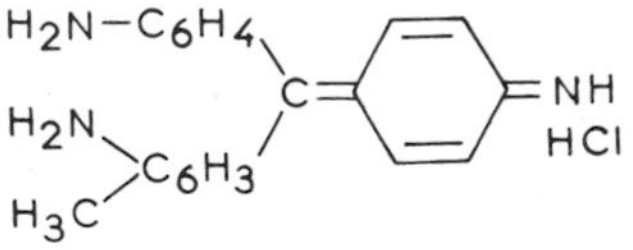

(I)

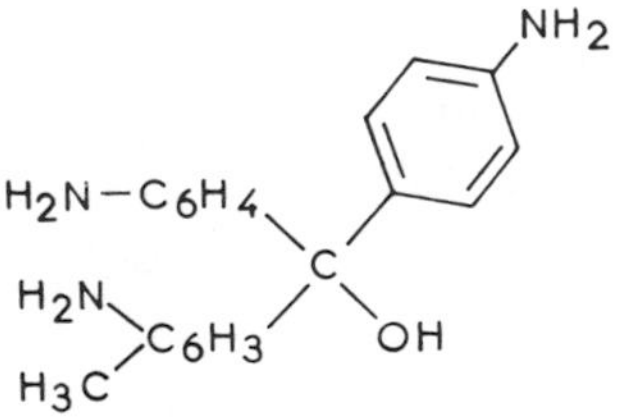

(II)

This base is colorless. Commercial samples often are violaceous red, but may be used for the above determination, whereas the hydrochloride (I) (metallic green, lustrous crystals) is not suited for this purpose.

2. The lower homolog of rosaniline, or triaminotriphenylcarbinol, is named pararosaniline. The salt corresponding to (I) is parafuchsine.

3. Schiff reagent is obtained by decolorizing fuchsine with sulfur dioxide (see Chapter 16, Section V).

IV. 2-HYDROXY-1-CARBOXYLIC ACIDS

A. Ferric Chloride

Principle (14)

Reaction with ferric chloride: yellow color.

Reagent

Dissolve 0.10 g of ferric chloride hexahydrate in 10 ml of water, add 0.1 ml of concentrated hydrochloric acid, and dilute to 100 ml with water.

Procedure

To 2 ml of aqueous solution of the acid, add 2 ml of reagent, and read.

	λ Max, nm	A = 0.3 (1 cm cell) Sample, μg
Glycolic acid	360	170
Lactic acid	360	350
Mandelic acid	365	130
Malic acid	360	200
Tartaric acid	360	125
Citric acid	370	360

V. 2,3-DIHYDROXY-1-CARBOXYLIC ACIDS

A. Sodium Metaperiodate and Resorcinol

Principle (15)

Oxidation with periodate ion, and reaction of the glyoxylic acid formed with resorcinol in acid medium to give the lactone of 2,2',4,4'-tetrahydroxydiphenylacetic acid, which is extracted and developed with potassium hydroxide: blue-violet color.

$$-CHOH-CHOH-COOH \xrightarrow{IO_4^-} OCH-COOH \xrightarrow{\text{resorcinol}} \text{lactone (HO, OH, CH, O=C–O, OH)}$$

Reagents

a. An 0.05 M aqueous solution of sodium metaperiodate adjusted to pH 10 with 1 N sodium hydroxide.

b. A 10% aqueous solution of ethylene glycol.

c. To 50 ml of 0.1 N aqueous silver nitrate, add 5 ml of 1 N sodium hydroxide. Let the precipitate settle, decant off the supernatant, wash the precipitate with water by decantation, dissolve it in 1 N sulfuric acid, and dilute to 100 ml with the acid. Titrate the silver with potassium thiocyanate.

d. A 1% solution of resorcinol in a 3:7 mixture of concentrated sulfuric acid and water.

e. A 1 N potassium hydroxide solution in ethanol.

Procedure

To 1 ml of neutralized aqueous solution of the acid, add 1 ml of reagent a, and let stand at 20-22° for 30 min. Add 0.1 ml of reagent b to destroy the excess periodate, and let stand for 5 min. Precipitate the iodate by adding the theoretical amount of reagent c (about 1 ml, calculated from the amount of periodate introduced), dilute to 5 ml with water, and filter. Pipet 1 ml of the filtrate, add to it 2 ml of reagent d, heat at 100° for 5 min, and allow to cool. Add 5 ml of water, and extract into 5 ml of ethyl acetate. To 1 ml of the organic layer, add 4 ml of ethanol and 0.5 ml of reagent e. Read at 550 nm within 10 min.

Results

	A = 0.3 (1 cm cell) Sample, μg
Tartaric acid	355
Mucic acid	420

Hexuronic acids can also be determined (Chapter 14, Section XI.B).

Notes

1. Formaldehyde, acetaldehyde, glycolic acid, oxalic acid, formic acid and malonic dialdehyde do not interfere. Glyoxal reacts weakly, but does not alter the results when present in amounts lower than glyoxylic acid.

2. Glyoxylic acid itself is oxidized by periodate ion, but it was established that in the presence of 0.05 M periodate at 20-22° it is stable at pH 10 (less than 3% oxidized after 1 hr). Between pH 3 and 7, it is completely oxidized within 15 min, whereas the reaction rate is very slow at a pH of about 2 (15).

3. The colorless 2,2',4,4'-tetrahydroxydiphenyl-acetic acid lactone obtained by condensing glyoxylic acid with resorcinol is insoluble in water and dilute acids, soluble in alkaline media. The blue-violet solutions so obtained are readily oxidized by atmospheric oxygen, the color rapidly turning to red-purple, then to orange, whereas a yellow-green fluorescence is displayed, thus allowing the fluorimetric determination of glyoxylic

acid (Section X.A). The quinone structure (III) was tentatively attributed to the oxidized species (16,17).

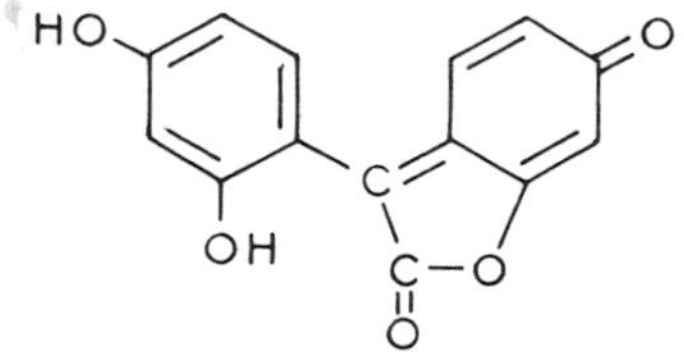

(I)

In a medium of ethanol, the blue-violet color is much more stable, thus permitting spectrophotometric measurements.

4. Periodates and periodate oxidation, see Chapter 16, Section IV.

VI. α-AMINO ACIDS

A. Cupric Phosphate, Oxalhydrazide, and Acetaldehyde

Principle (18,19)

Amino acids reacting with insoluble cupric phosphate give soluble cupric complexes. The copper so dissolved is developed with oxalhydrazide and acetaldehyde: violet-blue color.

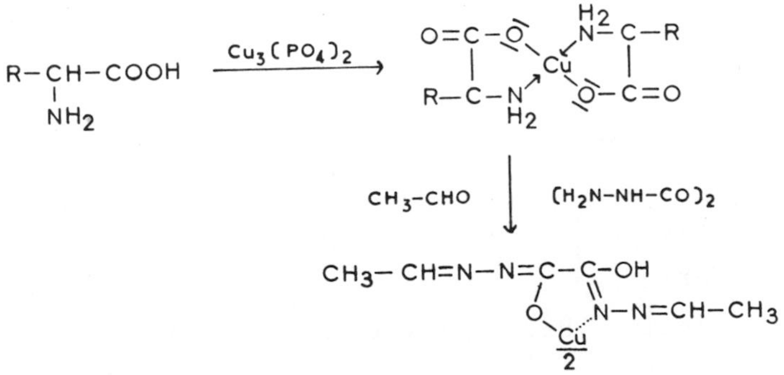

$$R-CH(NH_2)-COOH \xrightarrow{Cu_3(PO_4)_2} \text{Cu complex}$$

$$\xrightarrow{CH_3-CHO \quad (H_2N-NH-CO)_2} CH_3-CH=N-N=C-C(-OH)=N-N=CH-CH_3 \;\; (O \cdots Cu/2)$$

Reagents

a. To 80 ml of 6.85% aqueous solution of trisodium phosphate dodecahydrate, add with stirring 40 ml of 2.8% aqueous solution of cupric chloride dihydrate. Centrifuge, discard the supernatant, suspend the precipitate into 120 ml of 1.9% aqueous solution of sodium tetraborate decahydrate (pH 9.1), centrifuge, discard the supernatant, and repeat the washing once more. Suspend the precipitate into 200 ml of the borate solution, to which 6 g of sodium chloride has been added.

b. A 40% aqueous solution of acetaldehyde.

c. An 0.25% aqueous solution of oxalhydrazide.

Procedure

To 1 ml of aqueous solution of the amino acid adjusted to pH 7, add 1 ml of reagent a, shake for about 1 min, let stand for 5 min, and filter. To an 0.3 ml aliquot of the filtrate, add 0.7 ml of water, then 2 ml of reagent b, 2 ml of reagent c, and 1 ml of aqueous ammonium hydroxide (d = 0.92, 22° Bé). Let stand for 30 min in the dark, and read at 540 nm.

Results

	A = 0.3 (1 cm cell) Sample, μg
Glycine	96
α-Alanine	110
Phenylalanine	192
Serine	135
Threonine	135
Cysteine (hydrochloride)	260
Methionine	170
Glutamic acid	184
Asparagine	175

With lysine, Beer's law is not followed.

Notes

1. Dipeptides may also react. A = 0.3 is given by 76 μg of glycylglycine.

2. Sarcosine also reacts, but Beer's law is not followed.

3. Aliphatic and aromatic amines do not react.

4. Determination of copper with oxalhydrazide and an aldehyde, see Chapter 8, Section II.A, Notes 2 and 3.

B. p-Nitrobenzoyl Chloride

Principle (20,21)

Condensation with p-nitrobenzoyl chloride in pyridine medium to give a p-nitrophenyloxazolone: transient violaceous color turning to stable yellow or orange.

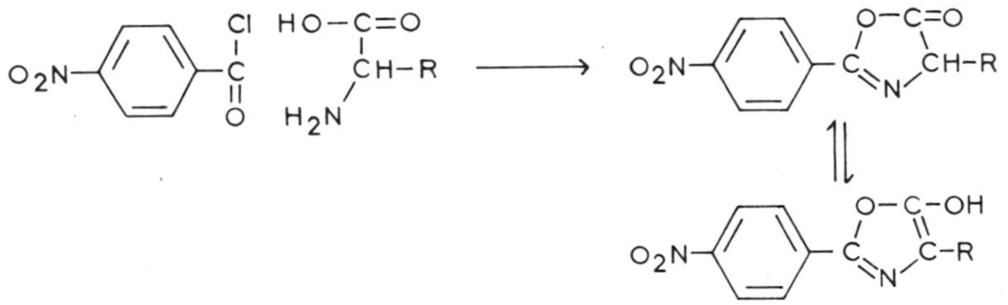

Reagent

A 5% solution of p-nitrobenzoyl chloride in pyridine. Prepare immediately before use.

Procedure

To 1 ml of aqueous solution of the amino acid, add 4 ml of pyridine, then dropwise with shaking, 2 ml of reagent. Let stand for 5 min, and read.

Results

	λ Max, nm	A = 0.3 (1 cm cell) Sample, μg
Glycine	480	7.5
Threonine[a]	410	280
Cysteine (hydrochloride)	405	380
Cystine	410	720
Asparagine	400	3000
Histidine	420	26
Tryptophan[a]	480	44

[a]Let stand for 15 min before reading.

Note

Colors developed with leucine, phenylalanine, tyrosine, methionine, arginine, and glutamic acid are too transient to allow colorimetric determinations.

C. p-Nitrobenzenediazonium Fluoborate

Principle and Procedure

See Chapter 4, Section VI.E, determination of primary aliphatic amines.

Results

	A = 0.3 (1 cm cell) Sample, μg
Glycine	5.0
α-Alanine	4.0
Methionine	10.3
Arginine (dihydrate)	20.5
Histidine	10.5

With leucine, Beer's law is not followed.

D. 2,5-Diethoxytetrahydrofuran and p-Dimethylaminobenzaldehyde

Principle and Procedure

See Chapter 4, Section VI.F, determination of primary aliphatic amines.

Results

	T_1, min	T_2, min	λ Max, nm	A = 0.3 (1 cm cell) Sample, μg
Glycine	10	5	558	1.8
Phenylalanine	10	20	558	5.7
Threonine	10	30	507	3.3
Cysteine	10	30	545	7.4
Arginine (dihydrate)	10	30	515	7.4

E. 2,5-Diethoxytetrahydrofuran and Flavylium Perchlorate

Principle and Procedure

See Chapter 4, Section VI.G, determination of primary aliphatic amines.

Results

	T_1, min	T_2, min	λ Max, nm	A = 0.3 (1 cm cell) Sample, μg
Glycine	10	15	500	6
Phenylalanine	10	20	455	12
Threonine	10	15	450	8
Cysteine	10	30	450	12

F. Ninhydrin and Potassium Cyanide

Principle (22)

Upon reaction with ninhydrin, formation of the anion of diketohydrindylidene-diketohydrindamine (Ruhemann's purple): blue-violet color.

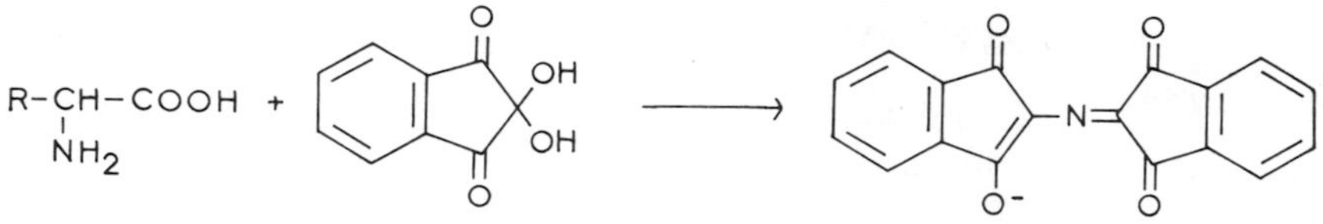

Reagents

_a_. Buffer for pH 5: Dissolve 21.00 g of citric acid in 200 ml of water, add 200 ml of 1 N sodium hydroxide, and dilute to 500 ml with water.

_b_. To 10 ml of 5% solution of ninhydrin in 2-methoxyethanol (Methyl Cellosolve), add 50 ml of 1:49 mixture of 0.01 M aqueous potassium cyanide and Methyl Cellosolve. The resulting solution is at first red, but turns rapidly to yellow. Let stand for at least 10 min before use.

Procedure

To 1 ml of aqueous solution of the amino acid, add 1 ml of buffer _a_ and 1 ml of reagent _b_. Heat at 100° for 15 min, chill in ice water, add 2 ml of ethanol, and let stand for 2 min. Read at 410 nm or 575 nm.

Results

	A = 0.3 (1 cm cell) Sample, μg	
	410 nm	575 nm
Glycine	4.4	5.0
α-Alanine	5.5	6.5
Leucine	8.1	8.9
Phenylalanine	10.3	11.8
Threonine	7.5	8.3
Cysteine[a]	30	63
Cystine (dihydrochloride)[a]	13.0	18.5
Methionine	8.8	9.6
Glutamic acid	7.5	9.6
Lysine (dihydrochloride)	10.0	12.9
Arginine (dihydrate)	12.9	14.0
Histidine	8.0	10.0
Tryptophan	12.6	15.0

[a]Irregular results.

Notes

1. At the heating stage, the blank becomes red. It turns to yellow on chilling.

2. When compared with the color developed by leucine, for which a 100% recovery was postulated, the recoveries for the other amino acids tested vary from 92 to 105% at 575 nm (116% for lysine). Cysteine and cystine were not taken into account.

3. Ninhydrin, and mechanism of the above reaction, see Chapter 16, Section X.

G. Ninhydrin and Ascorbic Acid

Principle

Upon reaction with ninhydrin, formation of the anion of diketohydrindylidene-diketohydrindamine (Ruhemann's purple): blue-violet color.

Reagents

a. Buffer for pH 5: Dissolve 21.00 g of citric acid in 200 ml of water, add 200 ml of 1 N sodium hydroxide, and dilute to 500 ml with water.

b. A 1% solution of ninhydrin in 2-methoxyethanol (Methyl Cellosolve).

c. An 0.1% aqueous solution of ascorbic acid.

Procedure

To 1 ml of aqueous solution of the amino acid, add 0.5 ml of reagent a, 1 ml of reagent b, and 0.5 ml of reagent c. Heat at 100° for 15 min, chill in ice water for 2 min, and add 2 ml of water. Read at 405 nm or 575 nm.

Results

	A = 0.3 (1 cm cell) Sample, µg	
	405 nm	575 nm
Glycine	4.6	4.8
α-Alanine	5.4	5.7

	A = 0.3 (1 cm cell) Sample, μg	
	405 nm	575 nm
Leucine	7.7	8.2
Phenylalanine	10.6	11.5
Threonine	7.8	8.3
Cysteine[a]	48.0	≃90
Cystine (dihydrochloride)[a]	15.2	20.0
Methionine	9.2	10.0
Glutamic acid	8.3	9.6
Lysine (dihydrochloride)	9.2	10.0
Arginine (dihydrate)	12.6	13.8
Histidine	8.8	10.0
Tryptophan	14.4	16.7

[a]Irregular results. With cysteine, another absorption maximum is observed at 450 nm.

Notes

1. At the heating stage, all the tubes are intensely red. On chilling, the blank turns to yellow, and the samples to blue-violet.

2. When compared with the color developed by leucine, for which a 100% recovery was postulated, the recoveries for the other amino acids tested vary from 89.6 to 97.7% (121% for lysine, 77% for tryptophan). Cysteine and cystine were not taken into account.

3. Ninhydrin, and mechanism of the above reaction, see Chapter 16, Section X.

H. Ascorbic Acid

Principle (23)

Reaction with ascorbic acid in aqueous dimethylformamide: red color. As a matter of fact, the true reagent is dehydroascorbic acid, formed by air oxidation of the ascorbic acid. The mechanism of the reaction was recently discussed (see Note 3).

Reagent

Dissolve 0.050 g of ascorbic acid in 0.5 ml of water, and dilute to 50 ml with dimethylformamide.

Procedure

To 0.25 ml of aqueous solution of the amino acid, add 1 ml of reagent and 3.75 ml of dimethylformamide. Heat at 100° for 10 min, and cool to 15-20° in a water bath. Read at 390 nm or 525 nm.

Results

	A = 0.3 (1 cm cell) Sample, μg	
	390 nm	525 nm
Glycine	8.5	30
α-Alanine	14	50
Threonine	20	77
Methionine	18	75
Glutamic acid	31	106
Lysine (hydrochloride)	23	86
Histidine	30	105
Tryptophan	35	106
Sarcosine	17	72

Notes

1. Application of ascorbic acid to the colorimetric determination of primary aliphatic amines, see Chapter 4, Section VI.J.

2. Inversely, ascorbic acid was determined with glycine as reagent (24).

3. Devaux and Mesnard (25) reasoned that the structure of dehydroascorbic acid (IV) can be compared with that of 1,2,3-indantrione (V), whose 2-monohydrate is ninhydrin. On the other hand, the spectra of the condensates obtained from an α-amino acid and (IV) or ninhydrin are closely related. They concluded therefrom that dehydroascorbic acid should behave like ninhydrin (see Chapter 16, Section X). They confirmed this standpoint

by establishing that an aldehyde is formed in the course of the reaction, and Pecherer (<u>26</u>) had already shown that carbon dioxide is evolved.

Therefore, they proposed a mechanism leading to compound (VI), which would be responsible for the color.

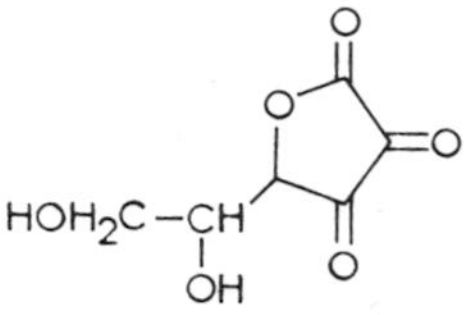

(IV) 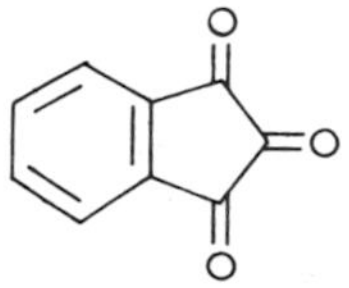 (V)

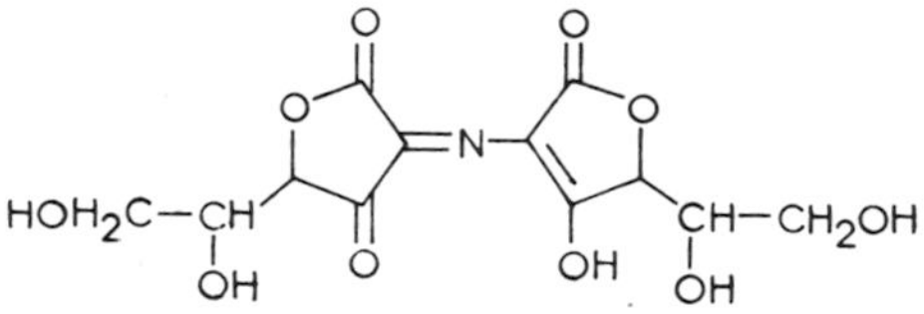

(VI)

VII. ANHYDRIDES AND LACTONES

A. Hydroxylamine Hydrochloride and Ferric Perchlorate

<u>Principle</u> (<u>27</u>)

Conversion of the anhydride or lactone to the corresponding hydroxamate, and development with ferric ion: red-violet color.

$$(R-CO)_2O + NH_2OH \longrightarrow R-COOH + R-CO-NHOH$$

$$>C(-O-)\cdots\cdots CO + NH_2OH \longrightarrow >C(OH)\cdots\cdots CO-NHOH \xrightarrow{Fe^{3+}}$$

See Section VIII.A, Note 1

Reagents

a. Neutralize (checking with phenolphthalein) a 12.5% solution of hydroxylamine hydrochloride in methanol with a 12.5% solution of sodium hydroxide in the same solvent, and filter. This reagent is stable for 4 hr.

b. Dissolve 5.0 g of ferric perchlorate in a mixture of 10 ml of 70% perchloric acid and 10 ml of water, and dilute to 100 ml with ethanol while cooling. Alternatively, dissolve with heating 0.8 g of iron filings in 10 ml of 70% perchloric acid, cool, add 10 ml of water, and dilute to 100 ml with ethanol.

To 4 ml of this stock solution, add 1.2 ml of 70% perchloric acid, and with cooling dilute to 100 ml with ethanol.

Procedure

To 0.5 ml of sample solution in benzene, add 0.3 ml of reagent a, and heat at 70° for 10 min, the tubes being fitted with air-condensers. Allow to cool, dilute to 5 ml with reagent b, and read.

Results

	λ Max, nm	A = 0.3 (1 cm cell) Sample, μg
Acetic anhydride	530	127
Propionic anhydride	530	427
Phthalic anhydride	540	1300
Butyrolactone	535	130
Santonin	535	1500

Phthalic anhydride and anhydrides which are insoluble in benzene can be dissolved in peroxide-free tetrahydrofuran.

Notes

1. According to this procedure, esters do not react. It can be applied to acid chlorides (27).

2. On the basis of kinetics studies (28), three mechanisms were proposed for the hydroxamation of lactones.

3. Hydroxamates, see following method, Notes.

VIII. CARBOXYLIC ESTERS

A. Hydroxylamine Hydrochloride and Ferric Perchlorate

Principle (27)

Conversion to the corresponding hydroxamate, and development with ferric ion: red-violet color.

$$R\text{—}COOR' + NH_2OH \longrightarrow R'\text{—}OH + R\text{—}CO\text{—}NHOH \xrightarrow{Fe^{3+}}$$

See Note 1

Reagents

a. Mix equal volumes of 12.5% solution of hydroxylamine hydrochloride in methanol and 12.5% solution of sodium hydroxide in the same solvent, and filter. This reagent is stable for 4 hr.

b. Prepare as described in the preceding method, under reagent b.

Procedure

To 0.5 ml of sample solution in ethanol, add 0.3 ml of reagent a, and heat at 70° for 5 min. Allow to cool, dilute to 5 ml with reagent b, and read.

Results

	λ Max, nm	A = 0.3 (1 cm cell) Sample, μg
Ethyl acetate	530	150
Butyl acetate	530	175
Benzyl benzoate	550	502
Ethyl malonate	520	135
Dimethyl phthalate	540	270

Notes

1. Hydroxamic acids were discovered in 1869 by Lossen (29). Two tautomeric forms may be taken into account, the hydroxyoxime (VII) and the N-hydroxyamide (VIII). Absorption spectra favor structure (VIII).

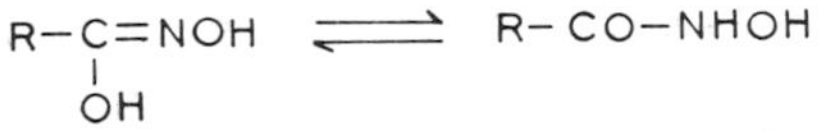

$$\underset{\text{(VII)}}{\text{R}-\underset{\substack{|\\ \text{OH}}}{\text{C}}=\text{NOH}} \rightleftharpoons \underset{\text{(VIII)}}{\text{R}-\text{CO}-\text{NHOH}}$$

They form highly colored ferric hydroxamates, to which the structure (IX) was attributed.

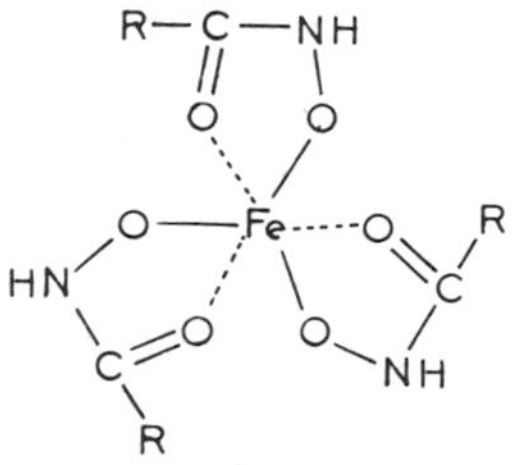

(IX)

The factors affecting the stability of the hydroxamic acid-iron complex have been studied in detail (30), as well as the kinetics and mechanisms of hydroxyaminolysis of succinamide (31), the kinetics of acetohydroxamic acid formation from ethyl acetate (32), and the mechanisms of hydroxamation of lactones (28). The stability of the color depends mainly on the ferric and hydrogen ion concentrations of the final solution (27).

2. Hydroxamation was applied to the characterization and determination of esters, lactones, amides, lactams, acid anhydrides, and chlorides, hence to the characterization and determination of compounds which, upon suitable reaction, may give rise to these species (33). These reactions are summarized in the following table.

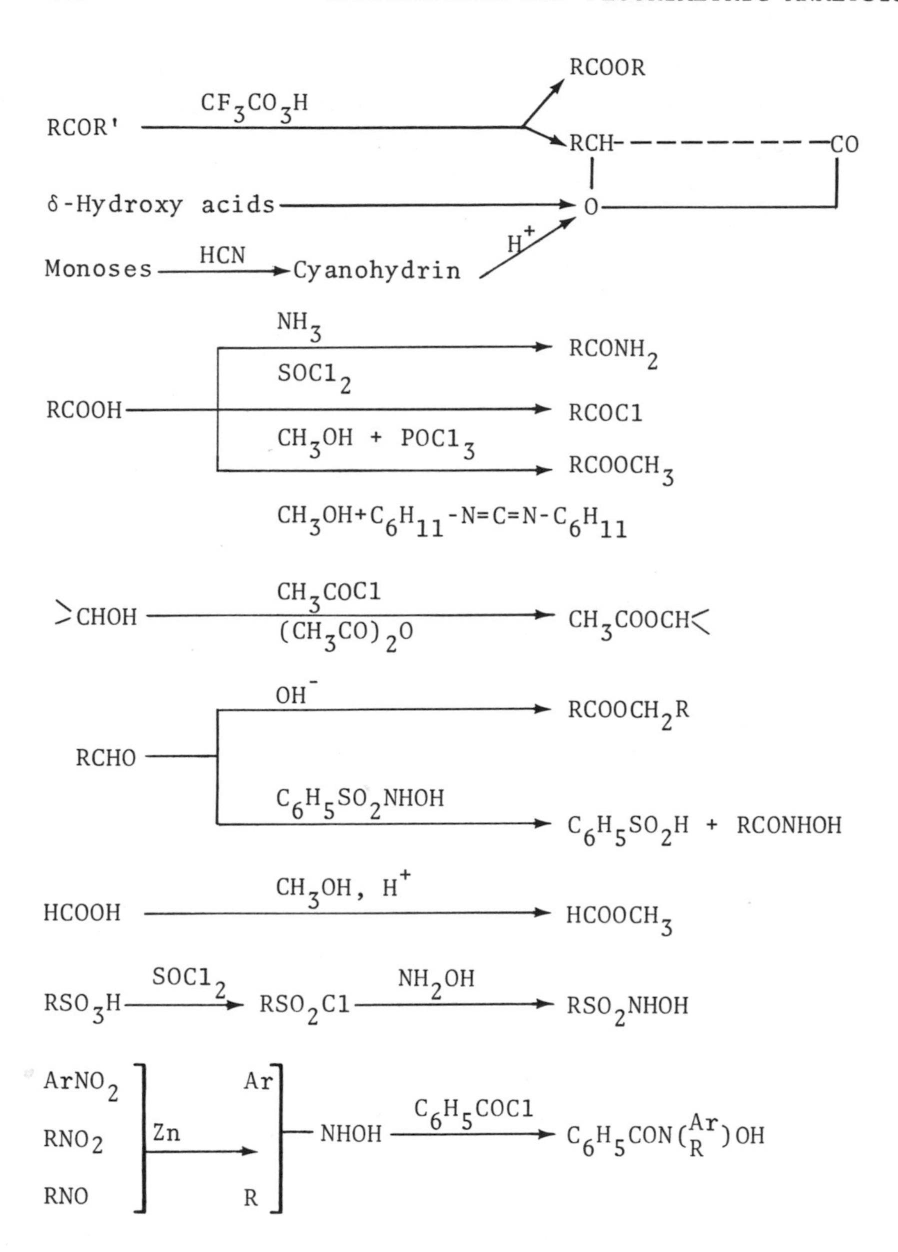
RCOOR
RCOR'
CF_3CO_3H
RCH
CO
δ-Hydroxy acids
O
Monoses
HCN
Cyanohydrin
H^+
NH_3
$RCONH_2$
$SOCl_2$
RCOOH
RCOCl
$CH_3OH + POCl_3$
$RCOOCH_3$
$CH_3OH+C_6H_{11}-N=C=N-C_6H_{11}$
>CHOH
CH_3COCl
$(CH_3CO)_2O$
$CH_3COOCH<$
OH^-
$RCOOCH_2R$
RCHO
$C_6H_5SO_2NHOH$
$C_6H_5SO_2H + RCONHOH$
CH_3OH, H^+
HCOOH
$HCOOCH_3$
RSO_3H
$SOCl_2$
RSO_2Cl
NH_2OH
RSO_2NHOH
$ArNO_2$
RNO_2
RNO
Zn
Ar
R
NHOH
C_6H_5COCl
$C_6H_5CON(^{Ar}_{R})OH$

In this book, hydroxamation was applied to the colorimetric determination of primary and secondary alcohols (Chapter 2, Section III.A), carboxylic acids (Sections II.D and II.E), anhydrides and lactones (preceding method), esters (herein-described method), and amides (following method).

IX. AMIDES

A. Hydroxylamine Sulfate and Ferric Chloride

Principle (34)

Conversion to the corresponding hydroxamate, and development with ferric ion: brown-violet color

$$R{-}CO{-}NH_2 + NH_2OH \longrightarrow NH_3 + R{-}CO{-}NHOH \xrightarrow{Fe^{3+}}$$

See Section VIII.A, Note 1

Reagents

a. Mix equal volumes of 16.4% aqueous solution of hydroxylamine sulfate and 3.5 N sodium hydroxide.

b. A 20% aqueous solution of ferric chloride hexahydrate in 0.1 N hydrochloric acid.

Procedure

To 3 ml of aqueous solution of the amide, add 2 ml of reagent *a*. Let stand at the given temperature for the given length of time (see Results), cool to room temperature, add 1 ml of 3.5 N hydrochloric acid and 1 ml of reagent *b*. Let stand for 5 min, and read at 500 nm.

Results

	Temperature, °C	Reaction time, hr	A = 0.3 (1 cm cell) Sample, μg
Acetamide	60	2	175
Acetamide	26	8	157
Dimethylformamide	26	4	375

	Temperature, °C	Reaction time, hr	A = 0.3 (1 cm cell) Sample, μg
Succinimide	60	2	300
Acetanilide	60	3	637
Nicotinamide	26	8	687

Note

Hydroxamates, see preceding method, Notes.

X. 2,3-DIHYDROXY-1-CARBOXYLIC ACIDS (F)

A. Sodium Metaperiodate and Resorcinol

Principle (35)

Oxidation with periodate ion, and reaction of the glyoxylic acid formed with resorcinol in acid medium to give the lactone of 2,2',4,4'-tetrahydroxydiphenylacetic acid, which is oxidized by atmospheric oxygen to give a quinone derivative: yellow-green fluorescence.

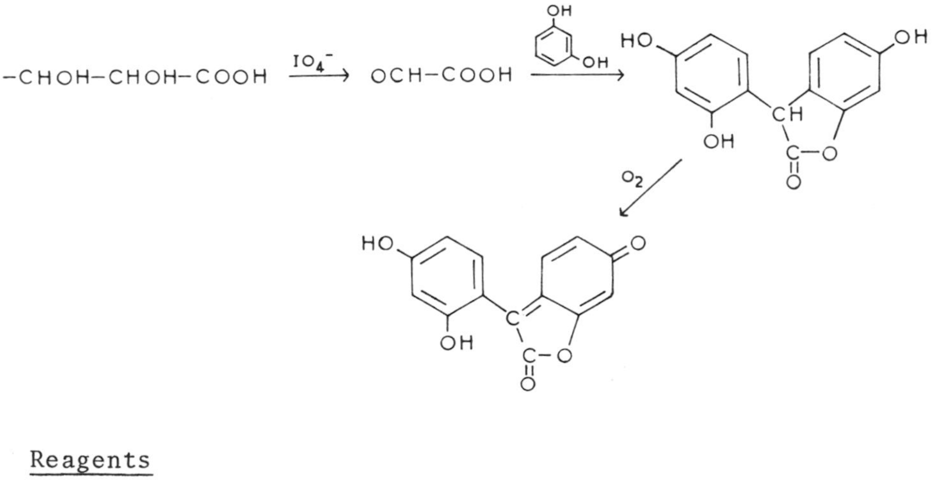

$$-CHOH-CHOH-COOH \xrightarrow{IO_4^-} OCH-COOH \xrightarrow{\text{resorcinol}} \text{lactone} \xrightarrow{O_2} \text{quinone derivative}$$

Reagents

a. An 0.05 M aqueous solution of sodium metaperiodate adjusted to pH 10 with 1 N sodium hydroxide.

b. An 0.1 N aqueous solution of potassium arsenite (1 liter of solution contains 4.945 g of arsenious oxide,

75 ml of 1 N potassium hydroxide, and 40.0 g of potassium bicarbonate).

c. To 50 ml of 0.1 N silver nitrate, add 5 ml of 1 N sodium hydroxide. Let the precipitate settle, decant off the supernatant, wash the precipitate with water by decantation, dissolve it in 1 N sulfuric acid, and dilute to 150 ml with the acid.

d. A 2% aqueous solution of resorcinol.

e. A saturated aqueous solution of potassium bicarbonate.

Procedure

To 1 ml of neutralized aqueous solution of the acid, add 0.2 ml of reagent a, and let stand at room temperature for 15 min. Add 0.5 ml of reagent b to destroy the excess periodate, let stand for 20 min at room temperature, add 0.3 ml of reagent c to precipitate the iodate, mix, add 1 ml of water, and filter. To 1 ml of the filtrate, add 1 ml of concentrated hydrochloric acid and 0.5 ml of reagent d. Heat at 60° for 20 min, cool to room temperature in a water bath, add slowly while shaking 5 ml of reagent e, and heat at 60° for 15 min. Allow to cool to room temperature, and read at exc: 436 nm; em: 530 nm.

Standard: An aqueous solution of sodium fluorescein.

Results

	Determination limits, μg	Reading 50 Sample, μg	Reading 50 Standard, μg/ml
Tartaric acid	1.0 - 5.0	2.1	0.13
Mucic acid	1.0 - 5.0	2.3	0.076

Hexuronic acids can also be determined (Chapter 14, Section XVII.A).

Notes

1. Several compounds afford glyoxylic acid upon suitable chemical reaction. It is selectively and readily

obtained from oxalic acid reduced by aluminum amalgam or a zinc-copper mixture (16), from dichloroacetic acid dehalogenated by hydrazine (17), and from acetic acid oxidized by periodate ion and hydrogen peroxide (36).

2. Colorimetric determination of 2,3-dihydroxy-1-carboxylic acids, see Section V.A.

3. Periodates and periodate oxidation, see Chapter 16, Section IV.

XI. α-AMINO ACIDS (F)

A. Acetylacetone and Formaldehyde

Principle (37,38)

See Chapter 4, Section XII.A, determination of primary aliphatic amines.

Reagent

To 10 ml of 1 M aqueous solution of sodium acetate, add 0.4 ml of acetylacetone and 1 ml of 30% aqueous formaldehyde, and dilute to 30 ml with water. Prepare immediately before use.

Procedure

To 1 ml of aqueous solution of the amino acid, add 1 ml of reagent, plug the tubes with cotton-wool, and heat at 100° for 10 min, protected against bright light. Chill in ice water, add 2 ml of water, and read at exc: 405 nm.

Standard: A solution of 2-methyl-5-carboxy-7-aminoquinoline in glacial acetic acid.

Results

	Em, nm	Determination limits, μg	Reading 50 Sample, μg	Reading 50 Standard, μg/ml
Glycine	485	2 - 10	4.5	0.042
Phenylalanine	490	8 - 40	17.2	0.044
Serine	485	5 - 25	10.0	0.032
Cysteine (hydrochloride)	500	20 - 100	38.0	0.040
Glutamic acid	485	20 - 100	43.0	0.035

B. o-Phthaldialdehyde and 2-Mercaptoethanol

Principle (39)

Condensation with o-phthaldialdehyde in the presence of 2-mercaptoethanol: blue fluorescence.

Reagent

A 1% solution of o-phthaldialdehyde in ethanol. Keep in the dark. Just before use, add 0.5 ml of this solution to 30 ml of buffer at pH 9.5, then add 0.5 ml of 0.5% solution of 2-mercaptoethanol in ethanol.

Buffer for pH 9.5: Adjust to this pH value an 0.05 M aqueous solution of sodium tetraborate decahydrate with 2 N sodium hydroxide.

Procedure

To 1 ml of aqueous solution of the amino acid, add 2 ml of reagent, mix, and read immediately at exc: 340 nm; em: 455 nm.

Standard: A solution of quinine sulfate in 0.1 N sulfuric acid.

Results

	Determination limits, μg	Reading 50	
		Sample, μg	Standard, μg/ml
Glycine	0.2 - 1.0	0.45	0.91
α-Alanine	0.2 - 1.0	0.45	1.02
Leucine	0.2 - 1.0	0.45	0.89
Serine	0.2 - 1.0	0.44	0.78
Threonine	0.2 - 1.0	0.42	0.72
Glutamine	0.2 - 1.0	0.45	0.85
Lysine (hydrochloride)[a]	0.4 - 2.0	0.88	0.83
Arginine	0.4 - 2.0	0.86	0.85
Histidine	0.3 - 1.5	0.64	0.96

[a]Use the Teorell and Stenhagen buffer for pH 7.0 (see Appendix, I.G) instead of the buffer for pH 9.5, and let stand for 5 min before reading.

The reaction is weakly sensitive with sarcosine and cysteine.

Notes

1. The mechanism of the reaction is as yet unknown. According to our experiments, 2-mercaptoethanol probably does not intervene only as a reducing agent.

2. Since the blank is almost nonfluorescent, more sensitive determinations are possible, but below the lower limit of 0.2 μg, the linear relationship does not hold.

C. Ninhydrin and Phenylacetaldehyde

Principle and Procedure

See Chapter 4, Section XII.B, fluorimetric determination of primary aliphatic amines.

Results

	Determination limits, μg	Reading 50	
		Sample, μg	Standard, μg/ml
Glycine	0.3 - 1.5	0.57	1.00
Leucine	1.6 - 8.0	3.0	1.56
Serine	1.0 - 5.0	1.9	0.71
Lysine (hydrochloride)	1.4 - 7.0	2.8	1.56
Tyrosine	1.4 - 7.0	2.65	1.07
Dopa	12 - 60	22.2	1.12

XII. α-KETO ACIDS (F)

A. o-Phenylenediamine

Principle (40)

Condensation with o-phenylenediamine in sulfuric acid medium to give a hydroxyquinoxaline: green to green-yellow fluorescence.

Reagent

Dissolve 0.010 g of o-phenylenediamine in 100 ml of 2 N sulfuric acid.

Procedure

To 1.5 ml of sample solution in 2 N sulfuric acid, add 1.5 ml of reagent. Fit the tubes with air-condensers, and heat at 100° for 1 hr. Chill for 5 min in ice water, and add 1.5 ml of concentrated sulfuric acid. Read at exc: 366 nm.

Standard: A solution of 2-hydroxy-3-methylquinoxaline in 50% sulfuric acid.

Results

	Em, nm	Determination limits, μg	Reading 50 Sample, μg	Reading 50 Standard, μg/ml
Glyoxylic acid (monohydrate)	520	1.0 - 5.0	1.9	0.27
Sodium pyruvate	490	0.4 - 2.0	0.85	0.30
α-Ketoglutaric acid	505	1.0 - 5.0	2.0	0.36
Oxaloacetic acid	500	1.0 - 5.0	2.2	0.56

Note

o-Phenylenediamine also allows the colorimetric determination of β-diketones (Chapter 8, Section IX.A).

REFERENCES

1. J. Bartos, *Mises au Point de Chimie Analytique*, 11e série, 55 (1963) (Masson Ed.).

2. Cf. Y. Nomura, *Bull. Chem. Soc. Japan*, 32, 536 (1959).

3. J. Bartos, *Talanta*, 8, 556 (1961).

4. M. Pesez and J. Bartos, *Talanta*, 12, 1049 (1965).

5. D.T. Elmore and J.J. Smyth, *Biochem. J.*, 94, 563 (1965).

6. L.K. Ramachandran, *Anal. Chem.*, 33, 1074 (1961).

7. S.W. Chaikin and W.G. Brown, *J. Amer. Chem. Soc.*, 71, 122 (1949).

8. T. Severin and R. Schmitz, *Chem. Ber.*, 95, 1417 (1962); T. Severin and M. Adam, *Chem. Ber.*, 96, 448 (1963); T. Severin, R. Schmitz, and H.L. Temme, *Chem. Ber.*, 96, 2499 (1963); T. Severin, R. Schmitz, and M. Adam, *Chem. Ber.*, 96, 3076 (1963); T. Severin, J. Loske, and D. Scheel, *Chem. Ber.*, 102, 3909 (1969).

9. J.C. Sheehan and G.P. Hess, *J. Amer. Chem. Soc.*, 77, 1067 (1955).

10. G. Amiard, R. Heymes, and L. Velluz, *Bull. Soc. Chim. Fr.*, 1464 (1955); L. Velluz, G. Amiard, J. Bartos, B. Goffinet, and R. Heymes, *Bull. Soc. Chim. Fr.*, 1464 (1956).

11. From H.A.C. Montgomery, J.F. Dymock, and N.S. Thom, *Analyst*, 87, 949 (1962).

12. From M. Pesez, *Ann. Pharm. Fr.*, 15, 173 (1957).

13. H.G. Krainick and F. Müller, *Mikrochem. Mikrochim. Acta*, 30, 7 (1941).

14. From A. Berg, *Bull. Soc. Chim. Paris*, [3]11, 882 (1894).

15. From M. Pesez and J. Bartos, *Bull. Soc. Chim. Fr.*, 481 (1960).

16. M. Pesez, *Bull. Soc. Chim. Fr.*, [5]3, 676, 2072 (1936); *J. Pharm. Chim.*, [9]2, 325 (1942).

17. M. Pesez, *Ann. Pharm. Fr.*, 9, 187 (1951).

18. J. Verdier, *Ann. Pharm. Fr.*, 25, 497 (1967).

19. Cf. B. Gauthier and M.C. Marechal, *Ann. Pharm. Fr.*, 20, 156 (1962).

20. Cf. E. Waser and E. Brauchli, *Helv. Chim. Acta*, 7, 757 (1924); E. Waser, *Mitt. Gebiete Lebensm. Hyg.*, 20, 260 (1929); S. Edlbacher and F. Litvan, *Z. Physiol. Chem.*, 265, 241 (1940); 267, 285 (1941).

21. Cf. P. Karrer and R. Keller, *Helv. Chim. Acta*, 26, 50 (1943).

22. E.W. Yemm and E.C. Cocking, *Analyst*, 80, 209 (1955).

23. J. Bartos, *Ann. Pharm. Fr.*, 22, 383 (1964).

24. M. Brunet, *Ann. Pharm. Fr.*, 26, 797 (1968).

25. G. Devaux and P. Mesnard, *Bull. Soc. Pharm. Bordeaux*, 110, 145 (1971).

26. B. Pecherer, *J. Amer. Chem. Soc.*, 73, 3827 (1951).

27. R.F. Goddu, N.F. Leblanc, and C.M. Wright, *Anal. Chem.*, 27, 1251 (1955).

28. T.C. Bruice and J.J. Bruno, *J. Amer. Chem. Soc.*, 83, 3494 (1961).

29. H. Lossen, *Justus Liebigs Ann. Chem.*, 150, 314 (1869).

30. R.E. Notari and J.W. Munson, *J. Pharm. Sci.*, 58, 1060 (1969).

31. R.E. Notari, *J. Pharm. Sci.*, 58, 1064 (1969).

32. R.E. Notari, *J. Pharm. Sci.*, 58, 1069 (1969).

33. Cf. F. Feigl, V. Anger, and O. Frehden, *Mikrochemie*, 15, 9, 23 (1934); F. Feigl, *Spot Tests in Organic Analysis*, Elsevier (Amsterdam), 7th Ed., 212, 214, 217 (1966); D. Davidson, *J. Chem. Educ.*, 17, 81 (1940); U.T. Hill, *Ind. Eng. Chem., Anal. Ed.*, 18, 317 (1946); M. Pesez and M. Legrand, *Bull. Soc. Chim. Fr.*, 453 (1960).

34. F. Bergmann, *Anal. Chem.*, 24, 1367 (1952).

35. M. Pesez and J. Bartos, *Talanta*, 14, 1097 (1967).

36. M. Pesez and J. Ferrero, *Ann. Pharm. Fr.*, 14, 558 (1956).

37. Cf. E. Sawicki and R.A. Carnes, *Anal. Chim. Acta*, 41, 178 (1968).

38. M. Pesez and J. Bartos, *Talanta*, 16, 331 (1969).

39. M. Roth, *Anal. Chem.*, 43, 880 (1971).

40. From J.E. Spikner and J.C. Towne, *Anal. Chem.*, 34, 1468 (1962).

Chapter 10

PEROXIDES

I. INTRODUCTION

The methods proposed for the colorimetric determination of peroxides offer some special features. Whereas the compounds which are dealt with in the other chapters are readily available in the pure state, thus allowing us to give the sample weights which afford an optical density of 0.3, peroxides are usually impure. They are more or less contaminated with unoxidized starting material, or with decomposition products originating from their instability.

We therefore consider that it is much better in every respect to give for each compound the amount of active oxygen affording A = 0.3. This result was calculated from the sample weight and the true concentration of peroxide in the material tested, as determined by the method recommended by the supplier.

On the other hand, the oxidizing efficiency of hydroperoxides, peracids, alkyl, acyl, and aroyl peroxides is often closely dependent upon the nature of the substituents. Therefore, the application of a given method to compounds other than those listed under Results may lead to low or negative results, and it can thence be necessary either to modify the procedure used, or to try another one. Inversely, each one of the procedures might give positive results with some peroxides which do not pertain to the classes mentioned at the head of the method.

Many peroxides oxidize iodide ion in acid medium, releasing elemental iodine which develops a yellow color.

Peracids, and acyl and aroyl peroxides oxidize dimethyl-p-phenylenediamine to give a colored species.

Hydroperoxides and peracids oxidize ferrous ion to ferric ion, which is then developed with p-aminobenzoic acid. In the course of this oxidation, the peroxide splits into a stable anion and a free radical. This latter reacts with ethanol, which is in turn converted to a radical which reduces Blue Tetrazolium to the corresponding strongly colored formazan.

II. MISCELLANEOUS PEROXIDES

A. Potassium Iodide

Principle (1)

Oxidation of iodide ion in acid medium, to give elemental iodine: yellow color.

$$R{-}O{-}O{-}R' + 2\,HI \longrightarrow ROH + R'OH + I_2$$

R and R' = Hydrogen, alkyl, acyl or aroyl

Reagent

A 10% aqueous solution of potassium iodide.

Procedure

To 2 ml of sample solution in water or chloroform, add 2 ml of the same solvent, 0.2 ml of 1 N hydrochloric acid, and 1 ml of reagent. Shake thoroughly, and let stand for 15 min in the dark. Then add 5 ml of water when the sample solvent is water, or 10 ml of ethanol when it is chloroform, and mix thoroughly to make homogeneous. Read at 360 nm.

Results

	Solvent[a]	A = 0.3 (1 cm cell) Active oxygen, μg
Hydrogen peroxide	W	1.3
tert-Butyl hydroperoxide	W	1.4
Cumene hydroperoxide	C	1.1
Monoperphthalic acid	W	2.2
m-Chloroperbenzoic acid	C	2.2
p-Nitroperbenzoic acid	C	5.4
Benzoyl peroxide	C	2.2

[a]W = Water, C = Chloroform.

Lauroyl peroxide and di-tert-butyl peroxide react but very weakly.

Note

The determination of peroxides through release of elemental iodine has been subjected to numerous studies, but mainly as a titrimetric method (2). It can be applied to a large variety of peroxides, but the reaction rate varies widely from one compound to the other, and it also depends upon the solvent and the acid used. Whereas an isolated double bond generally does not interfere, multiple, and particularly conjugated double bonds react with the released iodine.

III. PERACIDS, ACYL AND AROYL PEROXIDES

A. Dimethyl-p-phenylenediamine Oxalate

Principle (3)

Oxidation of dimethyl-p-phenylenediamine by the peroxide: pink color.

Reagent

Dissolve with gentle warming 0.3 g of dimethyl-p-phenylenediamine oxalate in 10 ml of glacial acetic acid, and dilute to 100 ml with methanol. Keep in the dark.

Procedure

To 2 ml of reagent, add 1 ml of sample solution in methanol and 1 ml of water. Mix, and keep in the dark for 3 min. Read at 560 nm.

Results

	A = 0.3 (1 cm cell) Active oxygen, μg
Monoperphthalic acid	1.0
m-Chloroperbenzoic acid	1.1
p-Nitroperbenzoic acid	1.1
Lauroyl peroxide[a]	2.9
Benzoyl peroxide	1.2

[a]Heat at 50° for 10 min in the dark, and chill in ice water for 2 min.

Notes

1. Hydroperoxides react but weakly. For instance, after 10 min at 50° in the dark, the amount of active oxygen affording A = 0.3 is of 170 μg for hydrogen peroxide and 450 μg for cumene hydroperoxide.

2. Dimethyl-p-phenylenediamine also allows the colorimetric determination of aliphatic (Chapter 8, Section IV.D) and aromatic (Chapter 8, Section VI.A) aldehydes, and of ketosteroids (Chapter 15, Sections VIII.A and XI.B).

3. Stability of dimethyl-p-phenylenediamine, see Chapter 3, Section II.O, Note 2.

IV. HYDROPEROXIDES AND PERACIDS

A. Ferrous Sulfate and p-Aminobenzoic Acid

Principle

Oxidation of the ferrous ion to ferric ion, and development with p-aminobenzoic acid: yellow color. The absorption maximum is in the ultraviolet range.

Reagent

Dissolve 0.40 g of ammonium ferrous sulfate hexahydrate in 40 ml of water, add 100 ml of methanol, and dissolve 0.50 g of p-aminobenzoic acid in this mixture. Prepare fresh prior to use.

Procedure

To 3 ml of reagent, add 1 ml of sample solution in methanol, and read at 340 nm.

Results

	A = 0.3 (1 cm cell) Active oxygen, μg
Hydrogen peroxide	1.65
tert-Butyl hydroperoxide	1.45
Cumene hydroperoxide	1.85
Monoperphthalic acid	2.35
m-Chloroperbenzoic acid	3.25
p-Nitroperbenzoic acid	2.8

Notes

1. Acyl and aroyl peroxides give a sensitive reaction only in the heat. The amount of active oxygen affording A = 0.3 is of 1 μg for lauroyl peroxide after 3 min at 50°, and of 3.5 μg for benzoyl peroxide after 7 min at 50°.

Di-tert-butyl peroxide develops only a very faint color.

2. The colorimetric determination of peroxides based on the oxidation of ferrous ion was subjected to numerous studies (2). Thiocyanate was generally used for the development of the ferric ion, but organic reagents such as salicylic or sulfosalicylic acid can be substituted for it. They develop a red color and can be used instead of p-aminobenzoic acid in the herein-described procedure, but we selected this latter reagent because it affords more sensitive and regular results.

3. The reaction of hydroperoxides and peracids with ferrous ion is not stoichiometric (4). The first step involves the reaction of one molecule of peroxide with one ferrous ion by a one-electron transfer, in the course of which the peroxide splits into a stable anion and a reactive free radical.

$$ROOH + Fe^{2+} \longrightarrow Fe^{3+} + OH^- + RO\cdot$$

The radical can then undergo various subsequent reactions with the excess of ferrous ion, the solvent R'H, or air oxygen, as summarized by the following equations:

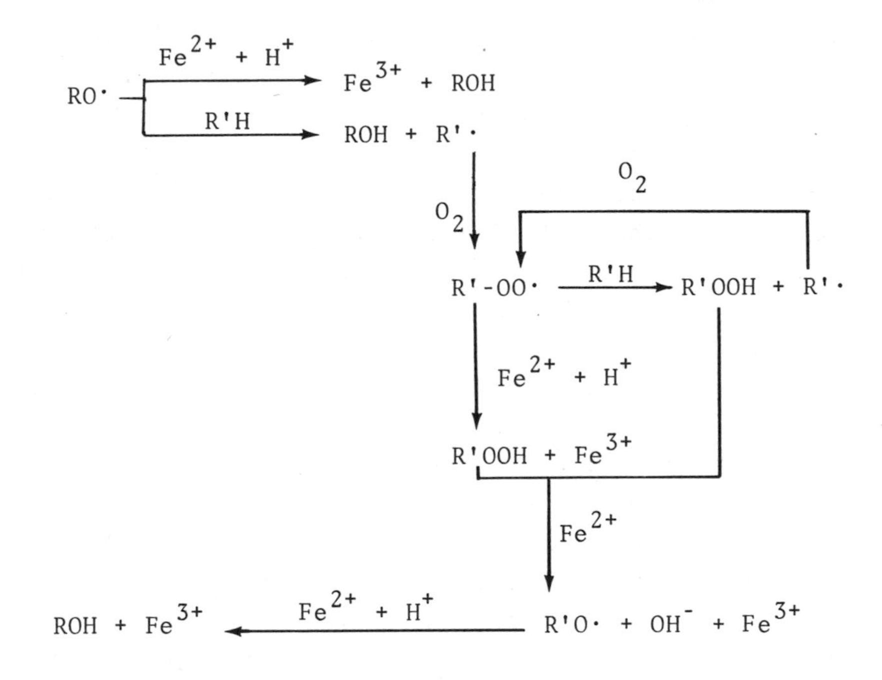

Under favorable circumstances, the overall course of the reaction may be determined almost entirely by the nature of the solvent radicals rather than of the peroxide radicals.

The above scheme shows that in comparison with the theoretical stoichiometry, high results are generally obtained in the presence of air, and low results in its absence.

B. Ferrous Sulfate, Ethanol, and Blue Tetrazolium

Principle (5,6)

Upon the action of a hydroperoxide in the presence of ferrous ion, ethanol is oxidized to a free radical, which in turn reduces Blue Tetrazolium to the corresponding formazan: violet color.

$$Fe^{2+} + ROOH \longrightarrow Fe^{3+} + OH^- + RO\cdot$$

$$RO\cdot + CH_3{-}CH_2OH \longrightarrow CH_3{-}\dot{C}HOH + ROH$$

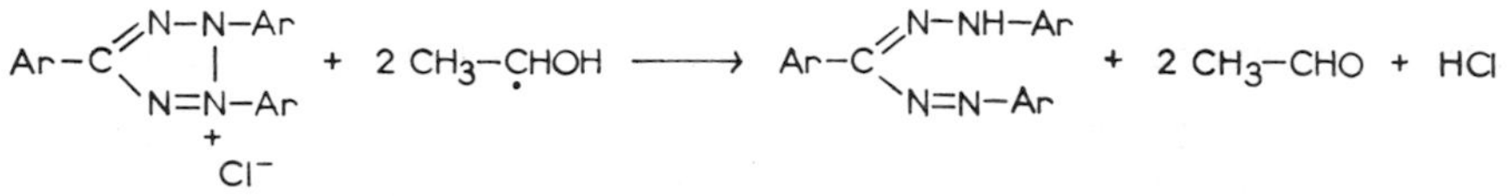

$$Ar{-}C\begin{matrix} \nearrow N{-}N{-}Ar \\ | \\ \searrow N{=}\overset{+}{N}{-}Ar \end{matrix}\ Cl^- + 2\,CH_3{-}\dot{C}HOH \longrightarrow Ar{-}C\begin{matrix} \nearrow N{-}NH{-}Ar \\ \searrow N{=}N{-}Ar \end{matrix} + 2\,CH_3{-}CHO + HCl$$

Sodium fluoride acts as masking agent for ferric ion, thus avoiding side reactions.

Reagents

a. A 4.2% solution of sodium fluoride in 0.1 N sulfuric acid.

b. A 7.84% solution of ammonium ferrous sulfate hexahydrate in 0.1 N sulfuric acid.

c. 1:1 mixture ethanol-water.

d. A saturated aqueous solution of Blue Tetrazolium.

These reagents are kept in ice water.

Procedure

With the tubes immersed in ice water, introduce successively 0.5 ml of reagent a, 0.1 ml of reagent b, 0.4 ml of reagent c, 1 ml of reagent d, then 0.5 ml of sample solution in water or 50% ethanol. Remove the tubes from the bath, and add 1.5 ml of 1 N sulfuric acid. Read at 560 nm.

Results

	A = 0.3 (1 cm cell) Active oxygen, μg
Hydrogen peroxide	4.2
tert-Butyl hydroperoxide	5.9
Cumene hydroperoxide	7.4
Monoperphthalic acid	7.8
m-Chloroperbenzoic acid	31
p-Nitroperbenzoic acid	156

Notes

1. Peroxides of the type R-O-O-R, acyl and aroyl peroxides do not react.

2. In the absence of sodium fluoride, ferric ion oxidizes the free radical much more rapidly than does the tetrazolium salt, thus making colorimetric determinations impossible.

3. For the more general reaction of ferrous ion with hydroperoxides and peracids, see Section IV.A, Note 3.

4. The mixture of hydrogen peroxide and ferrous sulfate is known as Fenton reagent and has been used as oxidant. It allows the characterization of tartaric acid, with which it develops a blue color (7).

5. Tetrazolium salts and formazans, see Chapter 16, Section VIII.

REFERENCES

1. Cf. A.M. Siddiqi and A.L. Tappel, *Chemist Analyst*, 44, 52 (1955); F.W. Heaton and N. Uri, *J. Sci. Food Agric.*, 9, 781 (1958); D.K. Banerjee and C.C. Budke, *Anal. Chem.*, 36, 792 (1964).

2. Cf. V.J. Karnojitzky, *Mises au Point de Chimie Analytique*, 12e série, 41 (1964) (Masson Ed.).

3. Cf. P.R. Dugan and R.D. O'Neill, *Anal. Chem.*, 35, 414 (1963).

4. I.M. Kolthoff and A.I. Medalia, *Anal. Chem.*, 23, 595 (1951).

5. J. Bartos, *Ann. Pharm. Fr.*, 30, 153 (1972).

6. Cf. I.M. Kolthoff and A.I. Medalia, *J. Amer. Chem. Soc.*, 71, 3777 (1949); D.J. Mackinnon and W.A. Waters, *J. Chem. Soc.*, 323 (1953).

7. H.J.H. Fenton, *Chem. News*, 33, 190 (1876).

Chapter 11

gem-POLYHALOGEN COMPOUNDS AND THIOLS

I. INTRODUCTION

A. Colorimetry

According to the Fujiwara reaction, gem-polyhalogen compounds open up the pyridine ring in alkaline medium to give a glutaconic dialdehyde derivative (determination levels of 15-36 μg for A = 0.3). Subsequent addition of benzidine affords a Schiff base of this aldehyde, making the sensitivity three- to fivefold greater.

Thiols are reducing compounds and can be determined as such with phosphotungstic acid or Blue Tetrazolium at levels of 17 to 140 μg, and with bis-(p-nitrophenyl) disulfide (6-15 μg of a thiol for A = 0.3).

The reactive hydrogen of the —SH group permits the condensation with 2,6-dichloroquinone-chlorimide to give a colored dichloroquinone sulfenimine. The sensitivity varies widely with the compound tested.

Thiols can also be determined through diazo coupling. The compound is converted to its S-nitroso derivative, which is then hydrolyzed in the presence of mercuric ion. The nitrous acid liberated diazotizes sulfanilamide, and the diazonium salt is developed with N-(1-naphthyl)ethylenediamine. A = 0.3 is given by 15-27 μg of a thiol.

According to an unknown mechanism, thiols condense with 2,3-dichloro-1,4-naphthoquinone. Depending on the compound, the color developed is extractable into chloroform or remains in the aqueous layer. Hence, some selective determinations are possible. The sensitivity varies widely with the compound tested.

B. Fluorimetry

The Fujiwara reaction also allows the fluorimetric determination of gem-polyhalogen compounds. The Schiff base obtained by reacting the glutaconic dialdehyde derivative with p-aminobenzoic acid is fluorescent, and determinations can be made with sample sizes of 1.1-3.6 μg for reading 50.

Thiols reduce thiamine disulfide, and the thiamine-thiol so formed is developed as thiochrome, allowing determination of 17-25 μg of compounds.

II. gem-POLYHALOGEN COMPOUNDS

A. Pyridine

Principle (1,2)

Upon action of the gem-polyhalogen compound in alkaline medium, opening of the pyridine ring: red color.

$$R{-}CCl_3 + \text{N} \longrightarrow \overset{+}{\text{N}}{-}CCl_2{-}R \;\; Cl^- \xrightarrow{OH^-} CH{=}N{-}CCl_2{-}R \;\; CHO^- \;\text{(I)} \longrightarrow CHO \;\; CHO^- \;\text{(II)}$$

Reagent

Pyridine, purified, Chapter 17, Section IV.

Procedure

To 5 ml of pyridine, add 1 ml of aqueous solution of the gem-polyhalogen compound, mix, and add 2 ml of 10 N potassium hydroxide. Mix well, heat at 100° for exactly 4 min, then immediately immerse the tubes in ice water for 3-5 min. To 3 ml of the supernatant pyridine layer, add 1 ml of water. Read immediately.

Results

	λ Max, nm	A = 0.3 (1 cm cell) Sample, μg
Trichloroethanol	440	36
Chloral hydrate	370	15
Trichloroacetic acid	370	15

Notes

1. The above determination is based on the Fujiwara reaction (1), which is bound up with determinations through the opening of pyridine ring (see Chapter 16, Section XI).

2. It may be assumed that the absorbing species which is spectrophotometrically estimated is (II) for chloral hydrate and trichloroacetic acid, (I) for trichloroethanol (3).

3. Determination of gem-polyhalogen compounds is used extensively in both toxicological screening and for quantitative work. A somewhat different procedure allows the determination of chloral hydrate, trichloroacetic acid, trichloroethanol, and urochloralic acid (2,2,2-trichloroethyl-β-D-glucosiduronic acid) in the presence of each other, thus permitting studies of the kinetics of chloral hydrate metabolism in mice (4).

B. Pyridine and Benzidine

Principle (5,6)

Upon action of the gem-polyhalogen compound in alkaline medium, opening of the pyridine ring, and development with benzidine in formic acid medium: red color.

>CCl_2 + N → N⁺ >CCl Cl^- $\xrightarrow{OH^-}$ CH=N–>CCl CHO⁻ $\xrightarrow{Ar-NH_2}$ CH=N–Ar CH–NH–Ar

Reagents

a. Pyridine, purified (Chapter 17, Section IV).

b. A 3% solution of benzidine in 88% formic acid. Prepare freshly.

Procedure

To 12 ml of sample solution in pyridine a, add 8 ml of 10 N potassium hydroxide. Mix well, heat at 100° for 3 min, chill in ice water, pipet 2 ml of the supernatant pyridine layer, and add to this aliquot 0.4 ml of reagent b. Let stand for 30 min, and read at 530 nm.

Results

	A = 0.3 (1 cm cell) Sample, μg
Chloroform	2.7
Iodoform	31
Trichloroethylene	4.2
sym-Tetrachloroethane	5.6
Chloral hydrate	3.7
Trichloroacetic acid	5.2

Notes

1. The reaction is negative with carbon tetrachloride, 1,2-dichloroethane, monochloroacetic acid, and chloralose.

2. Caution: benzidine is a potent carcinogenic agent.

3. Determinations through opening of the pyridine ring, see Chapter 16, Section XI.

III. THIOLS

A. Phosphotungstic Acid

Principle (7)

Reduction of phosphotungstic acid: blue color.

Reagents

a. To 100 g of sodium tungstate, add 200 ml of water and 50 ml of 85% phosphoric acid (d = 1.69). Reflux for 1 hr, add 5 drops of bromine, and boil gently for a few minutes, then vigorously without the condenser attached until complete disappearance of the excess bromine. Cool, and dilute to 1.25 liters with water.

b. Buffer for pH 5.2: To 32.8 g of sodium acetate trihydrate, add 7.2 g of acetic acid, and dilute to 100 ml with water.

c. A 35.7% aqueous solution of trisodium citrate dihydrate.

Procedure

To 1 ml of aqueous solution of the thiol, add 2 ml of reagent a, 2 ml of buffer b, and 1 ml of water or 1.5 ml of ethanol (for thiophenol). Let stand at room temperature for 10 min, and add 1 ml of reagent c. Read after 30 min.

Results

	λ Max, nm	A = 0.3 (1 cm cell) Sample, μg
Thiophenol[a]	730	76.5
Sodium thioglycolate	720	140
Cysteine	720	63

[a]Solubilized in water by adding the minimum amount of sodium hydroxide.

Note

Various analytical uses of heteropoly acids derived from molybdenum and tungsten have been described (8) (see also Chapter 3, Section II.C, Note 2).

B. Blue Tetrazolium

Principle (9)

Reduction of Blue Tetrazolium, or 3,3'-(3,3'-dimethoxy-4,4'-biphenylylene)bis[2,5-diphenyl-2H-tetrazolium chloride], into the corresponding formazan: pink-violet color.

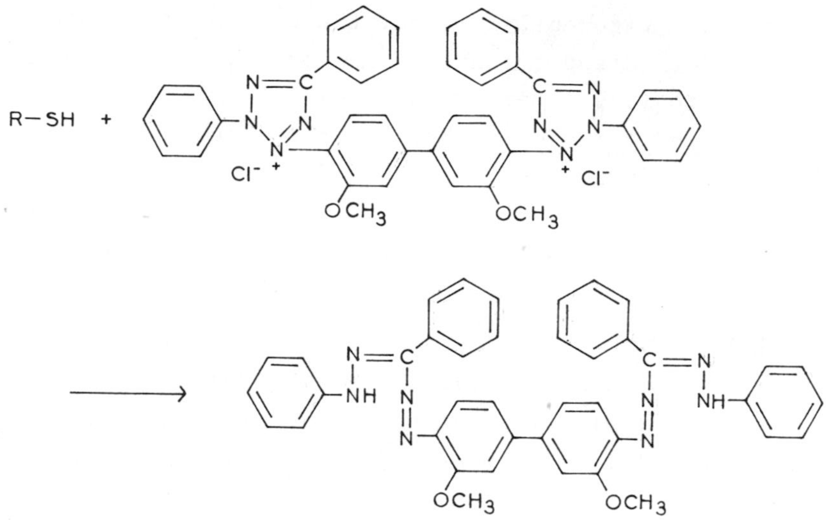

Reagents

a. An 0.25% solution of Blue Tetrazolium in ethanol.

b. Dilute 10 ml of 10% aqueous solution of tetramethylammonium hydroxide to 100 ml with ethanol.

Procedure

To 5 ml of sample solution in ethanol, add 1 ml of reagent a and 1 ml of reagent b. Incubate at 30° for 30 min, and read at 530 nm.

Results

	A = 0.3 (1 cm cell) Sample, μg
Dodecanethiol	64
Thioglycolic acid	17
Cysteine	23

Note

Tetrazolium salts and formazans, see Chapter 16, Section VIII.

C. Bis(p-Nitrophenyl) Disulfide

Principle (10)

Reduction of bis(p-nitrophenyl) disulfide to give the p-nitrothiophenate ion: yellow color.

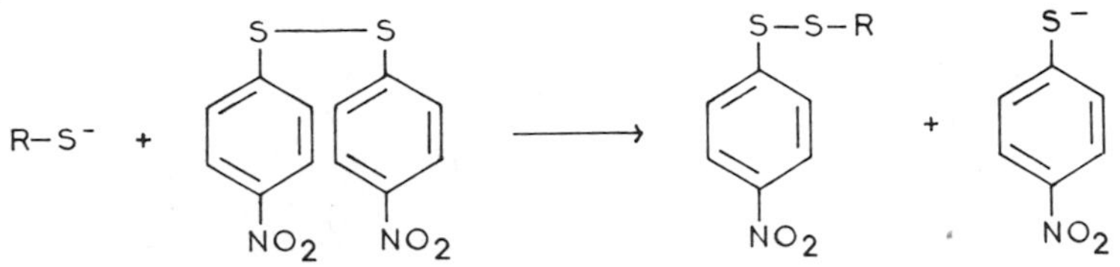

$$R{-}S^- + O_2N{-}C_6H_4{-}S{-}S{-}C_6H_4{-}NO_2 \longrightarrow O_2N{-}C_6H_4{-}S{-}S{-}R + {}^-S{-}C_6H_4{-}NO_2$$

Reagents

a. Buffer for pH 8.0: Adjust to this pH an 0.1 M aqueous solution of monosodium phosphate with 1 N sodium hydroxide.

b. An 0.06% solution of bis(p-nitrophenyl) disulfide in acetone. Keep at 0°, protected against light.

Procedure

Operate protected against light. To 1.5 ml of aqueous solution of the thiol, add 1 ml of buffer a, 1 ml of reagent b, 1 ml of acetone, and mix well. Read at 435 nm.

Results

	A = 0.3 (1 cm cell) Sample, μg
Dodecanethiol[a]	15
Thioglycolic acid	6.2
Cysteine	15

[a]Dissolve in 0.5 ml of acetone, and dilute to 1.5 ml with water.

Note

Dithiobis(2-nitrobenzoic) acid can be used instead of bis(p-nitrophenyl) disulfide (11).

D. 2,6-Dichloroquinone-chlorimide

Principle (12)

Reaction with 2,6-dichloroquinone-chlorimide in alkaline medium to give a dichloroquinone sulfenimine: yellow to brown color.

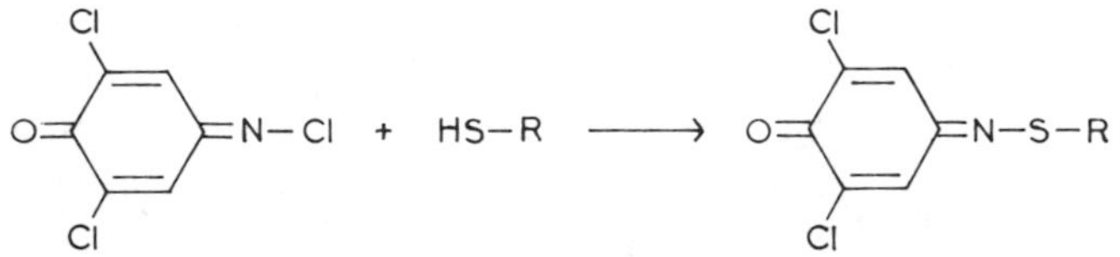

Reagents

a. An 0.5% aqueous solution of tetramethylammonium hydroxide.

b. An 0.5% solution of 2,6-dichloroquinone-chlorimide in dioxane.

Procedure

To 3 ml of aqueous solution of the thiol, add 0.2 ml of reagent a, keep for 1 min in ice water, and add 0.2 ml of reagent b. Let stand for 1 min, add 2 ml of dioxane, and read at 400 nm.

When the thiol is insoluble in water, dissolve the sample in 1 ml of dioxane, and dilute with 4 ml of the same solvent (instead of 2 ml) before reading at 370 nm.

Results

	λ Max, nm	A = 0.3 (1 cm cell) Sample, μg
Dodecanethiol	370	440
Sodium thioglycolate	400	118
Cysteine	400	33

Note

Reactions of 2,6-dihalogenoquinone-chlorimide, see Chapter 3, Section II.N, Notes 2, 3, and 4.

E. 2,3-Dichloro-1,4-naphthoquinone

Principle (13)

Reaction with 2,3-dichloro-1,4-naphthoquinone (unknown mechanism): yellow to orange color.

Reagents

a. An 0.092% solution of 2,3-dichloro-1,4-naphthoquinone in chloroform.

b. A 5% aqueous solution of potassium carbonate.

Procedure

Pipet 5 ml of reagent a, 2 ml of water, 1 ml of reagent b, and 1 ml of aqueous solution of the thiol into a separatory funnel. Shake vigorously for 2 min, allow the phases to separate, and collect the colored layer. Filter through filter paper if turbid, and read.

Results

	Colored phase[a]	Color	λ Max, nm	A = 0.3 (1 cm cell) Sample, μg
Dodecanethiol[b]	O	Yellow	440	970
Thiophenol[b]	O	Orange	440	145
Cysteine	A	Yellow	435	42
Glutathione	A	Pink-orange	450	137
Thiamine (hydrochloride)	O	Yellow	440	300
Cocarboxylase (chloride)	A	Pinkish	500	440

[a]A = aqueous, O = organic.

[b]These compounds being insoluble in water are dissolved in dioxane.

Note

In alkaline medium, the thiazole ring of vitamin B_1 and of cocarboxylase opens up to give a thiol derivative.

$$\text{Thiamine} \xrightarrow{OH^-} \text{thiol form (}CH_2-N(CHO)-C(CH_3)=C(SH)-CH_2-CH_2OH\text{)}$$

F. Sodium Nitrite, Sulfanilamide, and N-(1-Naphthyl)ethylenediamine Dihydrochloride

Principle (14,15)

Conversion of the thiol to its S-nitroso derivative, removal of the excess sodium nitrite, and mercuric-ion-assisted hydrolysis of the derivative. The nitrous acid liberated diazotizes sulfanilamide, and the diazonium salt is developed with N-(1-naphthyl)ethylenediamine (Bratton-Marshall reagent): pink color.

$$R-SH \xrightarrow{HNO_2} R-S-NO \xrightarrow{HgCl_2} R-S-HgCl + HNO_2 \xrightarrow{H_2N-C_6H_4-SO_2NH_2}$$

$$N\equiv\overset{+}{N}-C_6H_4-SO_2NH_2 \xrightarrow{C_{10}H_7-NH-CH_2-CH_2-NH_2} \text{(4-}(N{=}N-C_6H_4-SO_2NH_2)\text{-1-naphthyl)}-NH-CH_2-CH_2-NH_2$$

Reagents

a. To 5 ml of 0.01 M aqueous solution of sodium nitrite, add 40 ml of 1 N sulfuric acid, and dilute to 100 ml with water.

b. An 0.5% aqueous solution of ammonium sulfamate.

c. To 40 ml of 3.4% solution of sulfanilamide in 0.4 N hydrochloric acid, add 10 ml of 1% aqueous solution of mercuric chloride.

d. An 0.1% solution of N-(1-naphthyl)ethylenediamine dihydrochloride in 0.4 N hydrochloric acid.

Procedure

To 1 ml of sample solution in water or water-ethanol, add 5 ml of reagent a, let stand for 5 min, and add 1 ml of reagent b. Shake well for a few seconds, let stand for 2 min, rapidly add 10 ml of reagent c, mix, and dilute to 25 ml with reagent d. Read at 540 nm after 10 min.

Results

	A = 0.3 (1 cm cell) Sample, μg
Thiophenol[a]	27
Thioglycolic acid	15
Cysteine	21

[a]Can be solubilized in water by adding the minimum amount of sodium hydroxide.

Note

Diazo coupling, see Chapter 16, Section I.

IV. gem-POLYHALOGEN COMPOUNDS (F)

A. Pyridine and p-Aminobenzoic Acid

Principle (16)

Upon the action of the gem-polyhalogen compound in alkaline medium, opening of the pyridine ring, and development with p-aminobenzoic acid: The fluorescence is not visible under ultraviolet light.

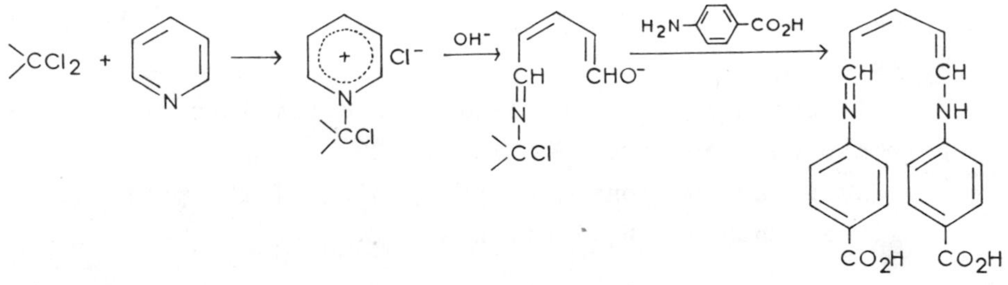

$>CCl_2$ + N → N⁺–C(Cl)< Cl^- $\xrightarrow{OH^-}$ CH=N–C(Cl)< ... CHO^- $\xrightarrow{H_2N-C_6H_4-CO_2H}$ CH=N–C_6H_4–CO_2H ... CH–NH–C_6H_4–CO_2H

Reagents

a. Pyridine, purified (Chapter 17, Section IV).

b. A 3% solution of p-aminobenzoic acid in ethanol.

Procedure

To 1 ml of sample solution in pyridine a, add 1 ml of 1 N sodium hydroxide. Mix well, heat at 100° for 3 min, chill in ice water, add 1.5 ml of 1 N hydrochloric acid and 1 ml of reagent b. Let stand at room temperature for the given length of time (see Results), and read at exc: 436 nm; em: 550 nm.

Standard: A solution of 2,6-dimethyl-3,5-diacetyl-1,4-dihydropyridine in ethanol.

Results

	Reaction time, min	Determination limits, μg	Reading 50 Sample, μg	Standard, μg/ml
Chloroform	20	0.5 - 2.5	1.1	0.92
Iodoform	30	1.2 - 6.0	2.5	0.85
Trichloro-ethylene[a]	40	1.0 - 5.0	1.6	0.94
sym-Tetrachloro-ethane[b]	30	1.0 - 5.0	2.7	1.00
Trichloroethanol	40	1.6 - 8.0	3.6	0.92
Chloral hydrate	30	0.6 - 3.0	1.32	0.91
Trichloroacetic acid	30	0.64 - 3.2	1.47	0.89

[a]Linear relationship is not satisfactorily followed.

[b]Linear relationship holds down to 2 μg.

Notes

1. Compare to the colorimetric determination of gem-polyhalogen compounds, Section II.B.

2. Determinations through opening of the pyridine ring, see Chapter 16, Section XI.

V. THIOLS (F)

A. Thiamine Disulfide and Potassium Ferricyanide

Principle (17,18)

Reduction of thiamine disulfide, and development of the formed thiaminethiol as thiochrome: blue fluorescence.

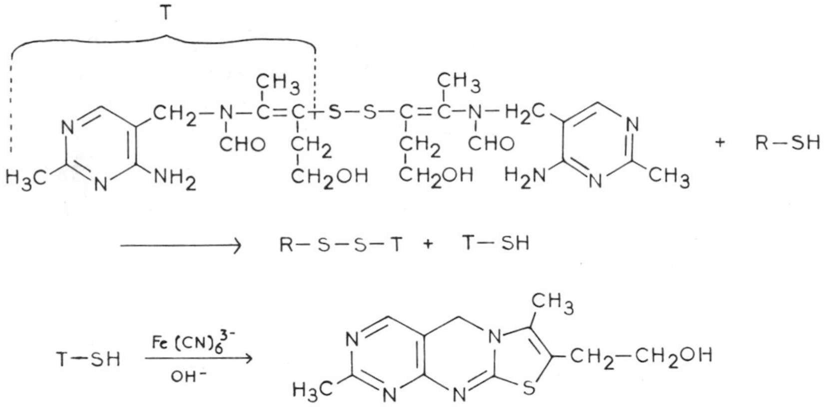

T

CH_3 CH_3

$CH_2-N-C=C-S-S-C=C-N-H_2C$

CHO CH_2 CH_2 CHO + R—SH

H_3C N NH_2 CH_2OH CH_2OH H_2N N CH_3

$\longrightarrow$ R—S—S—T + T—SH

T—SH $\xrightarrow[OH^-]{Fe(CN)_6^{3-}}$ CH_3, $-CH_2-CH_2OH$, H_3C (thiochrome)

Reagents

a. Buffer for pH 7.8-8.0: Mix 150 ml of 0.5 M aqueous dipotassium phosphate and 12 ml of 0.5 M aqueous monopotassium phosphate.

b. An 0.05 M aqueous solution of EDTA.

c. An 0.141% aqueous solution of thiamine disulfide.

d. A 25% solution of potassium chloride in 0.1 N hydrochloric acid.

e. Dilute 12.5 ml of 1% aqueous solution of potassium ferricyanide to 100 ml with 15% sodium hydroxide. Prepare fresh immediately before use.

Procedure

To 1 ml of aqueous solution of the thiol, add 2 ml of buffer a, 1 ml of reagent b, and 1 ml of reagent c. Incubate at 30° for 20 min, pipet a 1 ml aliquot of the mixture, and add to it 5 ml of reagent d and 3 ml of reagent e. Let stand for 2 min, add 13 ml of isobutanol, shake for 90 sec, and allow the phases to separate. Collect the organic layer, and if turbid dry it over anhydrous magnesium sulfate. Read at exc: 366 nm; em: 435 nm.

Standard: A solution of quinine sulfate in 0.1 N sulfuric acid.

Results

	Determination limits, μg	Reading 50 Sample, μg	Standard, μg/ml
Thiophenol	11 - 55	25	5.77
Thioglycolic acid	9 - 45	17	1.38
Cysteine	12 - 60	18	0.90

REFERENCES

1. Cf. K. Fujiwara, *Sitzber. Abhandl. Naturforsch. Ges. Rostock*, 6, 33 (1916).

2. P.J. Friedman and J.R. Cooper, *Anal. Chem.*, 30, 1674 (1958).

3. M.S. Moss and H.J. Rylance, *Nature*, 210, 945 (1966).

4. B.E. Cabana and P.K. Gessner, *Anal. Chem.*, 39, 1449 (1967).

5. Cf. F. Feigl, *Spot Tests in Organic Analysis*, Elsevier (Amsterdam), 5th Ed., 313 (1956).

6. From K.C. Leibman and J.D. Hindman, *Anal. Chem.*, 36, 348 (1964).

7. A.E. Mirsky and L.M. Anson, *J. Gen. Physiol.*, 18, 307 (1935).

8. L. Vignoli, B. Cristau, and A. Pfister, *Chim. Anal.*, 38, 392 (1956).

9. E.F. Salim, P.E. Manni, and J.E. Sinsheimer, *J. Pharm. Sci.*, 53, 391 (1964).

10. From G.L. Ellman, *Arch. Biochem. Biophys.*, 74, 443 (1958).

11. G.L. Ellman, *Arch. Biochem. Biophys.*, 82, 70 (1959).

12. Cf. D.N. Kramer and R.M. Gamson, *J. Org. Chem.*, 24, 1154 (1959).

13. From K. Hofmann, *Naturwiss.*, 52, 428 (1965).

14. B. Saville, *Analyst*, 83, 670 (1958).

15. Cf. H.F. Liddell and B. Saville, *Analyst*, 84, 188 (1959).

16. J. Bartos, *Ann. Pharm. Fr.*, 29, 221 (1971).

17. Cf. O. Zima, K. Ritsert, and T. Moll, *Z. Physiol. Chem.*, 267, 210 (1941).

18. K. Kohno, *J. Vitaminol.* (Japan), 12, 137 (1966).

Chapter 12

UNSATURATED COMPOUNDS

I. INTRODUCTION

A. Colorimetry

The determination of ethylenic compounds is based on the oxidation of the double bond under conditions which give rise to aldehydes. Therefore, only compounds of the type R—CH=C< (affording R—CHO), and R—CH=CH_2 (generating formaldehyde) can be taken into account.

Compounds of the type R—CH=C< are oxidized with ruthenium tetroxide, and the aldehyde formed is developed with 3-methylbenzothiazolin-2-one hydrazone. Sample sizes of 5-48 μg are necessary for A = 0.3.

Compounds of the type R—CH=CH_2 are oxidized with the Lemieux reagent (aqueous solution of sodium metaperiodate and potassium permanganate). The formaldehyde formed is developed either with phenylhydrazine and ferricyanide (Schryver reaction), or with chromotropic acid (Eegriwe reaction). The sensitivity varies widely with the compound tested.

As a rule, aromatic compounds are not oxidized under the above conditions. The dinitration of the benzene ring is hence turned to account for their determination. This reaction affords mainly m-dinitro derivatives, which are developed as Meisenheimer-like σ complexes with nitromethane in alkaline medium (26-300 μg for A = 0.3). A few percent of o- and p-dinitro compounds are also formed, which can be developed with ascorbic acid in alkaline medium, affording the dianionic form of o- and p-dinitrocyclohexadiene derivatives. This reaction is poorly sensitive (several hundred micrograms for A = 0.3).

A similar sensitivity is obtained with the procedure based on the Baudisch reaction: Upon the action of cupric ion, hydroxylamine, and hydrogen peroxide, the copper complex of an o-nitrosophenol is obtained.

The formation of colored cations when the benzene ring is reacted with a formaldehyde-sulfuric acid reagent was also taken into account, although more than 1 mg of sample may be necessary for A = 0.3.

B. Fluorimetry

Only one method is proposed, allowing the determination of ethylenic compounds of the type $R—CH=CH_2$. The formaldehyde obtained upon oxidation of the compound with the Lemieux reagent is developed through the Hantzsch reaction with ethyl acetoacetate and ammonia. Reading 50 is given by 3.5-16 µg of sample.

II. ETHYLENIC COMPOUNDS OF THE TYPE R—CH=C<

A. Ruthenium Tetroxide and 3-Methylbenzothiazolin-2-one Hydrazone

Principle (1)

Oxidation with ruthenium tetroxide, causing carbon-carbon double-bond cleavage, and development of the aldehydes formed with 3-methylbenzothiazolin-2-one hydrazone: blue color.

$$R-CH=C\langle \xrightarrow{RuO_4} R-CHO \longrightarrow \text{See Chapter 8, Section IV.B}$$

Reagents

a. Shake a suspension of 0.010 g of ruthenium dioxide (see Note 1) in 5 ml of carbon tetrachloride with 5 ml of 10% aqueous solution of anhydrous sodium metaperiodate until the organic phase turns to yellow-green. The solution of ruthenium tetroxide in carbon tetrachloride so obtained may be used for 2 hr when kept under the aqueous layer. Avoid contact with grease.

b. An 0.5% aqueous solution of 3-methylbenzothiazolin-2-one hydrazone hydrochloride.

c. An 0.3% aqueous solution of hydrogen peroxide (1 volume oxygen).

Procedure

To 3 ml of sample solution in glacial acetic acid, add 0.1 ml of reagent *a*, mix, let stand for 5 min, add 0.5 ml of reagent *b*, and let stand for 10 min. Add 0.2 ml of reagent *c*, let stand for 15 min, and dilute with 5 ml of acetone. After 5 min, read at 660 nm. During the whole procedure, avoid contact with grease.

Results

	Double bonds (nonaromatic)		A = 0.3 (1 cm cell) Sample, μg
	>C=CH–	>C=CH_2	
Styrene	0	1	5
Camphene	0	1	12
Secobarbital	0	1	24
Quinine	0	1	16
Testosterone	1	0	48
Hydrocortisone	1	0	48
Prednisolone	2	0	34
7-Dehydrocholesterol	2	0	41
Ergosterol	3	0	31
Calciferol	3	1	19

Notes

1. The ruthenium dioxide should be amorphous, and the diameter of the particles should not exceed 1 μm. Crystallized samples (as shown by Debye-Scherrer spectrum) are oxidized but poorly or not at all by sodium periodate.

2. Whereas osmium tetroxide has been widely used for the oxidation of organic compounds, ruthenium tetroxide was used practically not at all until 1958 (2). It is much more powerful an oxidizing agent than osmium tetroxide, the latter giving hydroxylation of the double bond, the former causing carbon-carbon double-bond cleavage, probably according to the following mechanism:

$$R-CH=C\langle \xrightarrow{RuO_4} R-\underset{H}{C}(O)-C(O)\langle \;[\text{cyclic }RuO_2\text{ ester}] \longrightarrow R-CHO + \rangle C=O + RuO_2$$

3. Under the above analytical conditions, aromatic double bonds practically do not react.

4. The periodate-permanganate reagent can be used instead of ruthenium tetroxide (3).

5. Primary alcohols are oxidized to aldehydes and can be so determined (Chapter 2, Section IV.A).

6. Properties of 3-methylbenzothiazolin-2-one hydrazone, see Chapter 16, Section III.

III. ETHYLENIC COMPOUNDS OF THE TYPE $R-CH=CH_2$

A. Potassium Permanganate, Sodium Metaperiodate, Phenylhydrazine Hydrochloride, and Potassium Ferricyanide

Principle (4,5)

Upon reaction with permanganate ion, the double bond is oxidized to a mixture of diol and ketol, which is in turn oxidized by periodate ion, and the liberated formaldehyde is developed with phenylhydrazine and ferricyanide ion: red color.

$$\begin{matrix} CH_2 \\ \| \\ CH \\ | \\ R \end{matrix} \xrightarrow[IO_4^-]{MnO_4^-} \left[\begin{matrix} CH_2OH \\ | \\ CHO \\ | \\ \text{and} \\ CH_2OH \\ | \\ CO \\ | \end{matrix} \right] \longrightarrow HCHO \xrightarrow[Fe(CN)_6^{3-}]{C_6H_5\text{-}NH\text{-}NH_2} \text{Formazan dye}$$

Reagents

a. An 0.05 N aqueous solution of sodium metaperiodate.

b. An 0.005 N aqueous solution of potassium permanganate.

c. An 0.1 N aqueous solution of potassium arsenite (1 liter of solution contains 4.945 g of arsenious oxide, 75 ml of 1 N potassium hydroxide, and 40.0 g of potassium bicarbonate).

d. A 1% aqueous solution of phenylhydrazine hydrochloride, freshly prepared.

e. A 2% aqueous solution of potassium ferricyanide, freshly prepared. Just before use, add 50 ml of concentrated hydrochloric acid to 20 ml of this solution.

Procedure

To 1 ml of sample solution in pyridine, add 4 ml of water, 1 ml of reagent a, and 1 ml of reagent b. Let stand for 30 min, add 2 ml of reagent c, and 1 ml of water. Pipet a 2 ml aliquot of this solution, add to it 2 ml of reagent d, let stand for 5 min, and add 7 ml of reagent e. Read at 510 nm.

Results

	A = 0.3 (1 cm cell) Sample, μg
Styrene	120
Allyl alcohol	67
Quinine	410

Notes

1. The reduced permanganate is reoxidized by the excess of periodate, so that only a catalytic amount of permanganate can be introduced in the reaction medium. According to our experiments, it can be replaced by an equivalent amount of a manganese(II) salt.

2. The formaldehyde formed can also be developed with chromotropic acid (see the following method).

3. Periodates and periodate oxidation, see Chapter 16, Section IV.

4. Colorimetric determination of formaldehyde, see Chapter 16, Section V.

5. Fluorimetric determination of ethylenic compounds of the type $R{-}CH{=}CH_2$, see Chapter 12, Section V.A.

B. Potassium Permanganate, Sodium Metaperiodate and Sodium Salt of Chromotropic Acid

Principle (4,6)

Formation of formaldehyde, see under Principle, preceding method; the formaldehyde is developed with chromotropic acid: violet color.

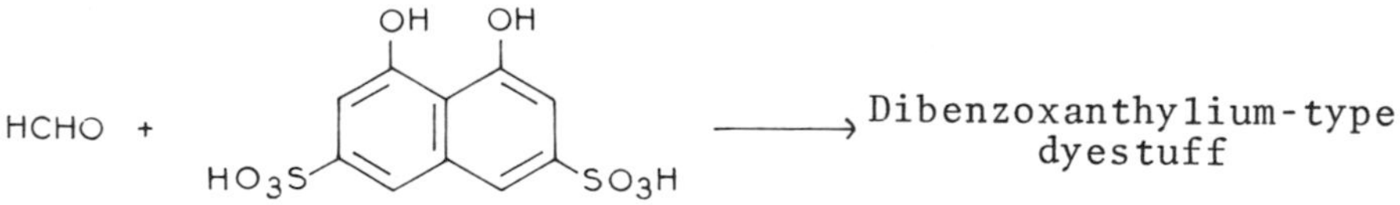

OH OH

HCHO + HO_3S SO_3H ⟶ Dibenzoxanthylium-type dyestuff

Reagents

a. An 0.05 N aqueous solution of sodium metaperiodate.

b. An 0.005 N aqueous solution of potassium permanganate.

c. An 0.1 N aqueous solution of potassium arsenite (1 liter of solution contains 4.945 g of arsenious oxide, 75 ml of 1 N potassium hydroxide, and 40.0 g of potassium bicarbonate).

d. Dissolve 1 g of sodium salt of chromotropic acid in 100 ml of water, filter, and dilute to 500 ml with 66% sulfuric acid.

Procedure

To 1 ml of sample solution in pyridine, add 5 ml of water, 1 ml of reagent a, and 1 ml of reagent b. Let stand for 30 min, and add 2 ml of reagent c. After 5 min, pipet a 2 ml aliquot of this solution, and add to it 5 ml of reagent d. Heat at 100° for 30 min, cool in a water bath, and read at 570 nm.

Results

	A = 0.3 (1 cm cell) Sample, μg
Styrene	100
Allyl alcohol	54
Quinine	340

Note

See Notes 1, 3, 4, and 5, preceding method.

IV. AROMATIC COMPOUNDS

A. Potassium Nitrate, Sulfuric Acid, Nitromethane and Benzyltrimethylammonium Hydroxide

Principle (7)

Dinitration of the benzene ring, and development of the m-dinitro derivative formed as a Meisenheimer-like σ complex with nitromethane in alkaline medium: pink to pink-violet color.

$$C_6H_6 \xrightarrow{HNO_3} C_6H_4(NO_2)_2 \xrightarrow[B^+\ OH^-]{CH_3\text{-}NO_2} [H(CH_2\text{-}NO_2)C_6H_4(NO_2)_2]^- \ B^+$$

Reagents

a. A 20% solution of potassium nitrate in concentrated sulfuric acid.

b. A 1:3 mixture of 40% solution of benzyltrimethylammonium hydroxide in methanol (see Chapter 17, Section III) and dimethylformamide. Prepare freshly.

Procedure

Evaporate the sample solution to dryness. To the residue, add 0.5 ml of reagent a, and heat at t° for T minutes. Chill to 0° in ice water, and add slowly 5 ml of prechilled water. Transfer the solution into a separatory funnel containing 2 ml of nitromethane, rinse the tube with 1 ml of water, and pour the rinsing into the funnel. Shake for 20 sec, let stand for a few minutes, and collect the lower organic layer. Pipet a 1 ml aliquot, add to it 4 ml of dimethylformamide and 0.1 ml of reagent b. Read immediately.

Results

	t, °C	T min	λ Max, nm	A = 0.3 (1 cm cell) Sample, μg
Benzoic acid	100	20	565	26.5
Phenylacetic acid[a]	70	5	555	23
Ephedrine	100	15	550	300
Phenobarbital	100	10	550	57
Atropine (sulfate)	70	15	550	105

[a]Let stand for 5 min before reading.

Notes

1. During the evaporation to dryness, some compounds may partly sublimate. This can be overcome by adding 0.01 to 0.05 g of anhydrous magnesium sulfate to the sample solution before evaporation. This salt does not hinder the nitration step, and clear solutions are obtained on dilution with water.

2. Methanol and methylene chloride were used as solvents for the compounds listed under Results. Ethyl ether

and such solvents which give rise easily to peroxides must be avoided, since an explosion may occur upon addition of the nitrating mixture.

3. This method is related to the Janovsky reaction, see Chapter 16, Section VI.

B. Potassium Nitrate, Sulfuric Acid, and Ascorbic Acid

Principle

Dinitration of the benzene ring, and reduction of the o- and p-dinitro derivatives formed with ascorbic acid in alkaline medium: red to violet color.

$$C_6H_6 \xrightarrow{HNO_3} C_6H_4(NO_2)_2 \xrightarrow{\text{Red.}} C_6H_6(NO_2)_2 \underset{}{\overset{OH^-}{\rightleftharpoons}} C_6H_4(=NO_2^-)_2$$

Reagents

a. A 20% solution of potassium nitrate in concentrated sulfuric acid.

b. A 1% aqueous solution of ascorbic acid. Prepare fresh immediately before use.

Procedure

Evaporate the sample solution to dryness. To the residue, add 1 ml of reagent a, and when the residue is dissolved heat at 100° for 15 min. Cool in a water bath, and add slowly 5 ml of ice water. Allow to cool, add 0.5 ml of reagent b, mix, add 5 ml of prechilled aqueous ammonium hydroxide (d = 0.92, 22° Bé), and read.

Results

	Color	λ Max, nm	A = 0.3 (1 cm cell) Sample, μg
Benzoic acid	Red	430	460
Phenylacetic acid	Red-violet	545	860

	Color	λ Max, nm	A = 0.3 (1 cm cell) Sample, μg
Phenylalanine	Red-violet	545	530
Phenobarbital	Violet	545	490
Atropine (sulfate)	Red-violet	545	980

Notes

1. Although the dinitration of benzene compounds mainly affords m-dinitro derivatives, small amounts of the o- and p-isomers are also formed, and are responsible for the color developed upon reduction. No color is displayed under the same conditions by pure m-dinitro compounds.

2. The structure of the colored species was particularly studied by Block and Bolling (8).

3. Other reductants were proposed for particular compounds: ammonium bisulfide (9) and hydroxylamine (10,11). Ascorbic acid was proposed in the case of phenylalanine (12,13).

4. Inversely, o-dinitrobenzene allows the characterization of miscellaneous reducing compounds (14).

C. Cupric Sulfate, Hydroxylamine Hydrochloride, and Hydrogen Peroxide

Principle (15)

Upon the action of cupric ion, hydroxylamine, and hydrogen peroxide, formation of a o-nitrosophenol developed as copper complex: pink-orange color.

Reagents

a. An 0.5% aqueous solution of cupric sulfate pentahydrate.

b. A 20% aqueous solution of hydroxylamine hydrochloride.

c. A 3% aqueous solution of hydrogen peroxide (10 volumes oxygen).

Procedure

To 1 ml of aqueous solution of the aromatic compound, add 0.45 ml of 0.1 N sodium hydroxide and 0.25 ml of reagent a. Mix well, and add 0.2 ml of reagent b and 0.1 ml of reagent c. Let stand at room temperature for the given length of time (see Results), add 1.5 ml of water, and read at 510 nm.

When the compound is insoluble in water, evaporate to dryness the sample solution in methylene chloride, then add 1 ml of water and 0.45 ml of 0.1 N sodium hydroxide to the residue, and proceed as above.

Results

	Reaction time, min	A = 0.3 (1 cm cell) Sample, μg
Benzoic acid	15	320
Phenylacetic acid	60	453
Phenylalanine	15	530
Phenobarbital	60	490
Atropine (sulfate)[a]	60	1170

[a]Beer's law is not followed.

Estrone and estradiol do not react.

Note

This method is based on the Baudisch reaction (16). Although it has been subjected to various studies, the mechanism of the formation of o-nitrosophenols through this reaction is as yet not well known (17).

D. Sulfuric Acid and Formaldehyde

Principle (18)

Reaction with a formaldehyde-sulfuric acid reagent: red color.

$$n\,C_6H_6 + CH_2OH^+ \xrightarrow{HCHO,\ H_2SO_4} n\,C_6H_5CH_2OH_2^+ \longrightarrow n\,C_6H_5CH_2^+$$

$$\left[C_6H_5-CH_2-(C_6H_4CH_2)_{n-2}-C_6H_4-CH_2\right]^+ \ HSO_4^-$$

Reagent

Dilute 2 ml of 35% aqueous solution of formaldehyde to 100 ml with concentrated sulfuric acid.

Procedure

Evaporate the sample solution to dryness. To the residue, add 5 ml of reagent, let stand for the given length of time (see Results), and read.

Results

	Reaction time, min	λ Max, nm	A = 0.3 (1 cm cell) Sample, μg
Phenylalanine	5 - 10	450	128
Ephedrine	30	335	2900
Phenobarbital	30	435	2600

Notes

1. The color developed varies with the reaction time, making results difficult to reproduce.

2. Compounds bearing Cl, SO_3H, NO_2, CO_2H, and $CH_2\overset{+}{N}R_3$ substituents on the benzene ring do not react.

3. More sensitive results can be obtained for each individual compound by a suitable selection of the formaldehyde concentration and of the reaction time and temperature (19).

V. ETHYLENIC COMPOUNDS OF THE TYPE R—CH=CH_2 (F)

A. Potassium Permanganate, Sodium Metaperiodate, Ethyl Acetoacetate, and Ammonium Acetate

Principle (20)

Oxidation with permanganate and periodate ions (see Section III.A, under Principle), and development of the formaldehyde formed with ethyl acetoacetate and ammonia: blue fluorescence.

$$R{-}CH{=}CH_2 \xrightarrow[IO_4^-]{MnO_4^-} HCHO \xrightarrow[NH_3]{CH_3{-}CO{-}CH_2{-}CO_2C_2H_5}$$

(2,6-dimethyl-3,5-dicarbethoxy-1,4-dihydropyridine: ring with $H_5C_2O_2C$ and $CO_2C_2H_5$ at 3,5; H_3C and CH_3 at 2,6; N–H)

Reagents

a. An 0.05 N aqueous solution of sodium metaperiodate.

b. An 0.005 N aqueous solution of potassium permanganate.

c. An 0.1 N aqueous solution of potassium arsenite (1 liter of solution contains 4.945 g of arsenious oxide, 75 ml of 1 N potassium hydroxide, and 40.0 g of potassium bicarbonate).

d. A 2% solution of ethyl acetoacetate in a 20% aqueous solution of ammonium acetate.

Procedure

To 1 ml of sample solution in pyridine, add 4 ml of water, 1 ml of reagent a, and 1 ml of reagent b. Let stand for 30 min at room temperature, add 2 ml of reagent c, mix, and add 2 ml of reagent d. Heat at 60° for 20 min, and cool to room temperature in ice water. Read at exc: 366 nm; em: 470 nm.

Standard: A solution of 2,6-dimethyl-3,5-dicarbethoxy-1,4-dihydropyridine in a 97:3 mixture of water and ethanol.

Results

	Determination limits, μg	Reading 50	
		Sample, μg	Standard, μg/ml
Styrene	4 - 20	7.2	2.55
Allyl alcohol	2 - 10	3.5	2.24
Quinine	10 - 50	16	1.97

Notes

1. See Note 1, Section III.A.

2. Periodates and periodate oxidation, see Chapter 16, Section IV.

3. Fluorimetric determination of formaldehyde, see Chapter 16, Section V.

REFERENCES

1. M. Pesez and J. Bartos, *Ann. Pharm. Fr.*, 22, 609 (1964).

2. L.M. Berkowitz and P.N. Rylander, *J. Amer. Chem. Soc.*, 80, 6682 (1958).

3. E. Sawicki, C.R. Engel, and M. Guyer, *Anal. Chim. Acta*, 39, 505 (1967).

4. Cf. R.U. Lemieux and E. von Rudloff, *Can. J. Chem.*, 33, 1710 (1955).

5. Cf. S.B. Schryver, *Proc. Roy. Soc. (London)*, B82, 226 (1910).

6. Cf. E. Eegriwe, *Z. Anal. Chem.*, 110, 22 (1937).

7. J. Bartos, *Chim. Anal.*, 53, 384 (1971).

8. R.J. Block and D. Bolling, *J. Biol. Chem.*, 129, 1 (1939).

9. M.E. Mohler, *Bull. Soc. Chim. Fr.*, [3]3, 414 (1890).

10. J. Grossfeld, *Z. Unters. Lebensm.*, 53, 467 (1927).

11. R. Kapeller-Adler, *Biochem. Z.*, 281, 175 (1935).

12. R. Kuhn and P. Desnuelle, *Chem. Ber.*, 70, 1907 (1937).

13. J. Roche, R. Michel, and M. Moutte, *Trav. Memb. Soc. Chim. Biol.*, 25, 1316 (1943).

14. F. Feigl, *Spot Tests in Organic Analysis*, Elsevier (Amsterdam), 6th Ed., 131 (1960).

15. J. Bartos, *Ann. Pharm. Fr.*, 27, 759 (1969).

16. O. Baudisch, *Naturwiss.*, 27, 768 (1939).

17. O. Baudisch and S.H. Smith, *Naturwiss.*, 27, 769 (1939); G. Cronheim, *J. Org. Chem.*, 12, 1 (1947); J.O. Konecny, *J. Amer. Chem. Soc.*, 77, 5748 (1955); K. Maruyama, I. Tanimoto, and R. Goto, *Tetrahedron Letters*, 5889 (1966).

18. Cf. G. Deniges, *Bull. Trav. Soc. Pharm. Bordeaux*, 39, 70 (1899); A.L. LeRosen, R.T. Moravek, and J.K. Carlton, *Anal. Chem.*, 24, 1335 (1952); M.J. Rosen, *Anal. Chem.*, 27, 111 (1955).

19. M.R.F. Ashworth, G. Cappel, and E. Hammer, *Anal. Chim. Acta*, 49, 301 (1970).

20. M. Pesez and J. Bartos, *Talanta*, 14, 1097 (1967).

Chapter 13

NITROGEN-CONTAINING HETERO-CYCLIC COMPOUNDS

I. INTRODUCTION

A. Colorimetry

Miscellaneous nitrogen-containing heterocyclic derivatives bearing an active hydrogen atom, including some alkaloids, can be determined with xanthydrol in acid medium with a fairly good sensitivity. Through an oxidative reaction, they also condense with β-naphthol, and the species formed is developed with 2,4-dinitrophenylhydrazine, but aliphatic amines behave the same.

The reducing properties of indole derivatives allow their determination with ferric ferricyanide at levels of 1.9-12 μg, but the reaction is far from specific.

The hydrogen atom in the α or β position to the nitrogen atom in the heterocyclic ring is active enough to permit condensations with p-benzoquinone through an oxido-reduction reaction (60-700 μg), with p-dimethylaminobenzaldehyde (3-120 μg) and with p-dimethylaminocinnamaldehyde (7-53 μg).

Pyridine derivatives form a complex with trisodium pentacyanoamminoferrate(III), allowing determinations at levels of 32-57 μg for A = 0.3.

The pyridine ring is opened up by cyanogen bromide or chloride, and the glutaconic dialdehyde derivative so obtained is developed either with p-aminobenzoic acid (1.9-27.5 μg) or with barbituric acid (8.2-820 μg for A = 0.3).

Imidazole derivatives couple with diazotized p-sulfanilic acid (Pauly reaction) when the heterocyclic nitrogen atom at position 1 is not substituted (26-29 μg for A = 0.3).

According to an unknown mechanism, imidazoline derivatives develop a pink color with sodium nitroprusside (80-110 μg samples).

2-Substituted pyrimidine derivatives condense with 2-thiobarbituric acid (unknown mechanism), and the reaction is highly sensitive: 3.7-5.6 μg for A = 0.3.

Barbiturates develop a violet color with cobaltous acetate and isopropylamine in methanol. However, the reaction is poorly sensitive, allowing accurate determinations only at levels higher than 1 mg.

Phenothiazine derivatives form a complex salt with palladous chloride (45-138 μg), and cationoid radicals through an oxidative reaction with p-benzoquinone (23-38 μg for A = 0.3). They also react with morpholine in the presence of bromine, presumably to give a Methylene Blue-like dyestuff. The sensitivity varies widely with the compound tested.

B. Fluorimetry

Methods are available only for indole, pyridine, and phenothiazine derivatives.

Through an unknown mechanism, indole derivatives react in acid medium with formaldehyde in the presence of trace amounts of hydrogen peroxide. The fluorescent species allows determinations at levels of 1.2-4.8 μg for reading 50.

The Schiff base which is formed by condensing p-aminobenzoic acid with the glutaconic dialdehyde obtained from pyridine derivatives (see above) is fluorescent under suitable conditions, and determinations are possible with 0.44-11 μg samples (reading 50).

The native fluorescence of phenothiazine derivatives is strongly increased upon reaction with permanganate ion, allowing determinations of 1.25-1.60 μg of the compound tested, but the maximum of the emission spectrum is located in the ultraviolet range.

II. MISCELLANEOUS N-HETEROCYCLIC DERIVATIVES

A. Xanthydrol

Principle (1)

Condensation with xanthydrol in acid medium: blue to violet color.

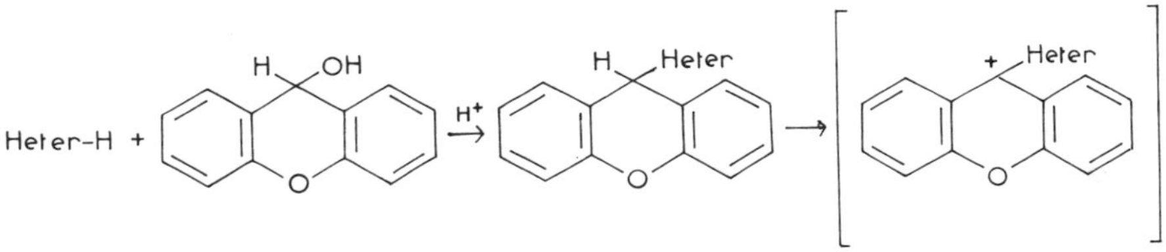

Reagent

Dissolve 0.020 g of xanthydrol in 100 ml of a 99:1 mixture of glacial acetic acid and concentrated hydrochloric acid.

Procedure

To 1 ml of sample solution in glacial acetic acid, add 5 ml of reagent, heat at 100° for exactly 10 min, then chill immediately in ice water, and read.

Results

	Color	λ Max, nm	A = 0.3 (1 cm cell) Sample, μg
Indole	Violet	530	20
Tryptophan	Violet	525	35
Carbazole	Blue	575	17
Ergotinine	Green-blue	650	157
Yohimbine	Violet	515	108
Reserpine	Pink-violet	515	360

Notes

1. Similar results are obtained when operating at -20° (frozen solution) instead of +100°, whereas the reaction is almost negative at +20°. This phenomenon is attributable to the existence of highly concentrated liquid regions present among the crystalline solvent. In these liquid regions, the reaction between the two solutes may be accelerated by the concentration effect (2).

2. Xanthydrol can be replaced by dixanthylurea, which liberates xanthydrol in the reaction medium. This reagent can easily be prepared in a highly pure state and was applied to the colorimetric determination of digitalis glycosides (3).

3. Xanthydrol also reacts with antipyrine (4), 1-phenylsemicarbazide (cryogenine) (5), some phenols and amines (6), and erythromycin (7). It allows the colorimetric determination of 2-deoxy sugars (Chapter 14, Section IX.C) and 2-amino-2-deoxy sugars condensed with acetylacetone (Chapter 14, Section X.D).

B. β-Naphthol, Hydrogen Peroxide, and 2,4-Dinitrophenylhydrazine

Principle and Procedure

See Chapter 4, Section III.D, determination of aliphatic amines.

Results

	A = 0.3 (1 cm cell) Sample, μg
Pyridine	300
Piperidine	13
Morpholine	21
Quinine	54

III. INDOLE DERIVATIVES

A. Ferric Ferricyanide

Principle and Procedure

See Chapter 3, Section II.B, determination of phenols.

Results

	A = 0.3 (1 cm cell) Sample, μg
Indole	2.6
Tryptophan	1.9
Carbazole	8.4
Reserpine	10.0
Deserpidine	6.9
Dehydroreserpine (perchlorate)	12.0

Note

Alternatively, uranyl ferricyanide can be used (Chapter 3, Section II.B, Note 3). A = 0.3 is given by 4.8 μg of indole.

B. p-Benzoquinone

Principle (8)

Condensation with p-benzoquinone in acid medium through an oxido-reduction reaction: red to violet color.

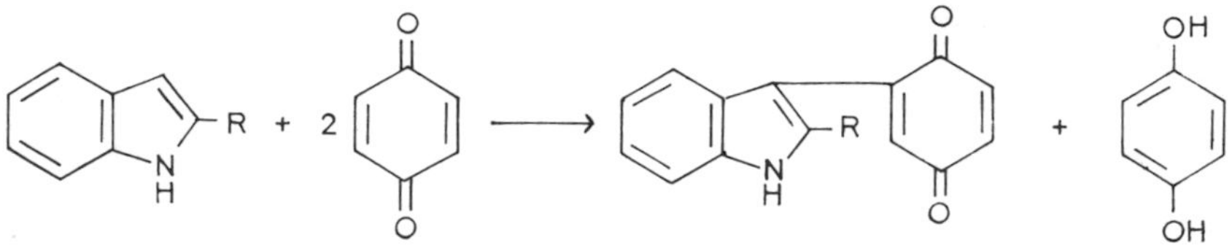

Reagents

a. A 2% solution of p-benzoquinone in ethanol. Prepare fresh immediately before used.

b. A 1:9 mixture of concentrated hydrochloric acid and ethanol.

Procedure

To 1 ml of sample solution in ethanol, add 1 ml of reagent a and 0.1 ml of reagent b. Let stand at the given temperature for the given length of time (see Results), add 2 ml of ethanol, and read immediately.

Results

	Reaction time, min	Temperature, °C	λ Max, nm	A = 0.3 (1 cm cell) Sample, μg
Indole	15	18 - 24	530	60
2-Methylindole	10	18 - 24	555	62
5-Ethoxyindole	5	18 - 24	530	60
Indole-3-acetic acid[a]	120	50	535	250
Indole-3-butyric acid	30	50	525	320
5-Methoxyindole-2-carboxylic acid	60	50	485	700
3-Hydroxyindole acetate	60	50	495	250
N-Acetyltryptophan	60	50	520	180

[a]Beer's law is not followed.

Pyrrole also reacts. A = 0.3 is given by 20 μg (565 nm) after 5 min at room temperature.

Notes

1. The sensitivity is less with compounds substituted at position 3.

2. Indole-2-carboxylic acid reacts but weakly, and the reaction is negative with the 3-isomer.

3. Carbazole, yohimbine, and reserpine are not detected.

4. Reaction of indole derivatives with quinones was particularly studied by Möhlau and Redlich (9), and by Bu'lock and Harley-Mason (10).

5. When, in the above procedure, the addition of reagent b is omitted, some other compounds display a red color after 5 to 10 min at room temperature. A = 0.3 is given by 35 μg of piperidine (500 nm), 22 μg of pyrrolidine (510 nm), and 60 μg of tryptamine (490 nm).

6. p-Benzoquinone also allows the colorimetric determination of phenothiazine derivatives (Section IX.B).

C. p-Dimethylaminobenzaldehyde

Principle (11)

Condensation with p-dimethylaminobenzaldehyde in acid medium to give species (I) or (II), or a mixture thereof: pink-violet color.

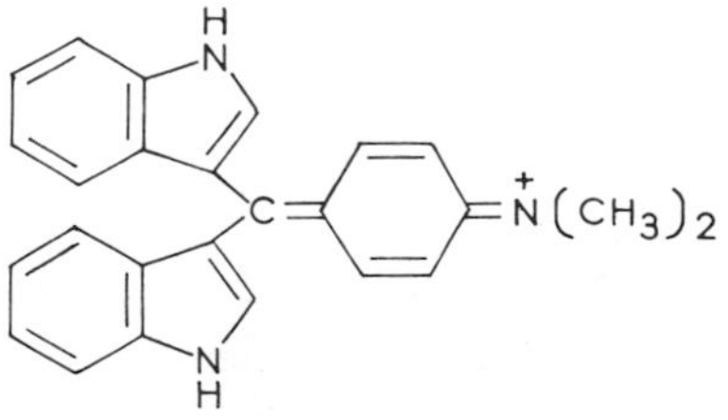

(I)

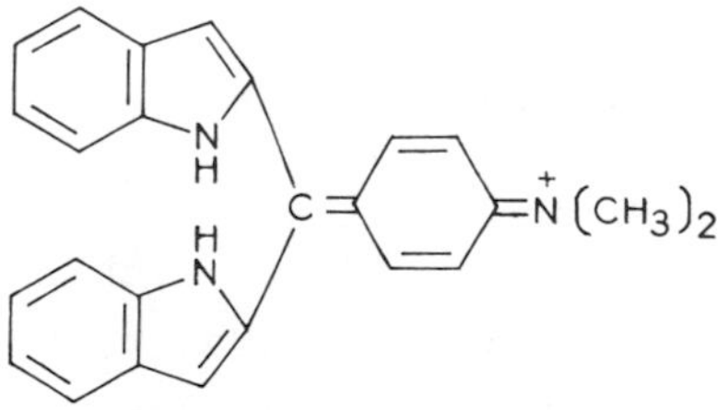

(II)

Reagent

A 6% solution of p-dimethylaminobenzaldehyde in 10% aqueous trifluoroacetic acid. Dissolve with slight warming.

Procedure

To 2 ml of aqueous solution of the indole derivative, add 2 ml of reagent, heat at 50° for the given length of time (see Results), cool to room temperature, and read.

Results

	Reaction time, min	λ Max, nm	A = 0.3 (1 cm cell) Sample, μg
Indole	0	563	3.0
Indole-3-butyric acid	30	575	14.5
N-Acetyltryptophan	90	570	120

Notes

1. The reaction is negative with derivatives bearing at position 3 substituents which deactivate the hydrogen at position 2, such as gramine, tryptophan, tryptamine, and serotonin.

2. It was established that benzhydrol interferes in the colorimetric assay of skatole. This interference is the result of an irreversible reaction of the indole nucleus with benzhydrol, taking place in competition with the reaction of the indole derivative with p-dimethylaminobenzaldehyde (12).

3. Uses of p-dimethylaminobenzaldehyde, see Chapter 5, Section IV.B, Notes 1 and 2.

D. p-Dimethylaminocinnamaldehyde

Principle (13)

Condensation with p-dimethylaminocinnamaldehyde in acid medium. Tentatively, formulas similar to those given for the reaction species obtained from p-dimethylaminobenzaldehyde (see preceding method) may be proposed: blue to blue-green color.

Reagent

A 1% solution of p-dimethylaminocinnamaldehyde in a 19:1 mixture of methanol and concentrated sulfuric acid. Prepare freshly.

Procedure

To 1 ml of sample solution in methanol, add 0.5 ml of concentrated sulfuric acid, the tubes being immersed in ice water. Allow the temperature to rise to 20°, then add 0.5 ml of reagent, and let stand at 20° for the given length of time (see Results). Add 2 ml of ethanol, 1 ml of water, and read.

Results

	Reaction time, min	λ Max, nm	A = 0.3 (1 cm cell) Sample, μg
Indole-3-acetic acid	3	600	7
Indole-3-butyric acid[a]	5	660	5
N-Acetyltryptophan	10	610	11
Reserpine	10	590	53

[a]Maintain at 0° instead of 20°.

Notes

1. Indole gives positive results, but Beer's law is not followed. Indole-2- and 3-carboxylic acids practically do not react.

2. p-Dimethylaminocinnamaldehyde also allows the colorimetric determination of m-diphenols (Chapter 3, Section V.B), and primary aliphatic (Chapter 4, Section VI.C) and aromatic (Chapter 5, Section IV.C) amines.

IV. PYRIDINE DERIVATIVES

A. Trisodium Pentacyanoamminoferrate(III)

Principle (14,15)

Formation of a 1:1 dissociable complex between the pyridine derivative and pentacyanoamminoferrate(III): yellow to purple color.

Reagents

a. Buffer for pH 6.5: Mix 3 volumes of 0.1 M aqueous citric acid and 7 volumes of 0.2 M disodium phosphate.

b. An 0.2% solution of trisodium pentacyanoammino-ferrate(III) in 0.02 N ammonium hydroxide. Prepare fresh immediately before use.

Procedure

To 2 ml of aqueous solution of the pyridine derivative, add 1.5 ml of buffer a and 0.5 ml of reagent b. Let stand at room temperature for 20 min, and read.

Results

	λ Max, nm	A = 0.3 (1 cm cell) Sample, μg
Pyridine	357	32
Nicotinic acid	378	57
Nicotinamide	385	52
Isoniazid	410	40

All these compounds display a yellow color.

Notes

1. The above procedure allows the determination of pyridine in dimethylformamide.

2. Relations between pentacyanoferrates, see Chapter 5, Section IV.A, Note 1.

B. Cyanogen Bromide and p-Aminobenzoic Acid

Principle (16)

Treatment with cyanogen bromide, and subsequent development with p-aminobenzoic acid to give an imino derivative of Schiff base type: yellow color.

BrCN → Br^- N–CN → $H_2N-C_6H_4-CO_2H$ → CH=CH–CH, CH–CH=N, CH=CH–NH, CO_2H CO_2H

Reagents

a. A 20% aqueous solution of sodium acetate trihydrate.

b. Cyanogen bromide solution: To 40 ml of 10% aqueous solution of potassium cyanide, add 5 ml of glacial acetic acid, then 40 ml of aqueous solution containing 3.0 g of potassium bromate and 4.0 g of potassium bromide (dissolved on slight warming). Maintain the mixture at 20° in a water bath, add dropwise with stirring 4 ml of concentrated sulfuric acid, and dilute to 100 ml with water. This reagent is stable for 1 day.

c. A 1% solution of p-aminobenzoic acid in ethanol.

Procedure

To 1 ml of aqueous solution of the pyridine derivative, add 2 ml of solution *a*, 0.3 ml of reagent *b*, 1 ml of acetone, and 1 ml of reagent *c*. Mix well after each addition, let stand at room temperature for the given length of time (see Results), and read.

Results

	Reaction time, min	λ Max, nm	A = 0.3 (1 cm cell) Sample, μg
Pyridine	30	470	1.9
3-Pyridinealdehyde	45	470	15.6
Nicotinic acid	10	455	10.4
Nicotinamide	30	465	21.6
N,N-Diethylnicotinamide	90	470	8.2
Isoniazid[a]	30	360	27.5

[a]No color is displayed.

3-Pyridinemethanol reacts but weakly (A = 0.3 for 817 μg, at 360 nm), and the reaction is negative with 4-hydroxypyridine.

Notes

1. Compounds substituted at α position to the heterocyclic nitrogen atom do not react.

2. A slightly different procedure allows the fluorimetric determination of pyridine derivatives (Chapter 13, Section XI.A).

3. Determinations through opening of the pyridine ring, see Chapter 16, Section XI.

C. Potassium Cyanide, Chloramine-T, and Barbituric Acid

Principle (17)

Treatment with cyanogen chloride generated in the reaction medium from a mixture of potassium cyanide and chloramine-T, and subsequent development with barbituric acid to give a glutaconic aldehyde derivative: blue to violet color.

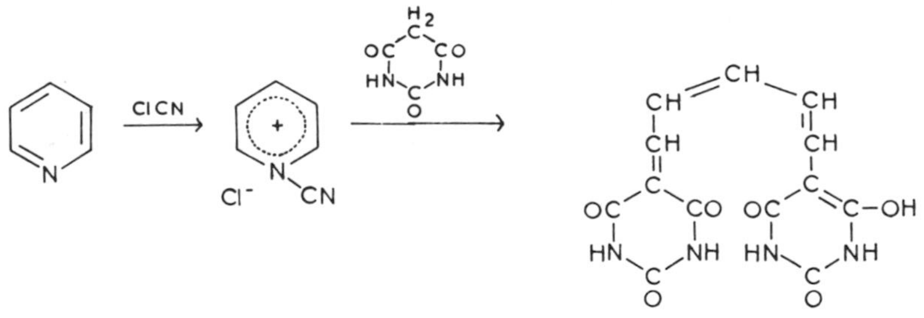

ClCN
Cl⁻ CN
H2 C OC CO HN NH C O
CH CH CH CH CH C C OC CO OC C-OH HN NH HN NH C C O O

Reagents

a. A 1% aqueous solution of potassium cyanide.

b. A 1% aqueous solution of chloramine-T.

c. A 1% aqueous solution of barbituric acid. Dissolve with slight warming.

Procedure

To 2 ml of 0.1 N hydrochloric acid, add 1 ml of reagent a and 5 ml of reagent b. Mix, and add 5 ml of aqueous solution of the pyridine derivative. Let stand for 5 min, add 10 ml of reagent c, let stand for 30 min, and read.

Results

	λ Max, nm	A = 0.3 (1 cm cell) Sample, μg
Pyridine	580	8.2
Nicotinic acid	486	820
Nicotinamide	580	144
N,N-Diethylnicotinamide	516	37
Isoniazid	636	260

Beer's law is not followed with 3-pyridinealdehyde. The reaction is negative with 4-hydroxypyridine.

Notes

1. Compounds substituted at the α position to the heterocyclic nitrogen atom do not react.

2. Chloramine-T is the sodium derivative of N-chloro-p-toluenesulfonamide.

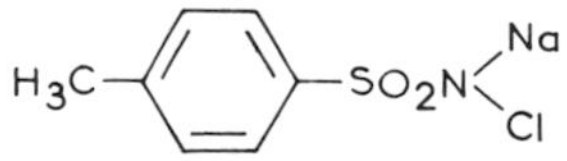

$$H_3C-C_6H_4-SO_2N(Na)Cl$$

3. Determinations through opening of the pyridine ring, see Chapter 16, Section XI.

V. IMIDAZOLE DERIVATIVES

A. Diazotized p-Sulfanilic Acid

Principle (18)

Coupling with the diazonium salt of sulfanilic acid in alkaline medium: red color.

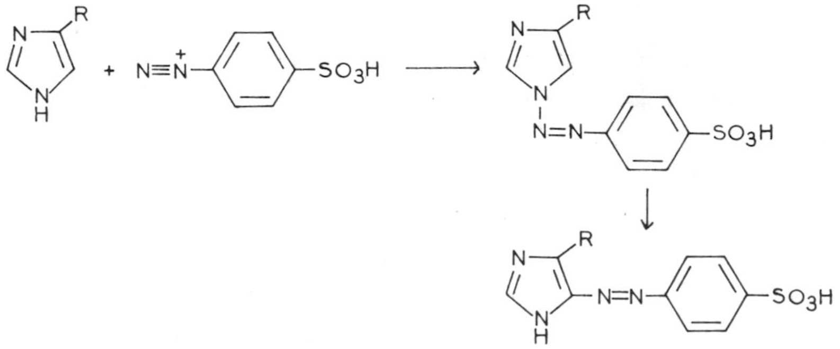

Reagents

a. A 20% aqueous solution of anhydrous sodium carbonate.

b. To 5 ml of 1% solution of p-sulfanilic acid in 1:9 hydrochloric acid, add dropwise 1 ml of 2% aqueous solution of sodium nitrite, and let stand for 3 min before use.

Procedure

To 2 ml of aqueous solution of the imidazole derivative, add 1 ml of reagent a, 0.5 ml of reagent b, mix, add immediately 1.5 ml of ethanol, and read.

Results

	λ Max, nm	A = 0.3 (1 cm cell) Sample, μg
Histamine (dihydrochloride)	490	26
Histidine (hydrochloride)	480	29

Notes

1. The addition of ethanol stabilizes the color, which otherwise is very unstable.

2. Compounds with substituted heterocyclic nitrogen atom do not react.

3. Diazo coupling, see Chapter 16, Section I.

VI. IMIDAZOLINE DERIVATIVES

A. Sodium Nitroprusside

Principle (19)

Reaction with nitroprusside ion in alkaline medium: pink color.

Reagents

a. A 5% aqueous solution of sodium nitroprusside.

b. A 1 M aqueous solution of sodium bicarbonate.

Procedure

To 1 ml of aqueous solution of the imidazoline derivative, add 0.5 ml of 1 N sodium hydroxide and 0.5 ml of reagent *a*. Let stand at room temperature for 10 min, and add 1 ml of reagent *b* and 2 ml of ethanol. Let stand for 5 min, and read.

Results

	λ Max, nm	A = 0.3 (1 cm cell) Sample, µg
Tolazoline	545	80
Naphazoline (nitrate)	545	100
Antazoline (hydrochloride)	535	110

Notes

1. Tolazoline, or 2-benzyl-2-imidazoline(III) is a smooth muscle relaxant and a peripheral vasodilator.

Naphazoline, or 2-(1-naphthylmethyl)imidazoline (IV), is a sympathomimetic.

Antazoline, or 2-(N-benzylanilinomethyl)imidazoline (V) is an antihistaminic.

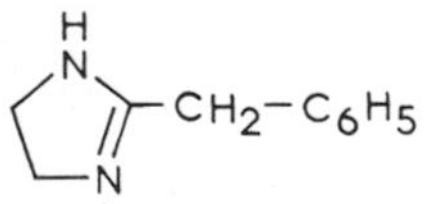

(III) 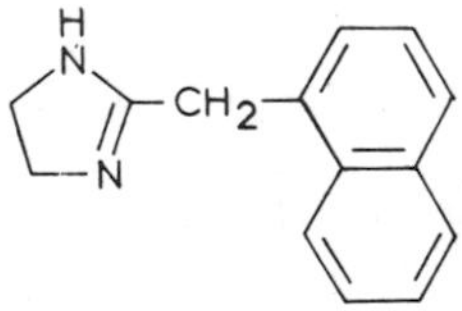 (IV) 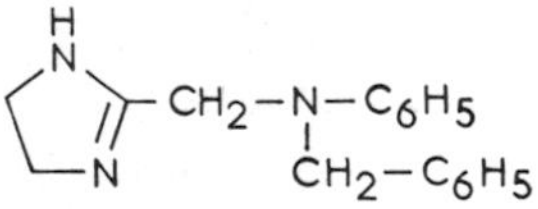 (V)

2. Nitroprusside also allows the colorimetric determination of primary (Chapter 4, Section VI.A), and secondary (Chapter 4, Section VII.A) aliphatic amines, α-methylene ketones (Chapter 8, Section VIII.A), and $\Delta^4$3-ketosteroids (Chapter 15, Section X.A).

3. Relations between nitroprusside and pentacyanoferrates, see Chapter 5, Section IV.A.

VII. 2-SUBSTITUTED PYRIMIDINE DERIVATIVES

A. 2-Thiobarbituric Acid

Principle (20)

Condensation with 2-thiobarbituric acid (unknown mechanism): red color.

Reagent

Dissolve on heating 0.5 g of 2-thiobarbituric acid in a mixture of 1 ml of 2 N sodium hydroxide and about 10 ml of water, and dilute to 50 ml with water.

Dissolve 3.7 g of trisodium citrate dihydrate in 3.2 ml of concentrated hydrochloric acid diluted to 25 ml with water.

Mix the two solutions, and adjust the pH to 2 if necessary.

Procedure

To 1 ml of aqueous solution of the pyrimidine derivative, add 4 ml of reagent, fit the tubes with air-condensers, and heat at 100° for the given length of time (see Results). Allow to cool, and read at 530 nm.

Results

	Reaction time	A = 0.3 (1 cm cell) Sample, μg
2-Aminopyrimidine	3 hr	5.6
Sulfadiazine	45 min	3.7

Note

Other analytical uses of 2-thiobarbituric acid, see Chapter 14, Section IX.D, Notes 3-5.

VIII. BARBITURATES

A. Cobaltous Acetate and Isopropylamine

Principle (21)

Formation of an adduct between the barbiturate, cobalt, and isopropylamine in methanol medium: violet color.

Reagents

a. A 1% solution of anhydrous cobaltous acetate (dried for 2 hr at 100°) in anhydrous methanol.

b. A 5% solution of anhydrous isopropylamine in anhydrous methanol.

Procedure

To 3 ml of sample solution in anhydrous chloroform, add 0.15 ml of reagent a and 0.9 ml of reagent b. Let stand for 15 min, and read at 565 nm.

Results

	A = 0.3 (1 cm cell) Sample, μg
Barbital	1700
Phenobarbital	1850

Notes

1. This condensation of barbiturates with a cobaltous salt and a base is sometimes named the Parri reaction (22). Besides isopropylamine (21,23), other bases such as barium or lithium hydroxide (24), barium methoxide (25), diethylamine (26), isobutylamine (27), and piperidine (28) were used.

2. The color fades in the presence of small amounts of water.

3. The reaction is also positive with various compounds with –CO–NH–CO– or –CO–NH–CS– groups, such as alloxan, cyanuric acid, allantoin, hydantoin, theobromine, theophylline, uracil, thiouracil, and propylthiouracil (29). Isatine and caffeine do not react. Since barbiturates are

often determined in pharmaceutical preparations, blood, urine, and other biological fluids, it may therefore be necessary to take these interferences into account, and to modify the above procedure. This is the reason why it is exemplified by only two determinations.

4. When operating with a mixture of cobalt and calcium salts, the addition of sodium hydroxide affords blue complexes which are insoluble in methanol and from which, upon acid hydrolysis, the barbiturate is liberated as shapely crystals which allow its characterization (30).

IX. PHENOTHIAZINE DERIVATIVES

A. Palladous Chloride

Principle (31)

Formation of a complex salt: red-orange color.

$$\text{Phenothiazine} \xrightarrow{Pd^{2+}} (\text{S})_2 \rightarrow Pd^{2+}$$

Reagent

Heat at 100° a mixture of 0.500 g of palladous chloride and 5 ml of concentrated hydrochloric acid, then add 200 ml of hot water in small increments with continued heating until solution is complete. Allow to cool, and dilute to 500 ml with water.

To 25 ml of this stock solution, add 50 ml of 8.2% aqueous solution of sodium acetate trihydrate, 48 ml of 1:9 hydrochloric acid, and dilute to 500 ml with water.

Procedure

To 3 ml of reagent, add 2 ml of aqueous solution of the phenothiazine derivative, and read immediately.

Results

	λ Max, nm	A = 0.3 (1 cm cell) Sample, μg
Phenothiazine[a]	550	45
Chlorpromazine (hydrochloride)	490	135
Promethazine (hydrochloride)	465	138

[a]Sample solution in ethanol: violet color.

Note

Upon addition of lauryl sulfate, the color turns to violet for chlorpromazine and promethazine, and the sensitivity is increased for the latter compound (31).

B. p-Benzoquinone

Principle (32)

Oxidative reaction with p-benzoquinone, to give cationoid radicals: pink color.

Reagent

An 0.05% aqueous solution of p-benzoquinone. Keep in the dark.

Procedure

To 1 ml of aqueous solution of the phenothiazine derivative, add 0.5 ml of reagent and 2.5 ml of 50% sulfuric acid. Read at 530 nm.

Results

	A = 0.3 (1 cm cell) Sample, μg
Phenothiazine[a]	23
Chlorpromazine (hydrochloride)	31
Promethazine (hydrochloride)	38

[a]Sample solution in ethanol.

Note

p-Benzoquinone allows also the colorimetric determination of indole derivatives (Section III.B).

C. Morpholine and Bromine

Principle

Formation of a phenazathionium perbromide and development with morpholine: blue color.

The following scheme may be tentatively proposed:

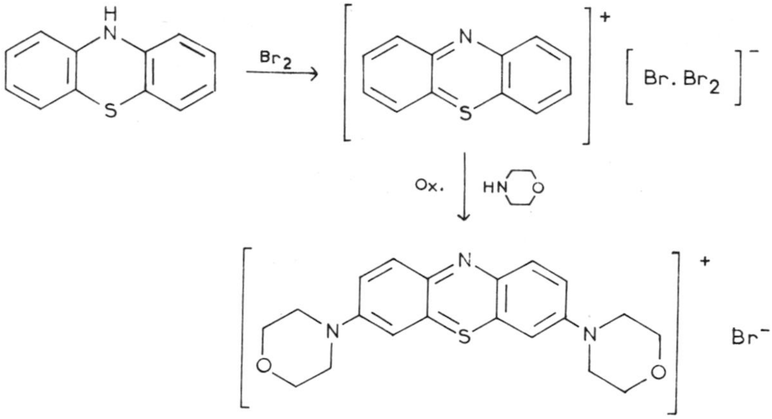

Br_2 → $[\ ]^+ [Br.Br_2]^-$

Ox. ↓ HN⟨ ⟩O

$[\ ]^+ Br^-$

Reagents

a. A 1% aqueous solution of morpholine.

b. Dilute 0.5 ml of bromine to 50 ml with methanol, and let stand for at least 24 hr in the dark, then dilute 5 ml of this stock solution to 20 ml with methanol. Keep in the dark.

Procedure

To 1 ml of sample solution in methanol, add 1 ml of reagent a and 0.2 ml of reagent b. Let stand for 15 min exposed to daylight, add 2 ml of water, and let stand for the given length of time (see Results). Read at 665 nm.

Results

	Reaction time, hr	A = 0.3 (1 cm cell) Sample, μg
Phenothiazine	1	10.5
Promethazine (hydrochloride)	5	40
10-(N,N-Dimethylcarbamoyl)phenothiazine	3	900
1-Acetyl-10-(2-carbomethoxyphenyl)phenothiazine	1.5	320

Chlorpromazine does not react.

Note

This procedure was derived from a paper of Bröll and Fischer (33), who described a spot test method for the detection of secondary amines based on the formation of Methylene Blue-like dyestuffs, using phenothiazine and bromine as reagents. Their method cannot be extended to the colorimetric determination of this class of compounds, since an excess of amine is necessary for color development.

X. INDOLE DERIVATIVES (F)

A. Hydrogen Peroxide and Formaldehyde

Principle (34)

Reaction in acid medium with formaldehyde in the presence of hydrogen peroxide: yellow to yellow-orange fluorescence.

Reagents

a. A 0.00015% aqueous solution of hydrogen peroxide (0.0005 volume of oxygen).

b. A 3% aqueous solution of formaldehyde.

Procedure

To 0.5 ml of sample solution in a 1:1 mixture of ethanol and glacial acetic acid, add 0.1 ml of reagent a, 0.1 ml of reagent b, and 0.6 ml of 65% perchloric acid.

Plug the tubes with cottonwool, heat at 60° for 1 hr, protected against light, and cool for 2 min in ice water. Add 4 ml of glacial acetic acid, and read at exc: 436 nm; em: 540 nm.

Standard: A solution of 2,6-dimethyl-3,5-diacetyl-1,4-dihydropyridine in a 1:49 mixture of ethanol and water.

Results

	Determination limits, μg	Reading 50 Sample, μg	Standard, μg/ml
Indole-3-carboxylic acid[a]	0.5 - 2.5	1.2	0.09
Indole-3-acetic acid	2 - 10	4.8	2.63
Tryptamine	1 - 5	2.3	0.74
Tryptophan	1 - 5	2.3	0.76
N-Acetyltryptophan	2 - 10	4.7	1.19

[a]Blue fluorescence. Read at exc: 366 nm; em: 490 nm. Standard: a solution of quinine sulfate in 0.1 N sulfuric acid.

Indole does not give satisfactory results.

XI. PYRIDINE DERIVATIVES (F)

A. Cyanogen Bromide and p-Aminobenzoic Acid

Principle (35)

See Section IV.B, colorimetric determination of pyridine derivatives: yellow-green fluorescence, visible under ultraviolet light only at high concentrations.

Reagents

a. A 10% aqueous solution of sodium acetate trihydrate.

b. Cyanogen bromide solution: To 40 ml of 10% aqueous solution of potassium cyanide, add 5 ml of glacial acetic acid, then 40 ml of aqueous solution containing 3.0 g of potassium bromate and 4.0 g of potassium bromide (dissolved on slight warming). Maintain the mixture at 20° in a water bath, add dropwise with stirring 4 ml of concentrated sulfuric acid, and dilute to 100 ml with water. This reagent is stable for 1 day.

c. An 0.2% solution of p-aminobenzoic acid in ethanol.

Procedure

To 1 ml of aqueous solution of the pyridine derivative, add 2 ml of solution a, 0.3 ml of reagent b, 1 ml of acetone, and 1 ml of reagent c. Mix well after each addition, let stand at room temperature for the given length of time (see Results), and read at exc: 436 nm.

Standard: A solution of 2,6-dimethyl-3,5-diacetyl-1,4-dihydropyridine in ethanol.

Results

	Reaction time, min	Em, nm	Determination limits, µg	Reading 50 Sample, µg	Standard, µg/ml
Pyridine	20	515	0.2 - 1.0	0.44	0.82
3-Pyridine-aldehyde	45	510	1.5 - 7.5	3.2	0.76
Nicotinic acid	10	505	2.5 - 12.5	5.5	0.75
Nicotinamide	45	510	2 - 10	4.5	0.80
N,N-Diethyl-nicotinamide	90	515	0.2 - 1.0	0.45	0.58
Isoniazid[a]	30	480	5 - 25	11.0	0.62

[a]Blue fluorescence. Read at exc: 366 nm. Standard: dissolve 2,6-dimethyl-3,5-dicarbethoxy-1,4-dihydropyridine in 0.7 ml of ethanol, and dilute to 100 ml with water.

3-Pyridinemethanol reacts but weakly, and the reaction is negative with 4-hydroxypyridine.

Notes

1. Compounds substituted at the α position to the heterocyclic nitrogen atom do not react.

2. p-Aminosalicylic acid may be used instead of p-aminobenzoic acid, but the sensitivity is weaker.

3. Determinations through opening of the pyridine ring, see Chapter 16, Section XI.

XII. PHENOTHIAZINE DERIVATIVES (F)

A. Potassium Permanganate

Principle (36)

Upon reaction with permanganate ion, development of an ultraviolet fluorescence.

Reagent

An 0.001% solution of potassium permanganate in 0.2 N sulfuric acid.

Procedure

To 0.5 ml of sample solution in ethanol, add 1 ml of reagent and 2.5 ml of 0.2 N sulfuric acid, mix, and let stand for 30 min. Read at exc: 310 nm; em: 380 nm.

Standard: A solution of protocatechuic acid in Clark and Lubs buffer for pH 8.9 (see Appendix I.C).

Results

	Determination limits, μg	Reading 50	
		Sample, μg	Standard, μg/ml
Phenothiazine	0.6 - 3.0	1.25	0.025
Chlorpromazine (hydrochloride)	0.8 - 4.0	1.60	0.020
Promethazine (hydrochloride)	0.8 - 4.0	1.60	0.020

Notes

1. Phenothiazine derivatives often show a native fluorescence emission in the range 450-475 nm. Upon oxidation, the fluorescence shifts to shorter wavelengths, whereas the sensitivity is increased.

2. Hydrogen peroxide was also proposed as oxidant (37).

REFERENCES

1. Cf. V. Arreguine, *Rev. Univ. Nacl. Cordoba (Argent.)*, 31, 1706, 1710 (1944); G. Gilbert, R.M. Stickel, and H.H. Morgan, Jr., *Anal. Chem.*, 31, 1981 (1959).

2. Cf. R.E. Pincock and T.E. Kiovsky, *J. Amer. Chem. Soc.*, 88, 51 (1966).

3. H. Poetter, *Pharmazie*, 20, 737 (1965).

4. M. Dantec, *Ann. Med. Pharm. Col.*, 32, 379 (1934).

5. V. Arreguine, *Rev. Univ. Nacl. Cordoba (Argent.)*, 29, 1740 (1942).

6. W. Lang, *Arzneim.-Forsch.*, 1, 230 (1951).

7. J.B. Tepe and C.V. St. John, *Anal. Chem.*, 27, 744 (1955).

8. M. Pesez and J. Bartos, *Ann. Pharm. Fr.*, 23, 783 (1965).

9. R. Möhlau and A. Redlich, *Ber.*, 44, 3605 (1911).

10. J.D. Bu'lock and J. Harley-Mason, *J. Chem. Soc.*, 703 (1951).

11. P. Byrom and J.H. Turnbull, *Talanta*, 10, 1217 (1963).

12. W.N. French, *J. Pharm. Sci.*, 54, 1726 (1965).

13. M. Pesez and J. Bartos, *Bull. Soc. Chim. Fr.*, 3802 (1966).

14. T.A. Larue, *Anal. Chim. Acta*, 40, 437 (1968).

15. Cf. V. Scardi, *Clin. Chim. Acta*, 2, 134 (1957); V. Scardi and V. Bonavita, *Clin. Chem.*, 3, 728 (1957).

16. Cf. D.K. Chaudhuri, *Indian J. Med. Res.*, 39, 491 (1951).

17. E. Asmus and H. Garschagen, *Z. Anal. Chem.*, 139, 81 (1953).

18. From A. Maciag and R. Schoental, *Mikrochemie*, 24, 243 (1938).

19. From S.C. Slack and W.J. Mader, *J. Amer. Pharm. Assoc., Sci. Ed.*, 46, 742 (1957); cf. H. Laubie, *Bull. Soc. Pharm. Bordeaux*, 88, 65 (1950); 91, 109 (1953); 93, 67 (1955).

20. R.G. Shepherd, *Anal. Chem.*, 20, 1150 (1948).

21. G.A. Levvy, *Biochem. J.*, 34, 73 (1940).

22. W. Parri, *Boll. Chim. Farm.*, 63, 401 (1924).

23. J.M. Dille and T. Koppanyi, *J. Amer. Pharm. Assoc.*, 23, 1079 (1934).

24. T. Koppanyi, J.M. Dille, W.S. Murphy, and S. Krop, *J. Amer. Pharm. Assoc.*, 23, 1074 (1934).

25. J.J.L. Zwikker, *Pharm. Weekblad*, 68, 975 (1931).

26. H. Griffon and R. Le Breton, *J. Pharm. Chim.*, [8]28, 49 (1938); *Ann. Pharm. Fr.*, 5, 393 (1947).

27. H. Baggersgaard-Rasmussen and B. Jerslev, *Dansk. Tidsk. Farm.*, 25, 29 (1951).

28. H. Gomahr and H. Kresbach, *Scientia Pharmaceutica*, 19, 148 (1951).

29. F.L. Kozelka and H.J. Tatum, *J. Pharmacol. Exptl. Therap.*, 59, 54 (1937); W.L. Holt and L.N. Mattson, *Anal. Chem.*, 21, 1389 (1949).

30. M. Pesez, *J. Pharm. Chim.*, [8]28, 69 (1938).

31. From J.A. Ryan, *J. Amer. Pharm. Assoc., Sci. Ed.*, 48, 240 (1959).

32. J. Meunier and F. Leterrier, *C. R. Acad. Sci., Paris*, 265 C, 1034 (1967).

33. H. Bröll and G. Fischer, *Mikrochim. Acta*, 249 (1962).

34. M. Pesez and J. Bartos, *Talanta*, 16, 331 (1969).

35. J. Bartos, *Ann. Pharm. Fr.*, 29, 71 (1971).

36. From T.J. Mellinger and C.E. Keeler, *Anal. Chem.*, 35, 554 (1963); 36, 1840 (1964).

37. J.B. Ragland, V.J. Kinross-Wright, and R.S. Ragland, *Anal. Biochem.*, 12, 60 (1965).

Chapter 14

SUGARS AND DERIVATIVES

I. INTRODUCTION

Although the classification of the stereoisomers of aldoses and ketoses is often attributed to Fischer, it is only fair to recall that the reference to glyceraldehyde and the conventional notation still in use were proposed by Rosannof (1).

According to this classification, an aldose or a ketose belongs to the D series when the hydroxyl group on the carbon atom directly attached to the end primary alcohol group is represented on the right in the stereochemical projection formula. It is well known that the symbols D and L refer to these configurations and not to the sign of rotation.

```
   CHO              CHO
    |                |
H—C—OH          HO—C—H
    |                |
   CH2OH            CH2OH
```

D-Glyceraldehyde   L-Glyceraldehyde

Aldotetroses

```
    CHO                 CHO
    |                   |
H - C - OH         HO - C - H
    |                   |
H - C - OH          H - C - OH
    |                   |
    CH2OH               CH2OH
```

D-Erythrose D-Threose

Aldopentoses

```
     CHO
     |
HO - C - H
     |
 H - C - OH   or
     |
 H - C - OH
     |
     CH2OH
```

D-Arabinose D-Lyxose D-Ribose D-Xylose

Aldohexoses

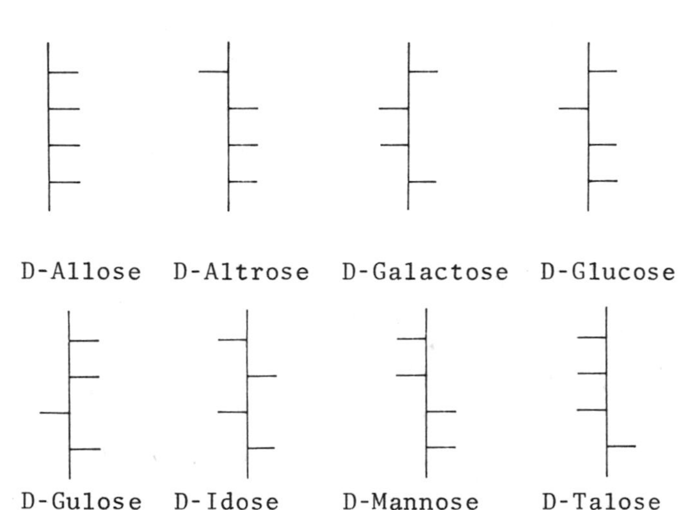

Ketohexoses

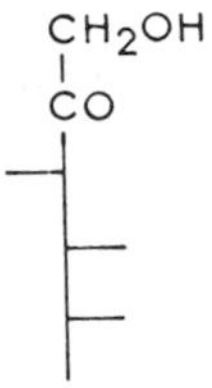

D-Fructose
(D-Levulose)

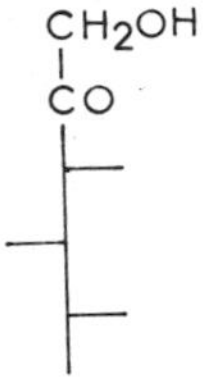

D-Sorbose

As a matter of fact, an equilibrium takes place in solution between the acyclic structure and inner cyclic acetals (I,II), and it is in favor of the pyranose form (I).

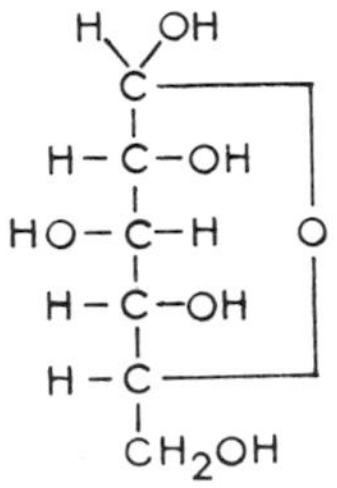

(I)

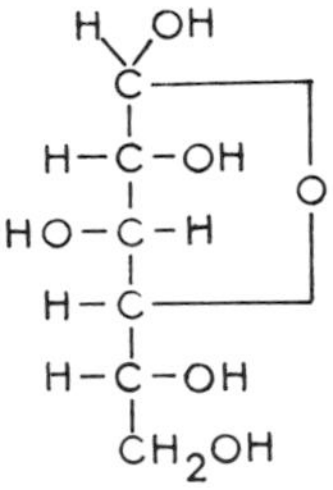

(II)

These pyranose and furanose forms account for the α,β isomerism and mutarotation. The concentration of the open-chain form is very low, but upon the action of a reagent of the carbonyl group, the equilibrium shifts toward the acyclic structure as the reaction proceeds.

A. Colorimetry

In the following, the numbers in parentheses are the sample weights which give A = 0.3.

Aldoses and ketoses are reductants and can therefore be determined with potassium ferricyanide (9.2-14 μg), molybdate ion (87-296 μg), 3,5-dinitrosalicylic acid (108-148 μg), and picric acid (240-390 μg).

Through their carbonyl group, they are determined with 2,4-dinitrophenylhydrazine (46-89 μg). The corresponding dyes have not been isolated, hence it cannot be stated whether the color is due to the hydrazone or the osazone. The condensates obtained with 2-hydrazinobenzothiazole or with 3-methylbenzothiazolin-2-one hydrazone are converted to the corresponding formazans with the diazonium salt generated in the reaction medium by oxidation of the excess reagent, affording highly sensitive determinations (9.5-16 μg and 3-4.5 μg, respectively).

In acid media, pentoses are dehydrated to furfural (III), methylpentoses to 5-methylfurfural(IV), and hexoses to 5-hydroxymethylfurfural(V).

CHOH—CHOH
R—CHOH CHOH—CHO ⟶ R—(furan ring)—CHO

(III) R = H
(IV) R = CH_3
(V) R = CH_2OH

These furan derivatives are then developed with miscellaneous reagents. The aptitude for dehydration depending on the sugar, suitable procedures allow selective determinations.

In this way, aldoses and ketoses are both determined with α-naphthol (12-28 μg), thymol (23-60 μg), or carbazole (72-264 μg); aldoses with benzidine (36-55 μg), and ketoses with diphenylamine (82-110 μg), indole-3-acetic acid (24-104 μg), resorcinol and ferric alum (23-24 μg), or 2-thiobarbituric acid (8-13 μg).

Pentoses only are determined with aniline (32-39 μg), p-bromoaniline (104-106 μg), or orcinol and ferric chloride

(5.8-9 μg), or cupric chloride (23-25 μg). The furfural obtained can also be determined as its p-nitrophenylhydrazone (115-325 μg).

Methylpentoses are determined with thioglycolic acid (17 μg).

Pentoses and hexoses both react with anthrone, but under somewhat different conditions which may allow selective determinations (21-51 μg and 46-80 μg, respectively).

In sulfuric acid medium, the methylol group of 5-hydroxymethylfurfural is split off to give formaldehyde, which is developed with chromotropic acid (150-220 μg).

Two reactions proceed through unknown mechanisms. Tetroses are determined with fructose and cysteine in sulfuric acid medium (12.6 μg), and pentoses with glucose and phloroglucinol in acetic-hydrochloric acid medium (14 μg).

2-Deoxyoses are developed through their methylene group with flavylium perchlorate (46-280 μg) or xanthydrol (9-20 μg). Upon periodate oxidation, they give rise to malonic dialdehyde, which is revealed with 2-thiobarbituric acid (2.8-3.8 μg) or 2-methylindole (2.3-6 μg).

In acid medium, 2-deoxyoses are at first converted to α-hydroxylevulinic aldehyde or a derivative thereof (substituted at C-5), which is then reacted with p-nitrophenylhydrazine to give a pyridazine derivative, and color is developed in alkaline medium upon opening of the ring (40-540 μg).

Under more acid conditions, the 2-deoxyose affords β-acetylacrolein or a derivative thereof (substituted at C-5), which then reacts with diphenylamine (Dische reaction) (80-220 μg).

2-Amino-2-deoxyhexoses react with acetylacetone in alkaline medium to give a pyrrole derivative, which can be developed either through its reducing properties with ferric ferricyanide (30-37 μg) or by condensing it with p-dimethylaminobenzaldehyde (46-64 μg), flavylium perchlorate (72-82 μg), or xanthydrol (305-315 μg). Phenylacetone can be substituted for acetylacetone, and the chromogen is then developed with p-dimethylaminobenzaldehyde (22.5-44 μg).

Upon the action of nitrous acid, 2-amino-2-deoxyhexoses are deaminated, and through an isomerization give rise to a 3,4-dihydroxy-5-hydroxymethyltetrahydrofuran-2-aldehyde, which is then developed with indole (23-25 μg) or with 3-methylbenzothiazolin-2-one hydrazone and ferric ion (8.0-8.5 μg).

Hexuronic acids react with naphthoresorcinol in acid medium to give a dibenzoxanthene-type dyestuff (10-25 μg). They can also be determined like other 2,3-dihydroxy-1-carboxylic acids through periodate oxidation and development of the glyoxylic acid formed with resorcinol (750-1000 μg).

B. Fluorimetry

Aldoses are converted to furfural or 5-hydroxymethylfurfural, which give fluorescent species upon reaction with resorcinol (2.7-8.6 μg for reading 50). Ketoses form a chelate with zirconyl chloride in acid medium (9.3-12.9 μg).

Furfural obtained from pentoses can be developed through its aldehyde group with cyclohexane-1,3-dione and ammonia (0.9-1.5 μg).

2-Deoxyoses react like α-methylene aldehydes with 3,5-diaminobenzoic acid (1.5-44 μg).

With pyridoxal and zinc ion in the presence of pyridine, 2-amino-2-deoxyhexoses give a zinc-N-pyridoxylidenehexosamine chelate (0.75-1.87 μg). They can also be condensed with ninhydrin and phenylacetaldehyde (2.5-4.8 μg), but the reaction is not specific; primary aliphatic amines and α-amino acids react also.

Hexuronic acids are determined like 2,3-dihydroxy-1-carboxylic acids through periodate oxidation and development of the glyoxylic acid formed with resorcinol (4.5-4.7 μg).

II. ALDOSES AND KETOSES

A. Potassium Ferricyanide

Principle (2)

Reduction of potassium ferricyanide in alkaline medium, and development as Prussian Blue.

Reagents

a. Adjust to 250 ml with water an aqueous solution containing 0.03 g of potassium ferrocyanide, 0.165 g of potassium ferricyanide, and 2 g of anhydrous sodium carbonate.

b. Dilute to 500 ml with water a mixture of 0.2 g of potassium ferricyanide and 25.6 g of 78% phosphoric acid (d = 1.61). Reflux until the color is discharged (30-45 min), and readjust to 500 ml with water if necessary.

Procedure

To 2 ml of aqueous solution of the sugar, add 2 ml of reagent a and 2 ml of water. Mix, heat at 100° for 15 min, and let stand for 10 min in a water bath at 20° (this temperature is critical). Add 4 ml of reagent b, mix, let stand for 15 min, and read at 690 nm.

Results

	A = 0.3 (1 cm cell) Sample, μg
Ribose	9.6
Xylose	9.2
Galactose	14.0

	A = 0.3 (1 cm cell) Sample, μg
Glucose	9.8
Fructose	10.8
Sorbose	11.6

Notes

1. Under the given conditions, Prussian Blue does not flocculate, and the addition of a surfactant is therefore avoided.

2. A small amount of ferrocyanide (reagent a) is necessary to initiate the reaction at low sugar concentrations. Otherwise, Beer's law is not followed.

B. Ammonium Molybdate

Principle (3)

Reduction of molybdate ion in acid medium: blue color.

Reagent

An 8% aqueous solution of ammonium molybdate.

Procedure

To 2 ml of aqueous solution of the sugar, add 2 ml of 1 N sulfuric acid and 4 ml of reagent. Heat at 100° for 30 min, cool in a water bath, and dilute to 10 ml with water. Read at 700 nm.

Results

	A = 0.3 (1 cm cell) Sample, μg
Ribose	268
Galactose	220
Glucose	296
Fructose	87
Sorbose	104

With xylose, Beer's law is not followed.

C. 3,5-Dinitrosalicylic Acid

Principle (4)

Reduction of 3,5-dinitrosalicylic acid in alkaline medium: orange color.

Reagents

a. Dissolve 1 g of 3,5-dinitrosalicylic acid, 0.2 g of phenol, and 0.05 g of anhydrous sodium sulfite in 0.25 N sodium hydroxide, and dilute to 100 ml with this alkaline solution.

b. A 40% aqueous solution of potassium sodium tartrate tetrahydrate (Rochelle or Seignette salt).

Procedure

To 3 ml of aqueous solution of the sugar, add 3 ml of reagent a, and heat at 100° for 15 min. Add 1 ml of reagent b, and cool to room temperature in a water bath. Read at 505 nm.

Results

	A = 0.3 (1 cm cell) Sample, μg
Ribose	108
Galactose	128
Fructose	148
Sorbose	140

With xylose, Beer's law is not followed.

D. Picric Acid

Principle (5)

Reduction of picric acid to picramic acid in alkaline medium: orange color.

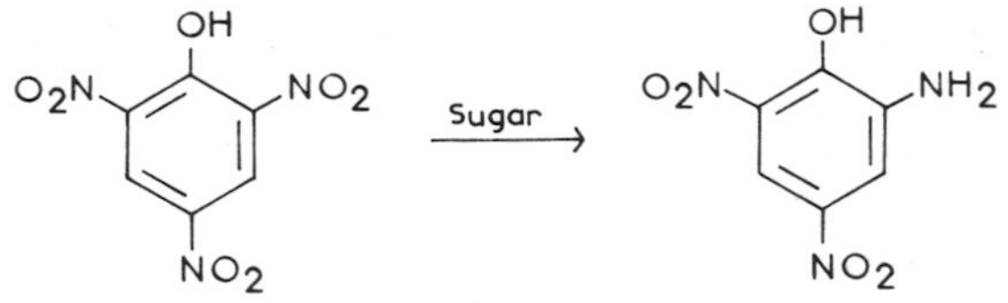

$$\text{picric acid (2,4,6-trinitrophenol)} \xrightarrow{\text{Sugar}} \text{picramic acid (2-amino-4,6-dinitrophenol)}$$

Reagent

Dissolve 0.2 g of picric acid and 0.4 g of anhydrous sodium carbonate in water, and dilute to 100 ml with water.

Procedure

To 1 ml of aqueous solution of the sugar, add 1 ml of reagent, and heat at 100° for 10 min. Allow to cool, dilute to 5 ml with water, and read at 460 nm.

Results

	A = 0.3 (1 cm cell) Sample, μg
Ribose	280
Xylose	240
Galactose	384
Fructose	390
Sorbose	270

Notes

1. Upon reaction with picric acid in alkaline medium, digitalis glycosides develop a red-orange color (6). The reaction is due to the active hydrogen of the lactone ring attached at position 17 of the steroidal skeleton.

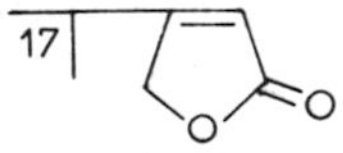

2. Polyphenols, various aldehydes and ketones, purine bases, and ascorbic acid also react.

3. Creatinine develops a red color (7), which is probably that of a tautomer of creatinine picrate.

E. 2,4-Dinitrophenylhydrazine in Acetic-Hydrochloric Acid

Principle and Procedure

See Chapter 8, Section II.B, colorimetric determination of aldehydes and ketones, but instead of 1 hr at room temperature, heat at 100° for 15 min. Read at 432 nm.

Results

	A = 0.3 (1 cm cell) Sample, μg
Ribose	58
Xylose	83
Galactose	89
Glucose	85
Fructose	46
Sorbose	47

F. 2-Hydrazinobenzothiazole and Hydrogen Peroxide

Principle (8)

Condensation with 2-hydrazinobenzothiazole in alkaline medium and development as formazan with the diazonium salt generated in the reaction medium by oxidation of the excess reagent (see mechanism, Chapter 8, Section IV.A): blue-violet color.

It may be postulated that, at least with ketoses, the hydrazone initially formed is converted to the corresponding osazone, thus giving rise to an aldehyde-hydrazone group, which then reacts according to the usual scheme.

Reagents

a. Dissolve 0.040 g of 2-hydrazinobenzothiazole in 2.5 ml of 0.1 N hydrochloric acid, and dilute to 10 ml with water.

b. An 0.75% aqueous solution of hydrogen peroxide (2.5 volumes oxygen).

Procedure

To 1 ml of aqueous solution of the sugar, add 0.5 ml of reagent a and 0.5 ml of 0.1 N sodium hydroxide. Heat at 100° for 10 min, cool for 5 min in a water bath at 15°, and add 2 ml of reagent b. Let stand for 10 min, add 2 ml of ethanol, let stand for 5 min, and read at 580 nm.

Results

	A = 0.3 (1 cm cell) Sample, μg
Ribose	9.5
Xylose	10
Galactose	12
Glucose	16
Fructose	15
Sorbose	12

Notes

1. Under the same conditions, A = 0.3 is given by 6.5 μg of glyceraldehyde, 20 μg of dihydroxyacetone, or 40 μg of glucosamine hydrochloride.

2. 2-Hydrazinobenzothiazole also allows the colorimetric determination of aldehydes (Chapter 8, Section IV.A) and 17-hydroxy-17-ketolsteroids (Chapter 15, Section XIV.B).

3. Tetrazolium salts and formazans, see Chapter 16, Section VIII.

G. 3-Methylbenzothiazolin-2-one Hydrazone and Ferric Chloride

Principle (8)

Condensation with 3-methylbenzothiazolin-2-one hydrazone in alkaline medium, and development as formazan with the diazonium salt generated in the reaction medium by oxidation of the excess reagent (see mechanism, Chapter 8, Section IV.B): blue color.

It may be postulated that, at least with ketoses, the hydrazone initially formed is converted to the corresponding osazone, thus giving rise to an aldehyde-hydrazone group, which then reacts according to the usual scheme.

Reagents

a. An 0.5% aqueous solution of 3-methylbenzothiazolin-2-one hydrazone hydrochloride.

b. An 0.25% aqueous solution of ferric chloride hexahydrate.

Procedure

To 1 ml of aqueous solution of the sugar, add 0.5 ml of reagent a and 0.5 ml of 0.1 N sodium hydroxide. Heat at 100° for 10 min, cool for 5 min in a water bath at 15°, add 0.5 ml of 1 N hydrochloric acid, and 2 ml of reagent b. Let stand for 30 min at room temperature, and read at 620 nm.

Results

	A = 0.3 (1 cm cell) Sample, μg
Ribose	3
Xylose	4
Galactose	3.5
Glucose	3.5
Fructose	4
Sorbose	4.5

Notes

1. Under the same conditions, A = 0.3 is given by 11 μg of glucosamine hydrochloride.

2. Properties of 3-methylbenzothiazolin-2-hydrazone, see Chapter 16, Section III.

H. α-Naphthol in Sulfuric Acid

Principle (9,10)

Conversion of pentoses to furfural, of hexoses to 5-hydroxymethylfurfural, and condensation with sulfonated α-naphthol: blue-violet color.

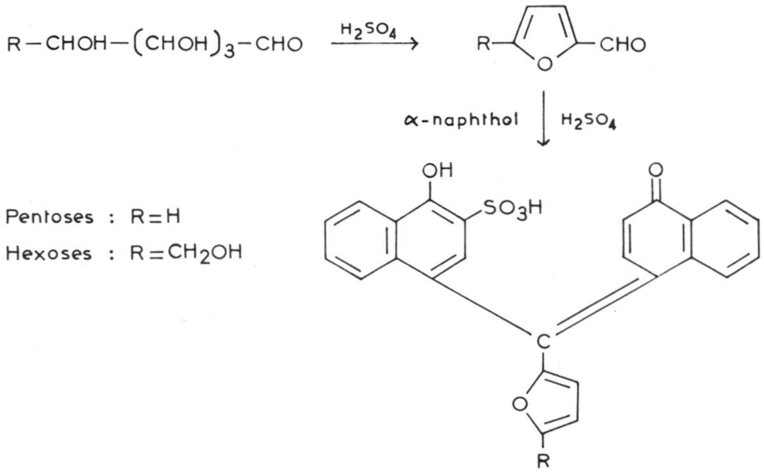

Reagent

Dissolve 0.40 g of α-naphthol in 100 ml of concentrated sulfuric acid, and let stand in the dark for 16 hr at room temperature.

Procedure

To 2 ml of aqueous solution of the sugar, add 5 ml of reagent. Agitate continually during the addition. Heat at 100° for 10 min, let stand in ice water for 30 min, and read.

Results

	λ Max, nm	A = 0.3 (1 cm cell) Sample, μg
Ribose	560	28
Xylose	550	12
Glucose	560	23
Fructose	560	18

I. Sulfuric Acid and Thymol

Principle (11)

Reaction with thymol in sulfuric acid medium: pink color.

Reagents

a. A 77% solution of sulfuric acid.

b. A 10% solution of thymol in ethanol.

Procedure

To 1 ml of aqueous solution of the sugar, add without mixing 7 ml of reagent a. Still without mixing, let stand for 10 min at room temperature, then 15 min in ice water. Add 0.1 ml of reagent b, 0.9 ml of water, and mix thoroughly. Heat at 100° for 20 min, chill for 5 min in ice water, let stand at room temperature for 25 min, and read.

Results

	λ Max, nm	A = 0.3 (1 cm cell) Sample, μg
Ribose	500	44
Xylose	485	34.5
Galactose	512	60
Glucose	512	55
Fructose	514	23.5
Sorbose	512	29.5

J. Sulfuric Acid and Carbazole

Principle (12)

Reaction with carbazole acid medium: pink color.

Reagents

a. An 8:1 mixture of concentrated sulfuric acid and water.

b. An 0.5% solution of carbazole in ethanol.

Procedure

To 1 ml of aqueous solution of the sugar, chilled in ice water, add 9 ml of reagent a prechilled to 0°. Mix, and, still maintaining at 0°, add 0.3 ml of reagent b. Heat for 10 min at 100°, let stand for 5 min in ice water, and read at 540 nm.

Results

	A = 0.3 (1 cm cell) Sample, μg
Ribose	264
Xylose	144
Galactose	164
Fructose	72
Sorbose	86

Notes

1. Carbazole in sulfuric acid medium was proposed as a reagent for the characterization of hexoses (13,14) and uronic acids (15).

2. The introduction of borates or sulfamates into the reaction medium makes the reaction somewhat selective, inhibiting the color development of glucose, galactose, and mannose (16).

III. ALDOSES

A. Benzidine

Principle (17)

Reaction with benzidine in acetic acid medium: orange-yellow color.

Reagent

Dissolve 0.20 g of benzidine and 0.10 g of stannous chloride dihydrate in 100 ml of glacial acetic acid, filter, and use immediately.

Procedure

To 1 ml of aqueous solution of the aldose, add 5 ml of reagent. Heat at 100° for 30 min, cool in ice water for 5 min, and read at 380 nm.

Results

	A = 0.3 (1 cm cell) Sample, μg
Arabinose	47
Ribose	43

Galactose	55
Glucose	36

Notes

1. No color is developed with fructose.

2. Although the wavelength of maximum absorption varies from one sugar to the other, Beer's law is followed only at 380 nm.

3. Benzidine is a potent carcinogenic agent.

IV. KETOSES

A. Diphenylamine

Principle (18)

Reaction with diphenylamine in hydrochloric acid medium: blue color.

Reagent

Mix 70 ml of 95% ethanol, 50 ml of concentrated hydrochloric acid, and 6 ml of 20% solution of diphenylamine in ethanol.

Procedure

To 1 ml of aqueous solution of the ketose, add 5 ml of reagent, heat at 100° for 15 min, cool in ice water for 5 min, and dilute to 25 ml with ethanol. Read at 640 nm.

Results

	A = 0.3 (1 cm cell) Sample, μg
Fructose	82
Sorbose	110

Notes

1. The reaction is negative with glucose.

2. The reaction of ketoses with diphenylamine was subjected to various studies. Formulas (VI) and (VII) were suggested for the colored species formed (19).

3. When operating in the presence of sulfuric and acetic acids, 2-deoxyribose develops a blue color (13) and glycolic aldehyde a green color (20).

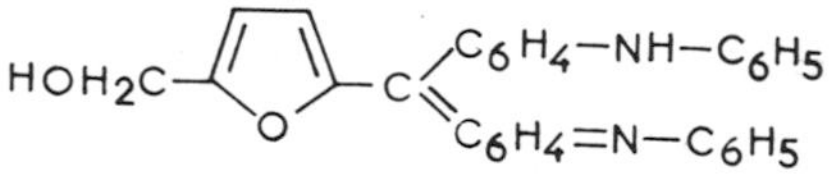

$$HOH_2C-C_4H_2O-C(=C_6H_4=N-C_6H_5)(-C_6H_4-NH-C_6H_5)$$

(VI)

$$\left[(C_6H_5)_2N-C(CH_2OH)=CH-CH=C(OH)-CH=\overset{+}{N}(C_6H_5)_2 \right] Cl^-$$

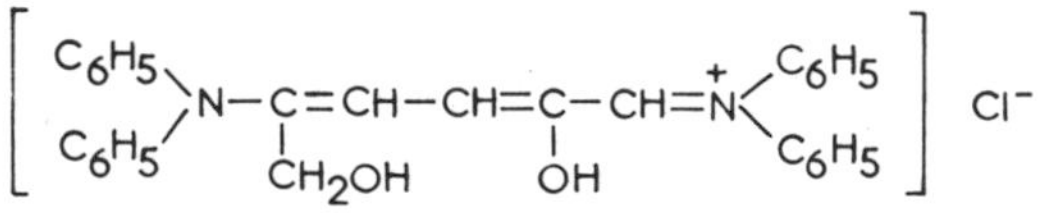

(VII)

4. Diphenylamine also allows the colorimetric determination of 2-deoxyoses (Section IX.F).

B. Indole-3-acetic Acid

Principle (21)

Reaction with indole-3-acetic acid in hydrochloric acid medium: blue-violet color.

Reagent

An 0.5% solution of indole-3-acetic acid in ethanol.

Procedure

To 1 ml of aqueous solution of the ketose, add 0.2 ml of reagent and 8 ml of concentrated hydrochloric acid. Incubate at 37° for 1 hr, and read at 520 nm.

Results

	A = 0.3 (1 cm cell) Sample, μg
Fructose	24
Sorbose	104

Note

The reaction is negative with glucose.

C. Resorcinol and Ferric Alum

Principle (22)

Reaction with resorcinol and ferric alum in hydrochloric acid medium: yellow-orange color.

Reagents

a. An 0.05% solution of resorcinol in ethanol.

b. Dissolve 0.216 g of ammonium ferric sulfate dodecahydrate in 1 liter of concentrated hydrochloric acid.

Procedure

To 2 ml of aqueous solution of the ketose, add 3 ml of reagent a and 3 ml of reagent b. Heat at 80° ± 3° for 1 hr, cool for 5 min in a water bath at room temperature, and read at 420 nm.

Results

	A = 0.3 (1 cm cell) Sample, μg
Fructose	23
Sorbose	24

Note

The reaction is negative with glucose.

D. 2-Thiobarbituric Acid

Principle (23)

Reaction with 2-thiobarbituric acid in hydrochloric acid medium: yellow color.

Reagent

Dissolve 0.288 g of 2-thiobarbituric acid in 100 ml of water.

Procedure

To 1 ml of aqueous solution of the ketose, add 1 ml of reagent and 1 ml of concentrated hydrochloric acid. Heat at 100° for 6 min, and cool for 5 min in a water bath at room temperature. Read at 432 nm.

Results

	A = 0.3 (1 cm cell) Sample, μg
Fructose	8
Sorbose	13

Notes

1. The reaction is almost negative with glucose (A = 0.02 given by 100 μg).

2. Other analytical uses of 2-thiobarbituric acid, see Section IX.D, Notes 3-5.

V. TETROSES

A. Fructose, Sulfuric Acid, and Cysteine Hydrochloride

Principle (24)

Reaction with fructose in sulfuric acid medium, then with cysteine: yellow color.

Reagents

a. An 0.1% aqueous solution of fructose.

b. A 6:1 mixture of concentrated sulfuric acid and water.

c. A 1% aqueous solution of cysteine hydrochloride.

Procedure

To 0.5 ml of prechilled aqueous solution of the tetrose, add 0.5 ml of reagent a prechilled to 0° and 4.5 ml of reagent b prechilled to 0°. Shake for 3 min, let stand for 3 min at room temperature, and heat at 100° for 3 min. Let stand for 3 min in ice water, add 1 ml of reagent c and 1 ml of water. Warm at 50° for 30 min, and allow to cool. Read at 458 and 480 nm. The difference between the two readings corresponds to the optical density of the tetrose.

Results

	$A_{458} - A_{480}$ = 0.3 (1 cm cell) Sample, μg
Erythrose	12.6

VI. PENTOSES

A. Aniline

Principle (25,26)

In acid medium, dehydration of the pentose into furfural, and condensation with aniline: orange color.

$$HOCH_2-(CHOH)_3-CHO \xrightarrow{H^+} \text{(furan)}-CHO \xrightarrow[H^+]{\text{Aniline}} \left[C_6H_5-NH-\overset{\displaystyle H \;\; H}{\underset{HO}{CH}}\!\!=\!\!C-C\!\!=\!\!C-CH=NH-C_6H_5 \right]^+$$

Reagent

To 100 ml of glacial acetic acid, add 10 ml of 5% aqueous solution of oxalic acid dihydrate, 24 ml of water, and 16 ml of colorless aniline.

Procedure

To 2 ml of aqueous solution of the pentose, add 6 ml of reagent, and let stand in the dark at room temperature for 24 hr. Read at 480 nm. Occasionally, the formation of a precipitate is observed. It settles out in the cell within 1 min and does not interfere with measurements.

Results

	A = 0.3 (1 cm cell) Sample, μg
Ribose	39
Xylose	32

Note

The reaction is negative with glucose.

B. p-Bromoaniline

Principle (27)

Reaction with p-bromoaniline in acetic acid medium: pink color.

Reagent

Dissolve 2 g of p-bromoaniline in 100 ml of glacial acetic acid saturated with thiourea (about 4 g).

Procedure

To 1 ml of aqueous solution of the pentose, add 5 ml of reagent. Heat at 70° for 10 min, cool for 5 min in a water bath at room temperature, and let stand for 70 min in the dark. Read at 520 nm.

Results

	A = 0.3 (1 cm cell) Sample, μg
Ribose	106
Xylose	104

Notes

1. The reaction is negative with glucose.

2. It may be assumed that the mechanism of the reaction is similar to that given in the preceding method.

C. Orcinol and Ferric Chloride

Principle (28)

Reaction with orcinol and ferric chloride in hydrochloric acid medium: blue-green color.

Reagent

Prepare an 0.166% solution of ferric chloride hexahydrate in concentrated hydrochloric acid. Just before use, add to this solution 0.010 g of orcinol per milliliter.

Procedure

To 1 ml of aqueous solution of the pentose, add 1 ml of reagent. Heat at 100° for 20 min, add 2 ml of water, and cool for 5 min in ice water. Read at 665 nm.

Results

	A = 0.3 (1 cm cell) Sample, μg
Ribose	9.0
Xylose	5.8

Notes

1. Glucose reacts weakly. A = 0.3 is given by 600 μg.

2. This reaction allows the characterization of ribonucleic acid (RNA) in the presence of deoxyribonucleic acid

(DNA) (29). It may be assumed that the ribose liberated from RNA in acid medium is at first converted into furfural, which then reacts with orcinol and ferric chloride, whereas 2-deoxyribose obtained under the same conditions from DNA gives rise to a levulinic acid derivative which does not react (see Section IX.F, Note 1).

D. Orcinol and Cupric Chloride

Principle (30)

Reaction with orcinol and cupric chloride in hydrochloric acid medium, and extraction of the colored species into isoamyl alcohol: green color.

Reagent

Dissolve 0.20 g of orcinol in concentrated hydrochloric acid, add 10 ml of 0.004 M cupric chloride solution in concentrated hydrochloric acid, and dilute to 100 ml with the same acid. This reagent is stable for several hours.

Procedure

To 5 ml of aqueous solution of the pentose, add 5 ml of reagent, heat at 100° for 40 min, cool for 5 min in ice water, and extract with 5 ml of isoamyl alcohol. After thorough decantation, collect the organic layer, and read at 670 nm.

Results

	A = 0.3 (1 cm cell) Sample, μg
Ribose	23
Xylose	25

Note

The reaction is almost negative with glucose.

E. Hydrochloric Acid, p-Nitrophenylhydrazine and Benzyltrimethylammonium Hydroxide

Principle

Conversion of the pentose into furfural, formation of the p-nitrophenylhydrazone, and development in alkaline medium (see Chapter 8, Section II.D): pink-violet color.

Reagents

a. An 0.04% solution of p-nitrophenylhydrazine in an 0.1% dilution of concentrated hydrochloric acid in ethanol.

b. Dilute 1 ml of 40% solution of benzyltrimethylammonium hydroxide (see Chapter 17, Section III) in methanol to 100 ml with dimethylformamide. Prepare fresh just before use.

Procedure

To 2 ml of aqueous solution of the pentose, add 1 ml of concentrated hydrochloric acid. Fit the tubes with air-condensers, and heat at 100° for 1 hr. Cool for 5 min in ice water, and neutralize: Add 1.15 ml of 10 N sodium hydroxide, then adjust to pH 6-7 by adding dropwise 1 N sodium hydroxide or 1 N hydrochloric acid. Pipet an 0.5 ml aliquot, add to it 0.5 ml of reagent *a*, heat at 70° for 30 min, cool to 20°, and add 4 ml of reagent *b*. Mix, let stand for 3 min, filter, and read immediately at 545 nm.

Results

	A = 0.3 (1 cm cell) Sample, μg
Arabinose	325
Ribose	115
Xylose	140

Note

The reaction is negative with glucose.

F. Glucose and Phloroglucinol

Principle (31)

Reaction with glucose and phloroglucinol in acetic-hydrochloric acid medium: red color.

Reagent

Mix 2 ml of concentrated hydrochloric acid, 110 ml of glacial acetic acid, 1 ml of 0.8% aqueous solution of glucose, and 5 ml of 5% solution of phloroglucinol in ethanol. Prepare fresh before use or keep at 0°.

Procedure

To 0.5 ml of aqueous solution of the pentose, add 0.5 ml of concentrated hydrochloric acid and 4.5 ml of reagent. Heat at 100° for 5 min, cool for 5 min in ice water, and read at 552 and 510 nm. The difference between the two readings corresponds to the optical density of the pentose.

Results

	$A_{552} - A_{510} = 0.3$ (1 cm cell) Sample, μg
Ribose	14
Xylose	14

G. Anthrone

Principle (32)

Reaction with anthrone in sulfuric acid medium: blue-green color.

Reagent

An 0.09% solution of anthrone in concentrated sulfuric acid.

Procedure

To 2 ml of aqueous solution of the pentose, add 4 ml of reagent prechilled to 0°, so as not to mix with the aqueous layer. Then mix quickly, immerse immediately in ice water for 5 min, and let stand at room temperature for 30 min. Read at 590 nm.

Results

	A = 0.3 (1 cm cell) Sample, μg
Ribose	51
Xylose	22

Notes

1. Glucose develops a green color, A = 0.1 for 40 μg, and Beer's law is not obeyed.

2. Under somewhat different conditions (diluted sulfuric acid), hexoses can also be determined (Section VIII.B).

3. It may be admitted that pentoses are at first dehydrated to furfural, which then condenses with anthrone to give various colored species. One of them was identified as 10-furfurylideneanthrone (VIII) (33).

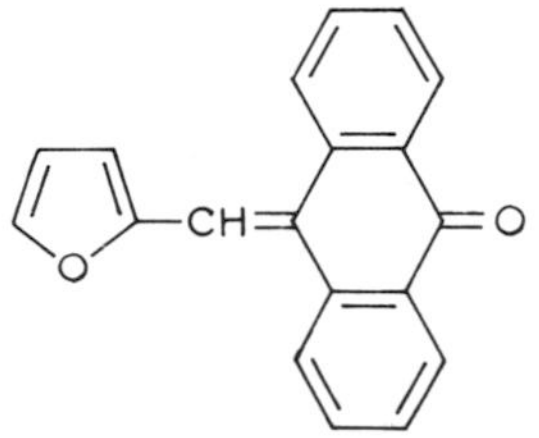

(VIII)

10-(5-Methylfurfurylidene)anthrone was isolated from the mixture obtained from rhamnose (34).

4. The reaction involving the formation of furfural, 2-deoxyoses do not react (35,36).

5. Anthrone in sulfuric acid allows the colorimetric determination of aliphatic aldehydes (Chapter 8, Section IV.C). It also reacts with some steroids (Chapter 8, Section IV.C, Note 2) and with ascorbic acid (red color) (37).

VII. METHYLPENTOSES

A. Sulfuric Acid and Thioglycolic Acid

Principle (38)

Reaction with sulfuric acid, then with thioglycolic acid: very light yellow color.

Reagents

a. A 6:1 mixture of concentrated sulfuric acid and water.

b. Dissolve 1 ml of thioglycolic acid in 29 ml of water.

Procedure

To 1 ml of aqueous solution of the methylpentose, add slowly with cooling in ice water 4.5 ml of reagent a. Mix, heat at 100° for 10 min, cool for 5 min in ice water, and add 0.1 ml of reagent b. Let stand for 3 hr in the dark, and read at 400 and 430 nm. The difference between the two readings corresponds to the optical density of the sugar.

Results

	$A_{400} - A_{430} = 0.3$ (1 cm cell) Sample, μg
Fucose	17
Rhamnose	17

Note

Glucose does not react.

VIII. HEXOSES

A. Sulfuric Acid and Sodium Salt of Chromotropic Acid

Principle (39)

Conversion of the hexose to 5-hydroxymethylfurfural and splitting of the methylol group to form formaldehyde, which is developed with chromotropic acid: violet color.

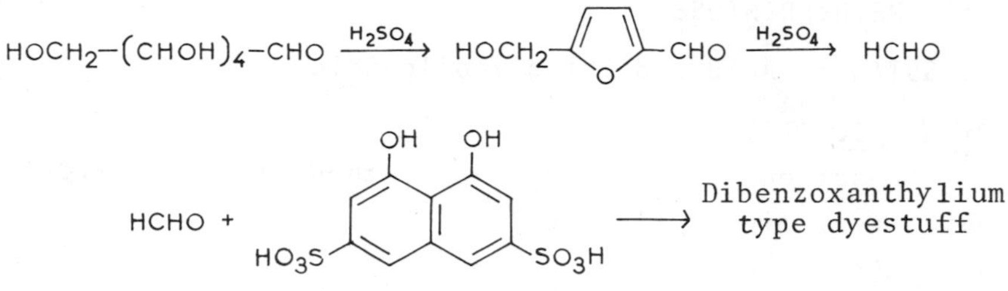

$$HOCH_2{-}(CHOH)_4{-}CHO \xrightarrow{H_2SO_4} HOCH_2{-}\text{(furan)}{-}CHO \xrightarrow{H_2SO_4} HCHO$$

$$HCHO + \text{(chromotropic acid: 1,8-(OH)}_2\text{-3,6-(SO}_3\text{H)}_2\text{-naphthalene)} \longrightarrow \text{Dibenzoxanthylium-type dyestuff}$$

Reagent

Dissolve 0.10 g of sodium salt of chromotropic acid in 1 ml of water, and dilute to 50 ml with 15 M sulfuric acid. Prepare fresh before use.

Procedure

To 1 ml of aqueous solution of the hexose, add 5 ml of reagent. Heat at 100° for 30 min, and cool for 5 min in ice water. Dilute to 10 ml with 9 M sulfuric acid, and read at 570 nm.

Results

	A = 0.3 (1 cm cell) Sample, μg
Galactose	220
Glucose	150
Fructose	200

Notes

1. Pentoses do not react.

2. Colorimetric determination of formaldehyde, see Chapter 16, Section V.

B. Anthrone

Principle (40)

Reaction with anthrone in sulfuric acid medium: blue-green color.

Reagent

Dissolve 0.20 g of anthrone in 100 ml of a 5:2 mixture of concentrated sulfuric acid and water.

Procedure

Chill 5 ml of reagent in ice water for 5 min, then layer on 1 ml of aqueous solution of the hexose. Let stand for 5 min at 0°, mix, and heat at 100° for 10 min. Cool for 5 min in a cold water bath, and read at 620 nm.

Results

	A = 0.3 (1 cm cell) Sample, μg
Glucose	80
Mannose	50
Fructose	46

Notes

1. With ribose, a yellow-green color is developed, and Beer's law is not followed. However, under somewhat different conditions (concentrated sulfuric acid), pentoses can also be determined (Section VI.G).

2. Compounds (IX) (41) and (X) (42) were identified in the reaction mixture.

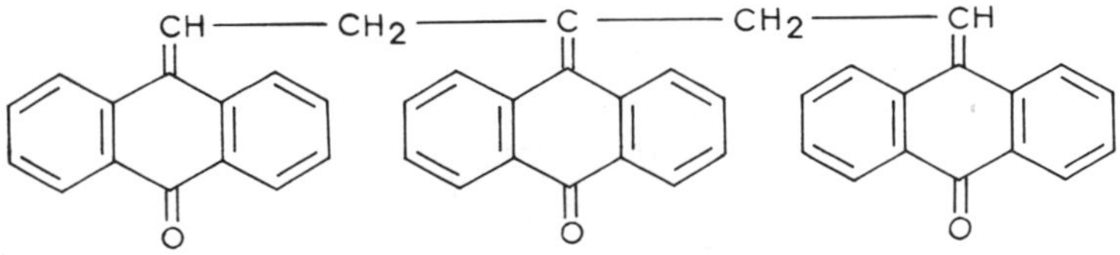

(IX)

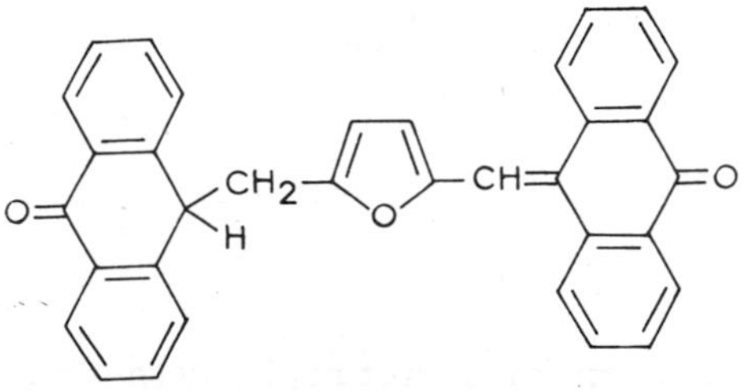

(X)

3. Other analytical uses of anthrone, see Chapter 8, Section IV.C, Notes 2 and 3, and Section VI.G, Note 5.

IX. 2-DEOXYOSES

A. p-Nitrophenylhydrazine

Principle (43)

The 2-deoxyose in acid medium is at first converted to α-hydroxylevulinic aldehyde, which then condenses with p-nitrophenylhydrazine. The acid causes the hydrazone to cyclize, and the pyridazine ring so formed is reopened in alkaline medium (44): violet color.

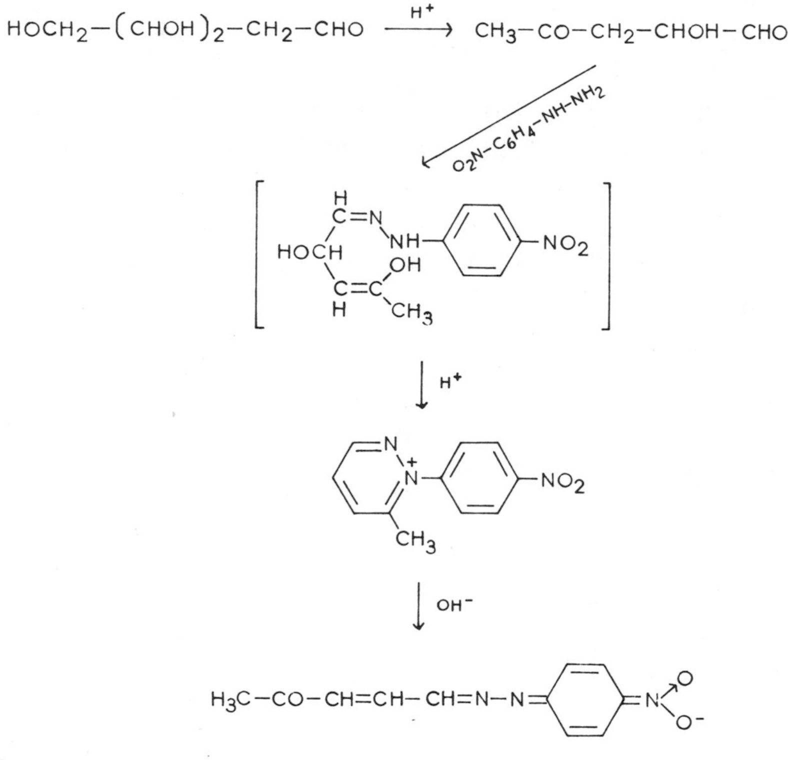

$$HOCH_2-(CHOH)_2-CH_2-CHO \xrightarrow{H^+} CH_3-CO-CH_2-CHOH-CHO$$

$$\xrightarrow{O_2N-C_6H_4-NH-NH_2}$$

$$\left[\begin{array}{l} H-C{=}N-NH-C_6H_4-NO_2 \\ HOCH \\ H-C{=}C(OH)-CH_3 \end{array}\right]$$

$$\downarrow H^+$$

$$\text{3-methyl-1-(4-nitrophenyl)pyridazinium} \; (N^+-C_6H_4-NO_2)$$

$$\downarrow OH^-$$

$$H_3C-CO-CH{=}CH-CH{=}N-N{=}C_6H_4{=}N(\rightarrow O)O^-$$

Reagent

An 0.5% solution of p-nitrophenylhydrazine in ethanol.

Procedure

To 4 ml of sample solution in 5% trichloroacetic acid, add 0.2 ml of reagent, fit the tubes with air-condensers, and heat at 100° for 20 min. Cool to room temperature in a water bath, add 10 ml of butyl acetate, shake for 3 min, and allow to decant. Collect a 3 ml aliquot of the aqueous layer, add to it 1 ml of 2 N sodium hydroxide and 1 ml of water, and read at 545 nm.

Results

	A = 0.3 (1 cm cell) Sample, μg
2-Deoxyribose	40
Digitoxose	93
2-Deoxyglucose	540

Notes

1. The reaction is negative with ribose, glucose, and glucosamine.

2. The reaction is specific for 2-deoxyoses with at least five carbon atoms (45,46).

B. Flavylium Perchlorate

Principle (47)

Reaction with flavylium perchlorate in acetic acid medium: yellow-green color.

Reagent

An 0.25% solution of flavylium perchlorate (Chapter 17, Section I.D) in 2% solution of anhydrous sodium acetate in glacial acetic acid. Prepare fresh just before use.

Procedure

To 5 ml of sample solution in glacial acetic acid, add 1 ml of reagent. Heat at 100° for 30 min, and let stand for 10 min at room temperature. Read at 450 nm.

Results

	A = 0.3 (1 cm cell) Sample, μg
2-Deoxyribose	66
Digitoxose	46
2-Deoxyglucose	280

Note

Analytical uses of flavylium perchlorate, see Chapter 5, Section II.B, Notes 1-5.

C. Xanthydrol and p-Toluenesulfonic Acid

Principle (48,49)

Condensation of xanthydrol in acid medium: pink color.

Reagents

a. An 0.04% solution of xanthydrol in glacial acetic acid.

b. A 1% solution of p-toluenesulfonic acid in glacial acetic acid.

Procedure

To 2 ml of sample solution in glacial acetic acid, add 1 ml of reagent a and 1 ml of reagent b. Heat at 100° for 5 min, let stand for 5 min in ice water, then 5 min at room temperature, and read.

Results

	λ Max, nm	A = 0.3 (1 cm cell) Sample, μg
2-Deoxyribose	520	16
Digitoxose	550	9
2-Deoxyglucose	550	20

Notes

1. Cardiotonic glycosides are characterized by the presence of a sugar linked glycosidically to a steroid nucleus. Upon suitable hydrolysis, they give an aglycone and one or more sugars. These latter are often characterized by the presence of deoxy groups on C-6 or C-2 and C-6. Apart from glucose and rhamnose, the sugars so obtained have not been found elsewhere. The above method may therefore be applied to the characterization and determination of this sugar moiety of cardiotonic glycosides.

For instance, the xanthydrol reaction can be applied to glycosides containing digitoxose (XI), cymarose (XII), or oleandrose (XIII) as illustrated by the accompanying table.

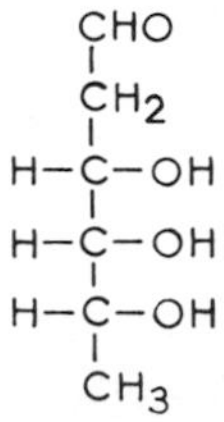

(XI)

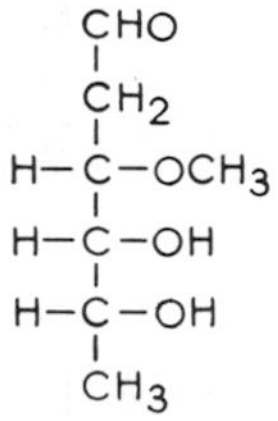

(XII)

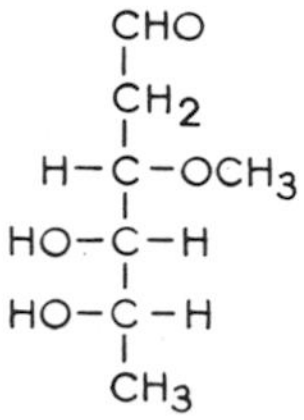

(XIII)

Aglycone	Substituents R	R'	Glycoside
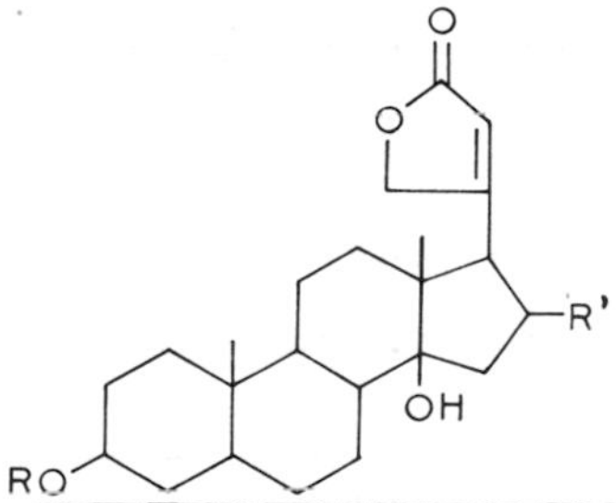	3 Digitoxose	H	Digitoxoside
	3 Digitoxose	OH	Gitoxoside
	Oleandrose	OAc	Oleandrin
	Cymarose	OAc	Hongheloside
	Cymarose	H	Somalin
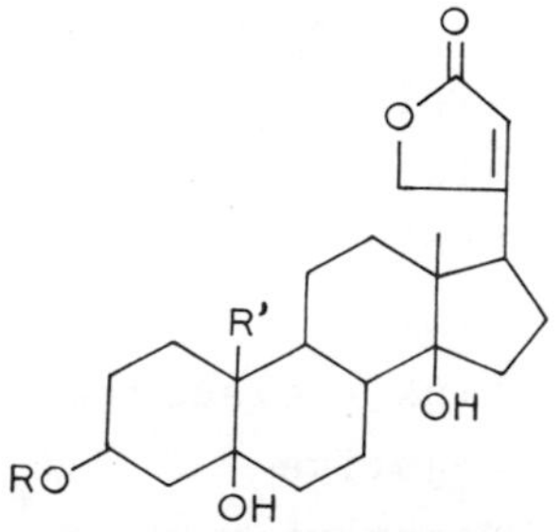	Cymarose	CH_3	Periplocymarin
	Cymarose	CH_2OH	Cymarol
	Cymarose	CHO	Cymarin
	Cymarose + glucose	CHO	K-Strophanthin β
	Cymarose + 2 glucose	CHO	K-Strophanthin

Ouabain (XIV), whose sugar moiety is rhamnose, does not react.

(a)

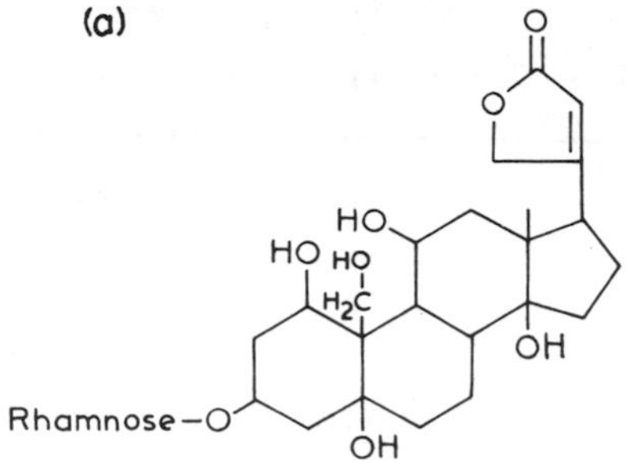

Rhamnose—O ... HO HO H_2C OH OH

(XIV)

2. It was also observed in our laboratory that digitoxose develops a violet color in acetone in the presence of perchloric acid, and the reaction can be extended to the characterization and determination of various cardiotonic glycosides (50).

Dissolve the compound in 1 ml of acetone, add 0.5 ml of 65% perchloric acid, and let stand at room temperature for 15 min. Add 3 ml of acetone, let stand for 10 min, and read at 570 nm.

Compounds which do not contain a 2-deoxyose do not react. (Thevetose is 3-methyl-6-deoxy-L-glucose.)

Glycoside	Sugar moiety	Color
Adynerin	Oleandrose	Violet
Cymarin	Cymarose	Violet
Digitoxoside	Digitoxose	Violet
Oleandrin	Oleandrose	Violet
K-Strophanthin	Cymarose	Pink-violaceous
Convallatoxin	Rhamnose	Negative
Honghelin	Thevetose	Negative
Neriifolin	Thevetose	Negative
Tanghiferin	Thevetose	Negative
Tanghinin	Thevetose	Negative

3. Other analytical uses of xanthydrol, see Chapter 13, Section II.A, Note 3.

D. Periodic Acid and 2-Thiobarbituric Acid

Principle (51)

Oxidation with periodic acid, and development of the malonic dialdehyde formed with 2-thiobarbituric acid: pink-violet color.

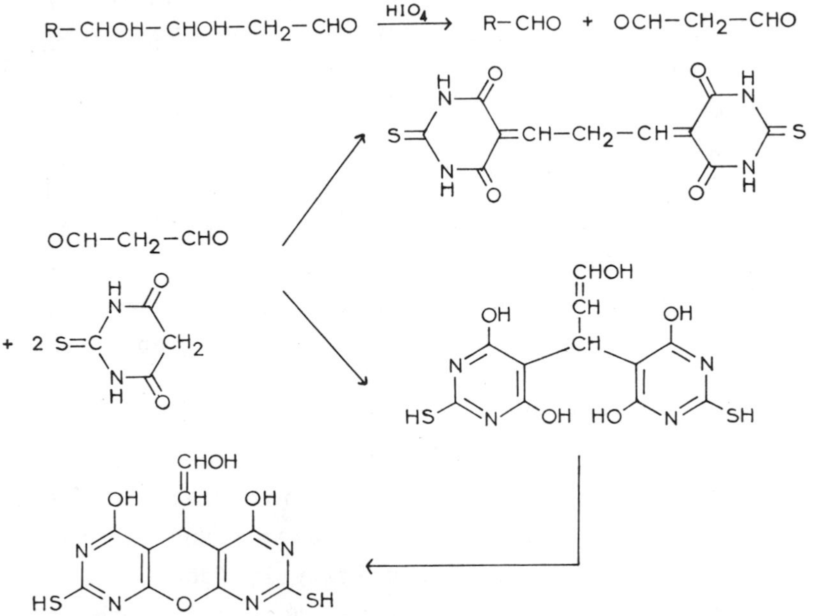

$$R-CHOH-CHOH-CH_2-CHO \xrightarrow{HIO_4} R-CHO + OCH-CH_2-CHO$$

Reagents

a. An 0.025 N solution of periodic acid in 0.1 N sulfuric acid.

b. A 2% solution of sodium arsenite in 0.5 N hydrochloric acid.

c. An 0.6% aqueous solution of 2-thiobarbituric acid, adjusted to pH 2.

Procedure

To 3.5 ml of neutral aqueous solution of the 2-deoxyose, add 0.5 ml of reagent a, and let stand at room temperature for 30 min. Add 1 ml of reagent b, let stand for 2 min, pipet a 2 ml aliquot, and add to it 2 ml of reagent c. Heat at 100° for 20 min, let stand for 5 min in a cold water bath, and add 1 ml of acetone. Read at 532 nm.

Results

	A = 0.3 (1 cm cell) Sample, μg
2-Deoxyribose	2.8
Digitoxose	3.0
2-Deoxyglucose	3.8

Notes

1. This reaction is not applicable to 2-deoxy-3-methoxy-5-methylpentose and related glycosides, such as cymarose, K-strophanthoside, neriifolioside, and thevetoside (51,52). It was applied to the determination of quinic acid and its caffeic esters (53), dihydrostreptomycin (54), and sialic acids (55).

2. The reaction was extended to the determination of sorbic acid, in which the unsaturated system is cleaved with acidic dichromate to yield malonic dialdehyde (56).

3. 2-Thiobarbituric acid has also found use in the determination of oxidative rancidity in food products (57). The intensity of the color can be increased by the addition of ferric ion (58).

4. 2-Thiobarbituric acid gives colored derivatives with α,β-unsaturated and aromatic aldehydes (59), and this reaction was applied to the analysis of drugs (54) and of α-amino alcohols previously oxidized by alkaline periodate (59).

5. 2-Thiobarbituric acid also allows the colorimetric determination of 2-substituted pyrimidine derivatives (Chapter 13, Section VII.A) and ketoses (Section IV.D).

6. Periodates and periodate oxidation, see Chapter 16, Section IV.

E. Periodic Acid and 2-Methylindole

Principle (60)

Oxidation with periodic acid, and development of the malonic dialdehyde formed with 2-methylindole: red color.

$$\text{R-CHOH-CHOH-CH}_2\text{-CHO} \xrightarrow{\text{HIO}_4} \text{R-CHO} + \text{OCH-CH}_2\text{-CHO}$$

$$\text{OCH-CH}_2\text{-CHO} + 2\ (\text{2-methylindole}) \longrightarrow [\text{2-methylindol-3-yl-CH=CH-CH=(2-methyl-3H-indolium)}]\ \text{X}^-$$

Reagents

a. An 0.025 N solution of periodic acid in 0.1 N sulfuric acid.

b. A 2% solution of sodium arsenite in 0.5 N hydrochloric acid.

c. Dissolve 0.10 g of 2-methylindole in 100 ml of ethanol. Chill to 0°, and just before use add 25 ml of concentrated hydrochloric acid.

Procedure

To 1 ml of aqueous solution of the 2-deoxyose, add 0.5 ml of reagent a, and let stand at room temperature for 10 min. Add 1 ml of reagent b, let stand for 2 min, add 2 ml of reagent c, and let stand at room temperature for 30 min. Read at 555 nm.

Results

	A = 0.3 (1 cm cell) Sample, μg
2-Deoxyribose	2.3
Digitoxose	2.4
2-Deoxyglucose	6.0

Note

Periodates and periodate oxidation, see Chapter 16, Section IV.

F. Diphenylamine

Principle (61)

Reaction with diphenylamine in acid medium: blue to pinkish color.

Reagent

Dissolve 1 g of diphenylamine in 100 ml of a 49:1 mixture of glacial acetic acid and concentrated sulfuric acid.

Procedure

To 2 ml of sample solution in 5% trichloroacetic acid, add 4 ml of reagent. Heat at 100° for 10 min, chill in ice water for 5 min, and read.

Results

	λ Max, nm	A = 0.3 (1 cm cell) Sample, μg
2-Deoxyribose	600	80
Digitoxose	490	275
2-Deoxyglucose	520	220

Notes

1. This reaction is often named the Dische reaction (13). Its mechanism has been subjected to numerous studies (62). It now seems well established (63) that the first step of the reaction is the dehydration of the 2-deoxyose to acetylacrolein in acid medium:

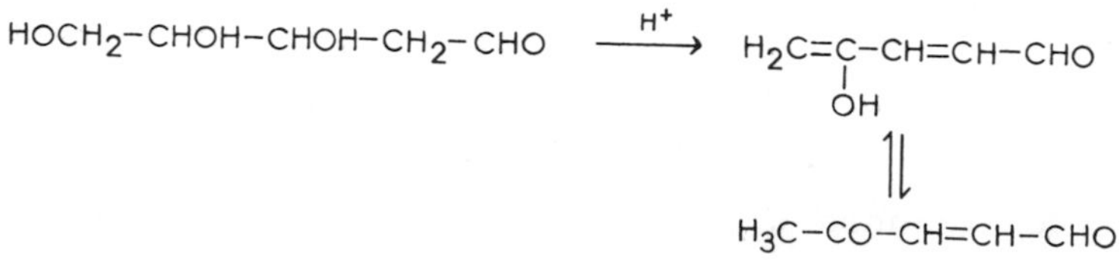

$$HOCH_2-CHOH-CHOH-CH_2-CHO \xrightarrow{H^+} H_2C=C(OH)-CH=CH-CHO \rightleftharpoons H_3C-CO-CH=CH-CHO$$

This compound reacts with 2 moles of diphenylamine to give compound (XV), which is oxidized to (XVI) by air oxygen, giving rise to the ionized species (XVII) in acid medium:

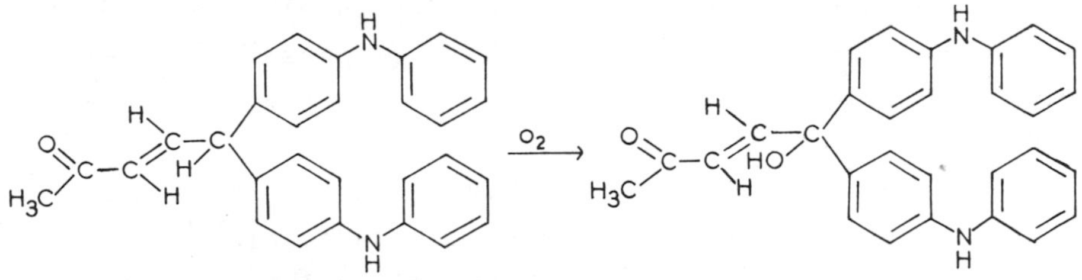

(XV)

(XVI)

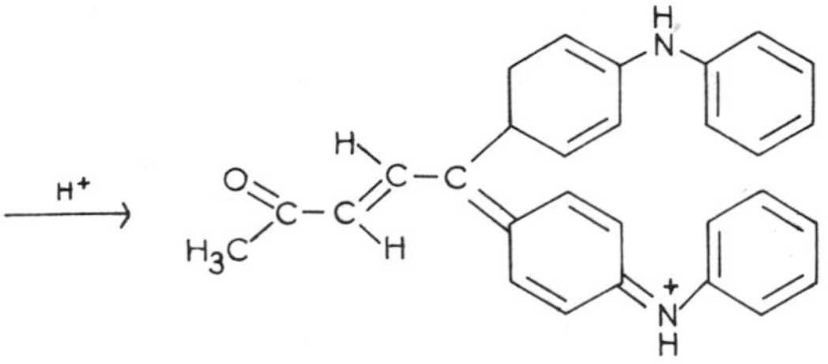

(XVII)

2. The Dische reaction has often been used for the colorimetric determination of deoxyribonucleic acids (DNA) in the presence of ribonucleic acid (RNA), the deoxyose being liberated by a suitable hydrolysis. However, under somewhat different conditions with a prolonged hydrolysis time, the ribose moiety of RNA can also react, and the reaction can be applied to the quantitative and qualitative analysis of RNA (64).

3. Diphenylamine also allows the colorimetric determination of ketoses (Section IV.A).

X. 2-AMINO-2-DEOXYHEXOSES

A. Acetylacetone and Ferric Ferricyanide

Principle (65)

Condensation with acetylacetone in alkaline medium to give a pyrrole derivative which reduces ferric ferricyanide into Prussian Blue.

Reagents

a. A 2% solution of freshly distilled acetylacetone in 0.5 M aqueous solution of sodium carbonate.

b. A 1:1 mixture of 0.6% aqueous solution of potassium ferricyanide and of 0.9% aqueous solution of ferric chloride hexahydrate. Prepare fresh immediately before use.

Procedure

To 0.6 ml of neutral aqueous solution of the 2-amino-2-deoxyose, add 0.6 ml of reagent a. Heat at 100° for 10 min, chill in ice water for 5 min, add 0.6 ml of 1 N hydrochloric acid, let stand for 2 min, and add 0.2 ml of reagent b. Let stand for 5 min at room temperature in the dark, add 2 ml of 1 N hydrochloric acid, heat at 100° for 5 min, and chill for 2 min in ice water. Read at 870 nm.

Results

	A = 0.3 (1 cm cell) Sample, μg
Galactosamine (hydrochloride)	37
Glucosamine (hydrochloride)	30

Notes

1. Formation of the pyrrole ring, see following method, Note 1.

2. Determinations with ferric ferricyanide, see Chapter 3, Section II.B, Note 1.

B. Acetylacetone and p-Dimethylaminobenzaldehyde

Principle (66)

Condensation with acetylacetone in alkaline medium to give a pyrrole derivative, and development with p-dimethylaminobenzaldehyde: red color.

Reagents

a. A 2% solution of freshly distilled acetylacetone in 0.5 M aqueous solution of sodium carbonate.

b. Dissolve 0.80 g of p-dimethylaminobenzaldehyde in 60 ml of concentrated hydrochloric acid.

Procedure

To 1 ml of neutral aqueous solution of the 2-amino-2-deoxyose, add 1 ml of reagent a and 2 ml of water. Heat at 100° for 10 min, allow to cool to room temperature, add 5 ml of ethanol, and heat at 75° for 5 min. Add 1 ml of reagent b, mix, and maintain heating at 75° for 30 min more. Allow to cool to room temperature, and dilute to 10 ml with ethanol. Read at 512 nm.

Results

	A = 0.3 (1 cm cell) Sample, μg
Galactosamine (hydrochloride)	46
Glucosamine (hydrochloride)	64

Notes

1. This reaction is usually named the Elson-Morgan reaction (67). These authors turned to account a work of Pauly and Ludwig (68), who had obtained the pyrrole derivative (XVIII) by reacting glucosamine with acetylacetone, and they postulated that the pyrrole ring could be revealed with p-dimethylaminobenzaldehyde (DMAB).

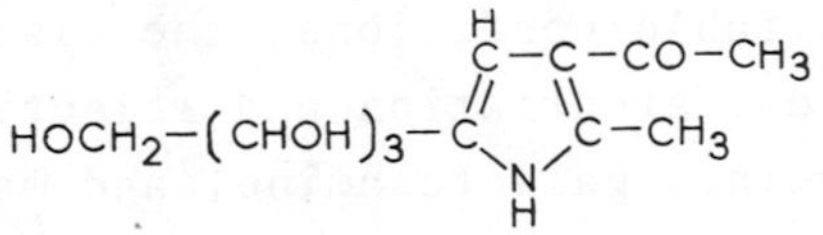

$$HOCH_2-(CHOH)_3-C \text{ pyrrole ring: } \overset{H}{C}-C-CO-CH_3,\ C-CH_3,\ \underset{H}{N}$$

(XVIII)

The Elson-Morgan reaction has been subjected to numerous studies.

Boyer and Furth (69) have admitted that under the action of DMAB the polyhydroxy group in compound (XVIII) is split off, and they have attributed formula (XIX) to the colored species.

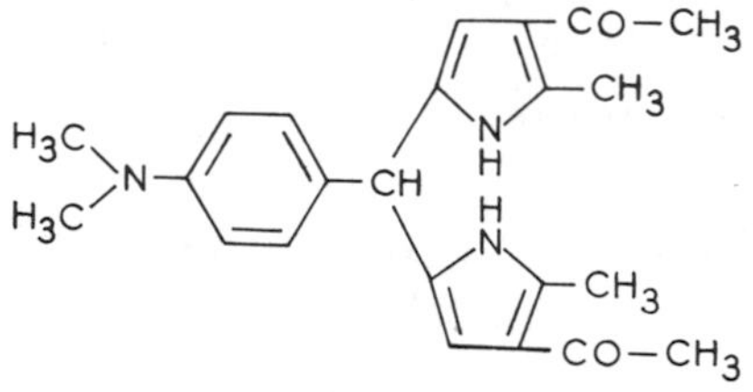

(XIX)

Schloss (70) has shown that the chromogen-forming reaction is much more complex than previously postulated. Two or three chromogens are formed, and one of them is volatile. Poorly reliable results are obtained when absorption measurements are made at the wavelength used by earlier workers (530 nm), because the compound absorbing maximally at 530 nm is not stable, and small variations in pH are critical. Absorption is greater at 512 nm, this wavelength corresponding to a stable colored species.

The volatility of one of the chromogens explains why some authors obtained more reproducible results when operating in sealed tubes or under conditions avoiding any losses (71).

Many authors have also dealt with interferences and determinations in the presence of sugars and amino acids (72).

2. Under suitable conditions, the Elson-Morgan reaction allows the differentiation and selective determinations of glucosamine, galactosamine, and mannosamine (73).

3. Caffeine develops a blue color when it reacts with acetylacetone in alkaline medium, then with DMAB in acid medium. Colorimetric determinations are possible; theobromine, theophylline, and other purine-containing compounds do not react (74).

4. Glucosamine, or chitosamine, is found in chitin, mucoproteins, and mucopolysaccharides. Its sulfamic derivative is a constituent of heparin.

Galactosamine, or chondrosamine, can be obtained by hydrolysis of chondroitinsulfuric acid, a mucopolysaccharide with N-acetylchondrosine (β-D-glucopyranosyluronic acid 2-deoxy-2-acetylamino-D-galactose) as a repeating unit and with one sulfate group per disaccharide unit.

5. Other analytical uses of p-dimethylaminobenzaldehyde, see Chapter 5, Section IV.B, Notes 1, 2.

C. Acetylacetone and Flavylium Perchlorate

Principle (75)

Condensation with acetylacetone in alkaline medium to give a pyrrole derivative, and development with flavylium perchlorate: orange color.

Reagents

a. A 1% solution of freshly distilled acetylacetone in 0.5 M aqueous solution of sodium carbonate.

b. Dissolve 0.10 g of flavylium perchlorate (Chapter 17, Section I.D) in a mixture of 100 ml of glacial acetic acid and 1 ml of concentrated hydrochloric acid.

Procedure

To 1 ml of neutral aqueous solution of the 2-amino-2-deoxyhexose, add 1 ml of reagent a. Heat at 100° for 10 min, cool for 5 min in ice water, and add 1 ml of reagent b. Heat at 100° for 5 min, cool for 5 min in ice water, and add 1 ml of glacial acetic acid. Read at 480 nm.

Results

	A = 0.3 (1 cm cell) Sample, μg
Galactosamine (hydrochloride)	82
Glucosamine (hydrochloride)	72

Notes

1. Formation of the pyrrole ring, see preceding method, Note 1.

2. Analytical uses of flavylium perchlorate, see Chapter 5, Section II.B, Notes 1-5.

D. Acetylacetone and Xanthydrol

Principle (75)

Condensation with acetylacetone in alkaline medium to give a pyrrole derivative, and development with xanthydrol: violet color.

Reagents

a. A 1% solution of freshly distilled acetylacetone in 0.5 M aqueous solution of sodium carbonate.

b. Dissolve 0.10 g of xanthydrol in a mixture of 100 ml of glacial acetic acid and 1 ml of concentrated hydrochloric acid.

Procedure

To 1 ml of neutral aqueous solution of the 2-amino-2-deoxyhexose, add 1 ml of reagent a. Heat at 100° for 10 min, cool for 5 min in ice water, and add 10 ml of reagent b. Fit the tubes with air-condensers, and heat at 100° for 15 min. Cool for 5 min in ice water, and read at 530 nm.

Results

	A = 0.3 (1 cm cell) Sample, μg
Galactosamine (hydrochloride)	315
Glucosamine (hydrochloride)	305

Notes

1. Formation of the pyrrole ring, see Section X.B, Note 1.

2. Analytical uses of xanthydrol, see Chapter 13, Section II.A, Note 3.

E. Phenylacetone and p-Dimethylaminobenzaldehyde

Principle (75)

Condensation with phenylacetone in alkaline medium, and development with p-dimethylaminobenzaldehyde: pink-violaceous color.

Reagents

a. A 5% solution of phenylacetone in a 3:1 mixture of propylene glycol and water.

b. A 5% solution of p-dimethylaminobenzaldehyde in 5 N hydrochloric acid.

Procedure

To 1 ml of neutral aqueous solution of the 2-amino-2-deoxyhexose, add 1 ml of reagent *a* and 0.5 ml of 5 N sodium hydroxide. Heat at 100° for 15 min, cool for 5 min in ice water, and add 0.5 ml of 5 N hydrochloric acid, 5 ml of ethanol, and 1 ml of reagent *b*. Let stand for 10 min at room temperature, and read at 560 nm.

Results

	A = 0.3 (1 cm cell) Sample, μg
Galactosamine (hydrochloride)	22.5
Glucosamine (hydrochloride)	44

Notes

1. It may be postulated that phenylacetone, or benzyl methyl ketone, $C_6H_5—CH_2—CO—CH_3$, behaves like acetylacetone to give a pyrrole ring upon condensation with the amino sugar (see Section X.B, Note 1).

2. This reaction allows the determination of glucosamine in the presence of its 3-methoxy derivative, which does not react (75).

3. Analytical uses of p-dimethylaminobenzaldehyde, see Chapter 5, Section IV.B, Notes 1, 2.

F. Sodium Nitrite and Indole

Principle (76)

Deamination and isomerization with nitrous acid, and development with indole: orange color.

CH_2OH O OH OH HO NH_2 $\xrightarrow{HNO_2}$ HOH_2C O CHO HO OH $\xrightarrow{\text{Indole}}$ Colored species

Reagents

a. A 5% aqueous solution of sodium nitrite.

b. 33% acetic acid.

c. A 12.5% aqueous solution of ammonium sulfamate.

d. A 1% solution of indole in ethanol.

Procedure

To 0.5 ml of neutral aqueous solution of the 2-amino-2-deoxyhexose, add 0.5 ml of reagent *a* and 0.5 ml of reagent *b*. Mix, let stand for 10 min, and add 0.5 ml of reagent *c*. Let stand for 30 min with occasional shaking. Add 2 ml of 1.5 N hydrochloric acid and 0.2 ml of reagent *d*. Heat at 100° for 5 min, add 2 ml of ethanol and cool for 5 min in ice water. Read at 492 and 520 nm. The difference between the two readings corresponds to the optical density of the sugar.

Results

	$A_{492} - A_{520} = 0.3$ (1 cm cell) Sample, μg
Galactosamine (hydrochloride)	25
Glucosamine (hydrochloride)	23

G. Sodium Nitrite, 3-Methylbenzothiazolin-2-one Hydrazone and Ferric Chloride

Principle (77)

Deamination and isomerization with nitrous acid, and development of the aldehyde formed with 3-methylbenzothiazolin-2-one hydrazone: blue color.

CH2OH O OH HO OH NH2 —HNO2→ HOH2C O CHO HO HO —(3-methylbenzothiazolin-2-one hydrazone: CH3, N, S, C=N-NH2)→ See Chapter 8, Section IV.B

Reagents

a. A 5% aqueous solution of sodium nitrite.

b. A 5% aqueous solution of potassium bisulfate.

c. A 12.5% aqueous solution of ammonium sulfamate.

d. An 0.5% aqueous solution of 3-methylbenzothiazolin-2-one hydrazone hydrochloride.

e. A 1% aqueous solution of ferric chloride hexahydrate.

Procedure

Operate protected against light. To 1 ml of neutral aqueous solution of the 2-amino-2-deoxyhexose, add 1 ml of reagent a and 1 ml of reagent b. Mix, let stand for 15 min, add 1 ml of reagent c, and let stand for 5 min with occasional shaking. Add 1 ml of reagent d, let stand for 1 hr, and add 1 ml of reagent e. Let stand for 30 min, and read at 655 nm.

Results

	A = 0.3 (1 cm cell) Sample, μg
Galactosamine (hydrochloride)	8.0
Glucosamine (hydrochloride)	8.5

Note

Properties of 3-methylbenzothiazolin-2-one hydrazone, see Chapter 16, Section III.

XI. HEXURONIC ACIDS

A. Naphthoresorcinol

Principle (78,79)

Condensation with naphthoresorcinol in acid medium to give a dibenzoxanthene-type dyestuff: blue color.

Reagent

Mix equal volumes of 0.2% aqueous solution of naphthoresorcinol (filtered) and of concentrated hydrochloric acid.

Procedure

To 2 ml of aqueous solution of the hexuronic acid, add 4 ml of reagent. Heat at 100° for 30 min, and cool in ice water for 5 min. Extract the color into 4 ml of n-butyl acetate, and read at 570 nm.

Results

	A = 0.3 (1 cm cell) Sample, μg
Galacturonic acid	10
Glucuronic acid	25

Note

This procedure is derived from the Tollens test for hexuronic acids (80).

B. Sodium Metaperiodate and Resorcinol

Principle and Procedure

See Chapter 9, Section V.A, colorimetric determination of 2,3-dihydroxy-1-carboxylic acids.

Results

	A = 0.3 (1 cm cell) Sample, μg
Galacturonic acid	750
Glucuronic acid	1000

XII. ALDOSES (F)

A. Hydrochloric Acid and Resorcinol

Principle (81)

Dehydration into furfural (pentoses) or 5-hydroxymethylfurfural (hexoses) in acid medium, and condensation with resorcinol to give presumably xanthenone derivatives: green (pentoses) or blue (hexoses) fluorescence.

Reagent

Dissolve 0.050 g of resorcinol in 20 ml of 2:1 mixture of concentrated hydrochloric acid and water.

Procedure

To 1 ml of aqueous solution of the aldose, add 1 ml of concentrated hydrochloric acid, fit the tubes with air-condensers, and heat at 100° for 1 hr. Cool for 5 min in ice water, add 0.5 ml of reagent and 2.5 ml of concentrated hydrochloric acid, heat at 100° for 10 min and cool for 5 min in ice water. Adjust the pH to about 8.6 with Thymol Blue as indicator: Add 3.5 ml of 10 N sodium hydroxide, then proceed dropwise with the addition, testing the pH on a spot plate. Dilute to 15 ml with water and read.

Pentoses: exc: 436 nm; em: 510 nm. Standard: A solution of 2,6-dimethyl-3,5-diacetyl-1,4-dihydropyridine in 50% ethanol.

Hexoses: exc: 366 nm; em: 427 nm. Standard: A solution of quinine sulfate in 0.1 N sulfuric acid.

Results

	Determination limits, μg	Reading 50	
		Sample, μg	Standard, μg/ml
Arabinose	4 - 12	5.3	1.0
Lyxose	2 - 6	2.7	1.12
Ribose	2 - 10	3.8	0.86
Xylose	2 - 10	4.0	1.17
Galactose	4 - 20	8.4	0.25
Glucose	2 - 10	3.8	0.25
Mannose	4 - 20	8.6	0.42

Note

No fluorescence is developed with ketoses (fructose and sorbose).

XIII. KETOSES (F)

A. Zirconyl Chloride

Principle (82)

Formation of a chelate: blue fluorescence.

$$\begin{array}{c} H \\ | \\ H-C-OH \\ | \\ C=O \\ | \\ (CHOH)_3 \\ | \\ CH_2OH \end{array} \rightleftharpoons \begin{array}{c} H-C-OH \\ \| \\ C-OH \\ | \\ (CHOH)_3 \\ | \\ CH_2OH \end{array} \xrightarrow[H_2O]{ZrOCl_2} \begin{array}{c} H \\ | \\ H-C-O \searrow \\ \| \qquad Zr-OH \\ C-O \nearrow \\ | \\ (CHOH)_3 \\ | \\ CH_2OH \end{array}$$

Reagent

An 0.1% aqueous solution of zirconyl chloride.

Procedure

To 1 ml of aqueous solution of the ketose, add 0.7 ml of 0.01 N hydrochloric acid and 0.5 ml of reagent, mix well, and add 2 ml of water. Heat at 70° for 1 hr, cool for 2 min in ice water, and read at exc: 345 nm; em: 400 nm.

Standard: A solution of quinine sulfate in 0.1 N sulfuric acid.

Results

	Determination limits, μg	Reading 50 Sample, μg	Reading 50 Standard, μg/ml
Fructose	6 - 24	9.3	0.115
Sorbose	6 - 30	12.9	0.071

Note

Under the above conditions, ribose, glucose, and 2-deoxyribose develop but a very faint fluorescence.

XIV. PENTOSES (F)

A. Hydrochloric Acid, Cyclohexane-1,3-dione, and Ammonium Acetate

Principle (83)

Dehydration into furfural in acid medium, and development of the aldehyde with cyclohexane-1,3-dione and ammonia (see Chapter 8, Section XI.A): yellow-green fluorescence.

Reagents

a. Prepare a buffer by diluting to 50 ml with water a mixture of 10 g of anhydrous sodium acetate and 10 ml of glacial acetic acid.

b. An 0.25% solution of cyclohexane-1,3-dione in 40% aqueous solution of ammonium acetate. Prepare freshly.

Procedure

To 1 ml of aqueous solution of the pentose, add 1 ml of concentrated hydrochloric acid, fit the tubes with air-condensers, and heat at 100° for 1 hr. Allow to cool, and adjust the pH with phenolphthalein as indicator: Add 1 ml of 10 N sodium hydroxide, then proceed dropwise with the addition, testing the pH on a spot plate. Add 1 ml of buffer a and 1 ml of reagent b, and heat at 60° for 30 min. Cool to 0° in ice water, add 2.5 ml of 10 N sodium hydroxide, mix, and read immediately at exc: 436 nm; em: 495 nm.

Standard: A solution of 2-methyl-5-carboxy-7-aminoquinoline in glacial acetic acid.

Results

	Determination limits, μg	Reading 50 Sample, μg	Standard, μg/ml
Arabinose	0.8 - 4.0	1.5	0.31
Lyxose	0.5 - 2.5	1.0	0.36
Ribose	0.5 - 2.5	1.0	0.40
Xylose	0.5 - 2.5	0.9	0.36

Notes

1. The fluorescence is rather unstable, and reading should be made immediately after the addition of alkali.

2. Aldohexoses are much less reactive. Determination limits are 20-100 μg for glucose and galactose, 50-500 μg for fructose, and no linear relationship is observed.

XV. 2-DEOXYOSES (F)

A. 3,5-Diaminobenzoic Acid Dihydrochloride

Principle (84)

See Chapter 8, Section XII.A, fluorimetric determination of α-methylene aldehydes: yellow-green fluorescence.

Reagent

Dissolve 2 g of 3,5-diaminobenzoic acid dihydrochloride in 57 ml of water, and add to this solution 43 ml of phosphoric acid (d = 1.71).

Procedure

To 2 ml of aqueous solution of the 2-deoxyose, add 2 ml of reagent, plug the tubes with cotton-wool, and heat at 100° for 15 min, protected against bright light. Cool for 2 min in ice water, and read at exc: 405 nm; em: 505 nm.

Standard: A solution of 2-methyl-5-carboxy-7-aminoquinoline in 10% acetic acid.

Results

	Determination limits, μg	Reading 50 Sample, μg	Reading 50 Standard, μg/ml
2-Deoxyribose	10 - 50	21	0.025
Digitoxose	20 - 100	44	0.030
2-Deoxyglucose	0.8 - 5.0	1.5	0.039

Notes

1. The reaction was applied to the determination of deoxyribonucleic acid (29,85) and of 2-deoxyglucose (86).

2. N-Acetylneuraminic acid (XX) (a sialic acid) is an α-keto acid which is readily deacetylated and decarboxylated in acid medium. The resulting 2-deoxy-4-aminooctose (XXI) then reacts with diaminobenzoic acid (87).

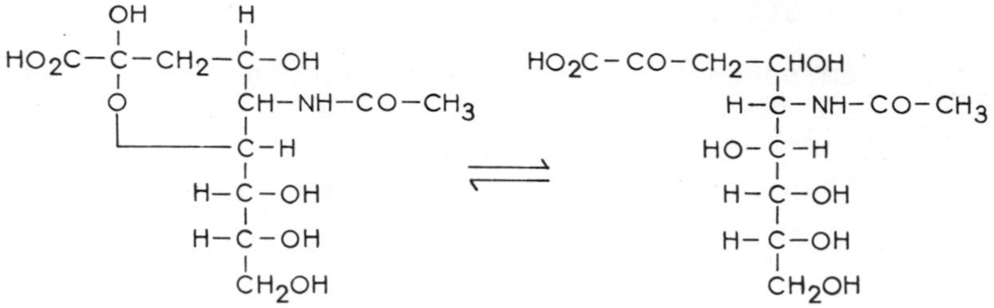

(XX)

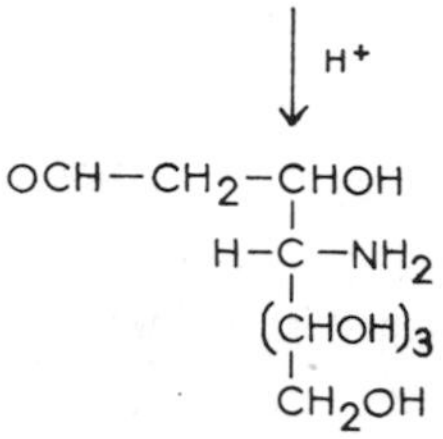

(XXI)

XVI. 2-AMINO-2-DEOXYHEXOSES (F)

A. Pyridoxal and Zinc Nitrate

Principle (88)

Reaction with pyridoxal and zinc ion in the presence of pyridine, to give a zinc-N-pyridoxylidenehexosamine chelate: blue fluorescence.

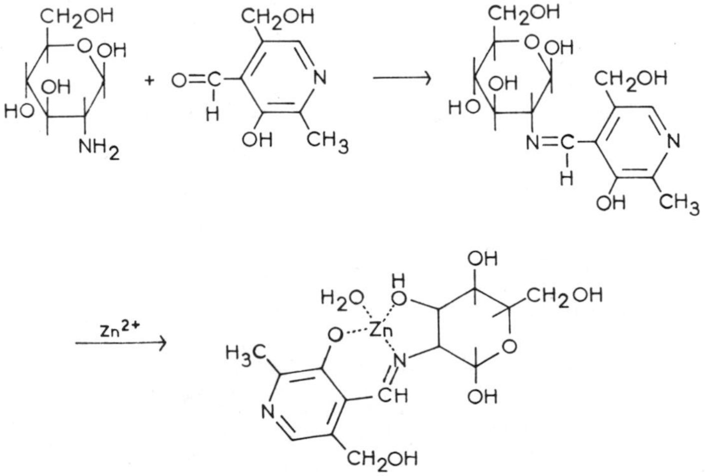

Reagents

a. A 1:4 mixture of pyridine and methanol.

b. An 0.05% solution of pyridoxal in methanol.

c. A 2% aqueous solution of zinc nitrate hexahydrate.

Procedure

To 0.5 ml of neutral aqueous solution of the 2-amino-2-deoxyhexose, add 0.5 ml of reagent *a* and 0.5 ml of reagent *b*. Let stand at room temperature for 30 min, add 0.5 ml of reagent *c*, let stand for 15 min, and add 3 ml of methanol. Read at exc: 410 nm; em: 470 nm.

Standard: To 2 ml of solution of esculin in ethanol, add 2.5 ml of 1 N sodium hydroxide.

Results

	Determination limits, μg	Reading 50	
		Sample, μg	Standard, μg/ml
Galactosamine (hydrochloride)	1.0 - 5.0	1.87	0.22
Glucosamine (hydrochloride)	0.4 - 2.0	0.75	0.08

B. Ninhydrin and Phenylacetaldehyde

Principle and Procedure

See Chapter 4, Section XII.B, fluorimetric determination of primary aliphatic amines.

Results

	Determination limits, μg	Reading 50	
		Sample, μg	Standard, μg/ml
Galactosamine (hydrochloride)	1.4 - 7.0	2.5	0.97
Glucosamine (hydrochloride)	3.0 - 15.0	4.8	1.00

XVII. HEXURONIC ACIDS (F)

A. Sodium Metaperiodate and Resorcinol

Principle and Procedure

See Chapter 9, Section X.A, fluorimetric determination of 2,3-dihydroxy-1-carboxylic acids: yellow-green fluorescence.

Results

	Determination limits, μg	Reading 50 Sample, μg	Standard, μg/ml
Galacturonic acid	2 - 10	4.7	0.077
Glucuronic acid	2 - 10	4.5	0.068

REFERENCES

1. M.A. Rosannof, *J. Amer. Chem. Soc.*, 28, 114 (1906).

2. M. Herbain, *Bull. Soc. Chim. Biol.*, 31, 1104 (1949).

3. Cf. M.Z. Barakat and M.E. Abd El-Wahab, *J. Pharm. Pharmacol.*, 3, 511 (1951).

4. G.L. Miller, *Anal. Chem.*, 31, 426 (1959).

5. Cf. W.M. Dehn and F.A. Hartman, *J. Amer. Chem. Soc.*, 36, 403 (1914).

6. H. Baljet, *Schweiz. Apotheker-Ztg.*, 56, 71, 84 (1918).

7. M. Jaffe, *Z. Physiol. Chem.*, 10, 391 (1886).

8. J. Bartos, *Ann. Pharm. Fr.*, 20, 650 (1962).

9. From A.W. Devor, *J. Amer. Chem. Soc.*, 72, 2008 (1950).

10. Cf. H. Bredereck, *Ber.*, 64, 2856 (1931); 65, 1110 (1932).

11. M.R. Shetlar and Y.F. Masters, *Anal. Chem.*, 29, 402 (1957).

12. F.B. Seibert and J. Atno, *J. Biol. Chem.*, 163, 511 (1946).

13. Z. Dische, *Mikrochemie*, 8, 4 (1930).

14. Z. Dische and E. Borenfreund, *J. Biol. Chem.*, 192, 583 (1951).

15. Z. Dische, *J. Biol. Chem.*, 167, 189 (1947).

16. J.T. Galambos, *Anal. Biochem.*, 19, 119 (1967).

17. J.K.N. Jones and J.B. Pridham, *Biochem. J.*, 58, 288 (1954).

18. A.C. Corcoran and I.H. Page, *J. Biol. Chem.*, 127, 601 (1939).

19. H. Thies and G. Kallinich, *Chem. Ber.*, 85, 438 (1952).

20. Z. Dische and E. Borenfreund, *J. Biol. Chem.*, 180, 1297 (1949).

21. A. Heyrovsky, *Chem. Listy*, 50, 1593 (1956).

22. R.G. Kulka, *Biochem. J.*, 63, 542 (1956).

23. F. Percheron, *Bull. Soc. Chim. Fr.*, 1684 (1963).

24. S. Pontremoli, E. Grazi, and A. de Flora, *Ital. J. Biochem.*, 9, 210 (1960).

25. M.V. Tracey, *Biochem. J.*, 47, 433 (1950).

26. Cf. W.F. Foley, Jr., G.E. Sanford, and H. McKennis, *J. Amer. Chem. Soc.*, 74, 5489 (1952).

27. J.H. Roe and E.W. Rice, *J. Biol. Chem.*, 173, 507 (1948).

28. From W. Mejbaum, *Z. Physiol. Chem.*, 258, 117 (1939).

29. M. Pesez, *Bull. Soc. Chim. Biol.*, 32, 701 (1950).

30. G. Ceriotti, *J. Biol. Chem.*, 214, 59 (1955).

31. L. Bolognani, G. Coppi, and V. Zambotti, *Experientia*, 17, 67 (1961).

32. R.R. Bridges, *Anal. Chem.*, 24, 2004 (1952).

33. R. Sawamura and T. Koyama, *Yakugaku Zasshi*, 84, 82 (1964).

34. T. Koyama and R. Sawamura, *Chem. Pharm. Bull. (Tokyo)*, 14, 1054 (1966).

35. L. Sattler and F.W. Zerban, *J. Amer. Chem. Soc.*, 72, 3814 (1950).

36. E. Voute, *Pharm. Weekblad*, 88, 144 (1953).

37. L. Sattler and F.W. Zerban, *Science*, 108, 207 (1948).

38. M. Gibbons, *Analyst*, 80, 268 (1955).

39. B. Klein and M. Weissman, *Anal. Chem.*, 25, 771 (1953).

40. W.E. Trevelyan and J.S. Harrison, *Biochem. J.*, 50, 298 (1952).

41. T. Momose, Y. Ueda, K. Sawada, and A. Sugi, *Pharm. Bull.*, 5, 31 (1957).

42. H. Hörmann and I.A. Siddiqui, *Justus Liebigs Ann. Chem.*, 714, 174 (1968).

43. J.M. Webb and H.B. Levy, *J. Biol. Chem.*, 213, 107 (1955).

44. K. Himmelspach and O. Westphal, *Justus Liebigs Ann. Chem.*, 668, 165 (1963).

45. I. Fromme, K. Himmelspach, O. Lüderitz, and O. Westphal, *Angew. Chem.*, 69, 643 (1957).

46. I. Fromme, O. Lüderitz, H. Stierlin, and O. Westphal, *Biochem. Z.*, 330, 53 (1958).

47. J. Bartos, *Ann. Pharm. Fr.*, 21, 603 (1963).

48. From J. Bartos, *Ann. Pharm. Fr.*, 21, 603 (1963).

49. Cf. M. Pesez, *Ann. Pharm. Fr.*, 10, 104 (1952).

50. J. Bartos, *Ann. Pharm. Fr.*, 22, 223 (1964).

51. P. Mesnard and G. Devaux, *Chim. Anal.*, 44, 287 (1962).

52. P. Mesnard and G. Devaux, *C. R. Acad. Sci., Paris*, 253, 497 (1961).

53. P. Mesnard and G. Devaux, *C. R. Acad. Sci., Paris*, 256, 1551 (1963).

54. P. Mesnard and G. Devaux, *Bull. Soc. Pharm. Bordeaux*, 103, 101 (1964).

55. D. Aminoff, *Biochem. J.*, 81, 384 (1961).

56. H. Schmidt, *Z. Anal. Chem.*, 178, 173 (1960).

57. B.G. Tarladgis, A.M. Pearson, and L.R. Dugart, Jr., *J. Amer. Oil Chemists' Soc.*, 39, 34 (1962).

58. E.D. Wills, *Biochim. Biophys. Acta*, 84, 475 (1964).

59. P. Mesnard, G. Devaux, and J. Fauquet, *Bull. Soc. Pharm. Bordeaux*, 104, 13 (1965).

60. From H. Scherz, G. Stehlik, E. Bancher, and K. Kaindl, *Mikrochim. Acta*, 915 (1967).

61. G. Delmon, R. Babin, and P. Blanquet, *Bull. Soc. Pharm. Bordeaux*, 91, 211 (1953).

62. Cf. R.E. Deriaz, M. Stacey, E.G. Teece, and L.F. Wiggins, *J. Chem. Soc.*, 1222 (1949); W.G. Overend, F. Shafizadeh, and M. Stacey, *J. Chem. Soc.*, 1027 (1950); K. Burton, *Biochem. J.*, 62, 315 (1956); L. Birkofer and R. Dutz, *Justus Liebigs Ann. Chem.*, 657, 94 (1962).

63. C. Izard-Verchere, P. Rumpf, and C. Viel, *Bull. Soc. Chim. Fr.*, 2134 (1971).

64. T. Pederson, *Anal. Biochem.*, 28, 35 (1969).

65. Cf. M. Pesez and J. Bartos, *Ann. Pharm. Fr.*, 23, 281 (1965).

66. R. Belcher, A.J. Nutten, and C.M. Sambrook, *Analyst*, 79, 201 (1953).

67. L.A. Elson and W.T.J. Morgan, *Biochem. J.*, 27, 1824 (1933).

68. H. Pauly and E. Ludwig, *Z. Physiol. Chem.*, 121, 176 (1922).

69. R. Boyer and O. Fürth, *Biochem. Z.*, 282, 242 (1935).

70. B. Schloss, *Anal. Chem.*, 23, 1321 (1951).

71. Cf. G. Blix, *Acta Chem. Scand.*, 2, 467 (1948); H.N. Horowitz, M. Ikawa, and M. Fling, *Arch. Biochem.*, 25, 226 (1950); J.P. Johnston, A.G. Ogston, and J.E. Stanier, *Analyst*, 76, 88 (1951); C.J.M. Rondle and W.T.J. Morgan, *Biochem. J.*, 61, 586 (1955).

72. J. Immers and E. Vasseur, *Nature*, 165, 898 (1950); Z. Stary, M. Yenson, S. Lisie, and M. Bilen, *Bull. Faculté Med. Istanbul*, 14, 392, 401 (1951); C. Nervi, *Boll. Soc. Ital. Biol. Sper.*, 30, 963 (1954); O.W. Neuhaus and M. Letzring, *Anal. Chem.*, 29, 1230 (1957).

73. J. Ludowieg and J.D. Benmaman, *Anal. Biochem.*, 19, 80 (1967).

74. H. Wachsmuth and L. Van Koeckhoven, *J. Pharm. Belg.*, 14, 79 (1959); F. Ordoveza and P.W. West, *Anal. Chim. Acta*, 30, 227 (1964).

75. M. Pesez, J. Bartos, and A. Sezerat, *Bull. Soc. Chim. Fr.*, 567 (1961).

76. Z. Dische and E. Borenfreund, *J. Biol. Chem.*, 184, 517 (1950).

77. A. Tsuji, T. Kinoshita, and M. Hoshino, *Chem. Pharm. Bull.*, 17, 217, 1505 (1969).

78. T. Miettinen, V. Ryhännen, and H. Salomaa, *Ann. Med. Exptl. Biol. Fenniae*, 35, 173 (1957).

79. Cf. A.H. Guerrero and R.T. Williams, *Nature*, 161, 930 (1948).

80. B. Tollens, *Ber.*, 41, 1788 (1908).

81. From C.J. Rogers, C.W. Chambers, and N.A. Clarke, *Anal. Chem.*, 38, 1851 (1966).

82. Cf. H. Trapmann and V.S. Sethi, *Z. Anal. Chem.*, 248, 314 (1969).

83. M. Pesez and J. Bartos, *Talanta*, 16, 331 (1969).

84. M. Pesez and J. Bartos, *Talanta*, 14, 1097 (1967).

85. J.M. Kissane and E. Robins, *J. Biol. Chem.*, 233, 184 (1958); R.T. Hinegardner, *Anal. Biochem.*, 39, 197 (1971).

86. F.B. Cramer and G.A. Neville, *J. Franklin Inst.*, 256, 379 (1953); M. Blecher, *Anal. Biochem.*, 2, 30 (1961).

87. H.H. Hess and E. Rolde, *J. Biol. Chem.*, 239, 3215 (1964).

88. M. Maeda, T. Kinoshita, and A. Tsuji, *Anal. Biochem.*, 38, 121 (1970).

Chapter 15

STEROIDS

I. INTRODUCTION

Rules of steroid nomenclature were published in 1960 (1), and more recently revised definitive rules were published (2). Only those rules which may be necessary for the understanding the methods included in this chapter are recalled here.

Steroids are numbered and rings are lettered as in formula (I).

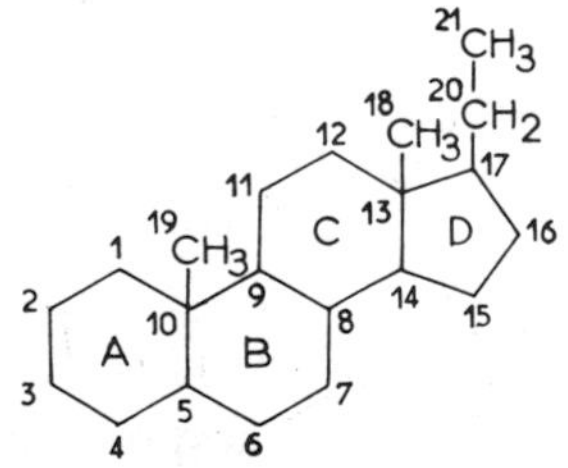

(I) Pregnane

If one or more of the carbon atoms shown in (I) are not present and a steroid name is used, the numbering of the remainder is undisturbed.

The parent tetracyclic hydrocarbon without the methyl groups at C-10 and C-13, and without a side chain at C-17, is named gonane (II). Formulas (III) and (IV) correspond respectively to parent hydrocarbons estrane and androstane. Formula (I) corresponds to pregnane.

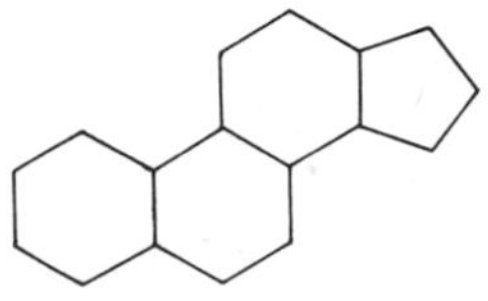

(II) Gonane 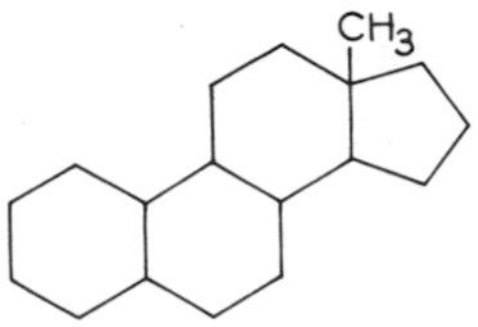 (III) Estrane 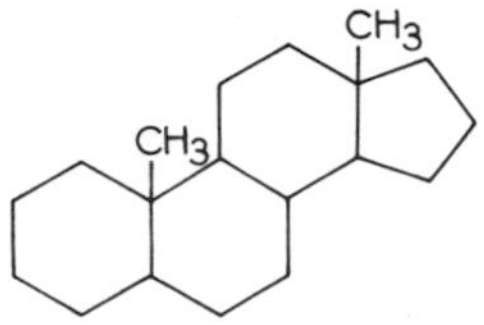 (IV) Androstane

Elimination of a methylene group from an alkyl group is indicated by the prefix "nor-." This prefix is in all cases preceded by the number of the carbon atom which disappears. For instance, estrane might be considered as 19-norandrostane and gonane is 18-norestrane.

Two other parent hydrocarbons are often encountered:

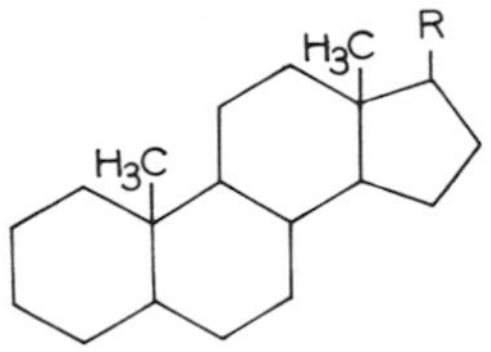

R= $CH(CH_3)-CH_2-CH_2-CH_3$
Cholane

R= $CH(CH_3)-CH_2-CH_2-CH_2-CH(CH_3)_2$
Cholestane

An atom or a group attached to a ring is termed α if it lies below the plane of the paper, and the bond is shown as a broken line. It is termed β if it lies above the plane of the paper, and the bond is shown as a solid line. When the configuration is unknown, it is represented by ξ and the bond is a waved line.

Unless stated to the contrary, use of a steroid name implies that atoms or groups attached at the ring-junction positions 8, 9, 10, 13, and 14 are oriented 8β, 9α, 10β, 13β, 14α. A carbon chain attached at position 17 is assumed to be β-oriented. The configuration of hydrogen or a substituent at the ring-junction position 5 can be α or β,

5α corresponding to A/B trans junction, 5β to A/B cis junction.

For a long time, the term "etiocholane" has been used to designate 5β-androstane, the term "androstane" thus implying the α configuration. Likewise, "allopregnane" has designated 5α-pregnane, the term "pregnane" implying the β orientation. This explains why the orientation of the hydrogen at position 5 is sometimes omitted for androstane and pregnane derivatives in papers dealing with steroids.

Unsaturation is indicated by changing terminal "-ane" to "-ene," "-adiene," "-atriene," etc. Although the use of Δ to designate unsaturation is not recommended by IUPAC, we have sometimes maintained it for the sake of brevity.

Most substituents can be designated either as suffixes or as prefixes. Halogen, alkyl, and nitro groups can be named only as prefixes. When possible, one type of substituent must be designated as suffix. When more than one type of substituent is present that could be designated as suffix, one type only may be so expressed and the other types must be designated as prefixes. Choice for suffix is made according to an order of preference that is laid down in the IUPAC Rules (2).

Many compounds listed in this chapter under Results sections are designated according to these rules. However, we have retained the trivial names for various important derivatives, either natural or synthetic, when they are commonly used by clinical chemists and pharmacists. The following table gives their corresponding systematic names.

Adrenosterone	4-Androstene-3,11,17-trione
Chlormadinone	6-Chloro-17-hydroxy-4,6-pregnadiene-3,20-dione
Cholesterol	5-Cholesten-3β-ol
Corticosterone	11β,21-Dihydroxy-4-pregnene-3,20-dione
Cortisone	17α,21-Dihydroxy-4-pregnene-3,11,20-trione
Dehydrocholic acid	3,7,12-Trioxo-5β-cholan-24-oic acid
Dehydroepiandrosterone	3β-Hydroxy-5-androsten-17-one
Deoxycorticosterone	21-Hydroxy-4-pregnene-3,20-dione
Dexamethasone	9α-Fluoro-16-methyl-11β,17α,21-trihydroxy-1,4-pregnadiene-3,20-dione (9α-Fluoro-16α-methylprednisolone)
Epiandrosterone	3β-Hydroxyandrostan-17-one
Ergosterol	5,7,22-Ergostatrien-3β-ol
Estrone	3-Hydroxy-1,3,5(10)-estratrien-17-one
Ethisterone	17α-Ethynyl-17β-hydroxy-4-androsten-3-one (17α-Ethynyltestosterone)
Fluocinolone	6α,9α-Difluoro-11β,16α,17α,21-tetrahydroxy-1,4-pregnadiene-3,20-dione
Fluocinolone acetonide	Fluocinolone cyclic 16,17-acetal with acetone
Hydrocortisone	11β,17α,21-Trihydroxy-4-pregnene-3,20-dione
Norethandrolone	17α-Ethyl-19-nortestosterone
Norethindrone	19-Norethisterone
Norethynodrel	17α-Ethynyl-17β-hydroxy-5(10)-estren-3-one
Oxymetholone	17α-Methyl-17β-hydroxy-2-(hydroxymethylene)androstan-3-one
Prednisolone	11β,17α,21-Trihydroxy-1,4-pregnadiene-3,20-dione
Prednisone	17α,21-Dihydroxy-1,4-pregnadiene-3,11,20-trione

Progesterone	4-Pregnene-3,20-dione
Testosterone	17β-Hydroxy-4-androsten-3-one

Strictly speaking, some of these names, such as 19-norethisterone, and others unlisted above do not obey the rules, but we felt that, for a better understanding, some "semitrivial" names should be kept up. It may be worth comparing results obtained with two closely related compounds, such as testosterone and 19-nortestosterone. This latter name certainly makes the comparison easier than the systematic: 17β-hydroxy-4-estren-3-one.

A. Colorimetry

The methods herein proposed are mainly based on reactions allowed by the presence of hydroxy or keto groups on the steroidal skeleton. Some properties peculiar to the 17-ketol grouping of corticosteroids are also turned to account. The sample size for A = 0.3 depending often mainly on the compound tested, the corresponding weights are not given here.

Three of the procedures described for alcohols can readily be extended to hydroxysteroids: reaction with vanadium oxinate, whose sensitivity can be enhanced by subsequent development with p-nitrobenzenediazonium ion of the 8-hydroxyquinoline liberated from the colored species, or esterification with 3,5-dinitrobenzoyl chloride and development with piperazine in nonaqueous medium, thus avoiding elimination of the excess reagent.

The hydroxyl group can also be esterified with 5-(p-sulfamoylphenylazo)salicylic acid (Salazosulfamide) in the presence of p-toluenesulfonyl chloride; the ester is then extracted into chloroform and developed with monoethanolamine, but aliphatic alcohols react also.

3-Hydroxysteroids develop a pink to yellow-brown color with the Deniges mercuric sulfoacetic reagent, and the reaction is specific for this class of compounds.

Like other carbonyl compounds, 3-, 17-, or 20-ketosteroids can be determined through the formation of hydrazones. They condense with p-nitrophenylhydrazine or 2,4-dinitrophenylhydrazine in acetic-hydrochloric acid, or with p-nitrophenylhydrazine in nonaqueous medium, the color then being developed with benzyltrimethylammonium hydroxide. The 2,4-dinitrophenylhydrazone formed under suitable conditions can also be developed as a Meisenheimer-like complex with nitromethane and a base. For a given procedure, the sensitivity depends on the location of the carbonyl group, and with 3-ketosteroids on the unsaturation of ring A.

Isoniazid gives colored hydrazones only with Δ^4- or $\Delta^{1,4}$-3-ketosteroids.

Formation of Schiff bases with dimethyl-p-phenylenediamine allows selective determination of Δ^1- or Δ^4- or $\Delta^{4,6}$-3-ketosteroids. $\Delta^{5(10)}$-3-keto-19-norsteroids can also be determined after isomerization in acid medium to the corresponding Δ^4-3-ketosteroids.

The Zimmermann reaction is based on oxidation by the excess of reagent of the Meisenheimer-like complex formed upon condensation of an α-methylene ketone with m-dinitrobenzene in alkaline medium. Suitable selection of the operating conditions and of the m-dinitro reagent allows more or less sensitive and selective determination of 3- and 17-ketosteroids.

m-Dinitrobenzene in ethanol-pyridine gives much more sensitive results with 17- than with 3-ketosteroids. For these latter, however, the sensitivity is greatly enhanced when passing from the normal to the 19-nor series, which behave like 17-keto derivatives. These differences are much less important when the reaction is performed with the same reagent in dimethylformamide solution, or in pyridine in the presence of benzyltrimethylammonium hydroxide.

The color developed with 1,3,5-trinitrobenzene in buffered medium is unusually stable (loss lower than 10% within 48 hr). For 17-ketosteroids, the sensitivity depends upon the orientation of the hydroxyl group at position 3. For 3-ketosteroids, it depends on the unsaturation of ring A. 20-Ketosteroids react but weakly.

Picric acid in dimethylformamide solution also allows sensitive but rather poorly selective determinations.

Through an unknown mechanism, 3- or 17-ketosteroids react with o-nitrobenzaldehyde in alkaline medium, the order of sensitivities being 17-keto > 3-keto with saturated ring A > 3-keto with one unsaturation.

Only 3-ketosteroids react with 2,6-di-tert-butyl-p-cresol in alkaline medium in the presence of hydrogen peroxide. The mechanism of the reaction is unknown. The color and sensitivity vary with the unsaturation of ring A, and with the character of the substituent at position 11.

3-Ketosteroids with unsaturated ring A can be determined with 4-aminoantipyrine in acid medium (the formation of a Schiff-like base may be postulated), and with Blue Tetrazolium in alkaline medium, but this latter reaction is far from specific.

Only Δ^4-ketosteroids react with sodium nitroprusside in alkaline medium (poorly sensitive results), with monoethanolamine in the presence of potassium hydroxide, and with tetramethylammonium hydroxide.

In alkaline medium, the pink color of thionine is partly discharged upon the action of a $\Delta^{5(10)}$-3-keto-19-norsteroid, and the reaction is almost specific for this class of compounds.

The 17-ketol grouping of corticosteroids is a reductant and can be determined as such. It reduces Blue Tetrazolium into the corresponding formazan, and p-nitroso-N,N-dimethylaniline into dimethyl-p-phenylenediamine, which is then developed with phenol in the presence of an oxidant. These reactions being performed in alkaline medium, readily hydrolyzable 21-esters react also.

Only 21-unesterified corticosteroids reduce molybdate ion in acid medium. They can also be determined through periodate oxidation of the 17-ketol grouping and development of the formaldehyde formed with phenylhydrazine and ferricyanide (Schryver reaction).

Some reactions can be applied only to compounds bearing both a hydroxyl group and the α-ketol grouping at position 17. These derivatives react with phenylhydrazine in sulfuric acid medium (Porter-Silber reaction), and with 2-hydrazinobenzothiazole or 3-methylbenzothiazolin-2-one hydrazone in the presence of an oxidant. These two latter reactions give also positive results with sugars.

B. Fluorimetry

Only a few methods are available.

The method of determination of alcohols with vanadium oxinate and magnesium ion can be extended to hydroxysteroids, but the reaction is poorly sensitive.

The species obtained by reacting Δ^4- or $\Delta^{1,4}$-3,11-diketosteroids with 2,6-di-tert-butyl-p-cresol in alkaline medium in the presence of hydrogen peroxide (see preceding section, Colorimetry) are also fluorescent.

In alkaline medium, the fluorescence of Pyronine G (also named Pyronine Y) is partly discharged upon the action of a $\Delta^{5(10)}$-3-keto-19-norsteroid (compare with Colorimetry).

The periodate oxidation of the 17-ketol grouping of corticosteroids gives rise to formaldehyde, which is then developed with ethyl acetoacetate and ammonia.

II. HYDROXYSTEROIDS

A. Vanadium Oxinate

Principle and Procedure

See Chapter 2, Section II.B, colorimetric determination of alcohols.

Results

	A = 0.3 (1 cm cell) Sample, μg
Cholesterol	1000
Testosterone	485
17α-Hydroxyprogesterone	2300
Cortisone acetate	670
20β-Hydroxypregnan-3-one	800
Hydrocortisone acetate	675
Pregnane-3α,20α-diol	375

Note

The primary hydroxyl group at position 21 of ketolsteroids does not react. No color is developed with 11-dehydrocorticosterone and 11-deoxycorticosterone. With 11-deoxy-17-hydroxycorticosterone, the color is too transient to allow measurements.

B. Vanadium Oxinate and p-Nitrobenzenediazonium Fluoborate

Principle and Procedure

See Chapter 2, Section II.C, colorimetric determination of alcohols.

Results

	A = 0.3 (1 cm cell) Sample, μg
Cholesterol	68
Dehydroepiandrosterone	30
Testosterone	40

C. 5-(p-Sulfamoylphenylazo)salicylic Acid, p-Toluenesulfonyl Chloride, and Monoethanolamine

Principle (3)

Esterification with 5-(sulfamoylphenylazo)salicylic acid (salazosulfamide) in pyridine medium in the presence of p-toluenesulfonyl chloride, extraction of the ester into chloroform, and development with monoethanolamine: yellow color.

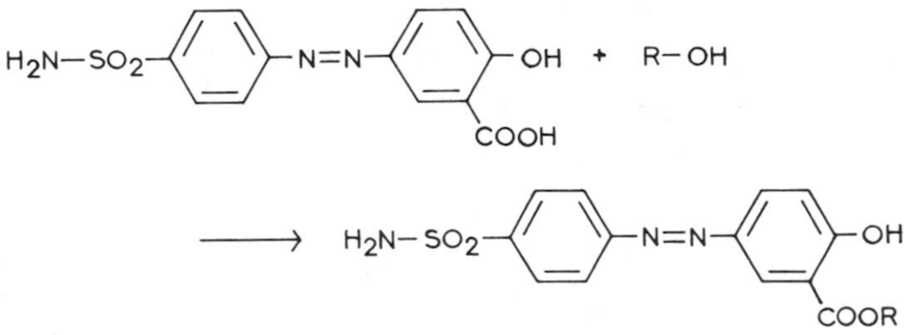

H_2N-SO_2 — N=N — OH, COOH + R—OH

⟶ H_2N-SO_2 — N=N — OH, COOR

Reagent

The reagent should be prepared in each tube separately. To 0.2 g of salazosulfamide (Chapter 17, Section I.G), add 0.04 g of p-toluenesulfonyl chloride and 1 ml of pyridine. Stopper the tube, mix, and let stand for 30 min before use.

Procedure

To the reagent, add 0.5 ml of sample solution in pyridine, and heat at 100° for 10 min. Add 0.5 ml of water, heat at 100° for 5 min more, and allow to cool. Add 1 ml of monoethanolamine, and transfer the solution to a 25 ml separatory funnel. Rinse the tube with 2 ml of chloroform, then 5 ml of water, pour the rinsings into the funnel, shake, and let stand until decantation is achieved. Collect the organic layer and repeat the extraction twice more with 2 ml portions of chloroform. Combine the chloroform extracts, wash with 5 ml of water, allow the phases to separate, and collect the organic layer into a 10 ml volumetric flask containing 1 ml of ethanol. Dilute to the mark with chloroform, add 1 ml of monoethanolamine, mix, and read at 425 nm.

Results

	A = 0.3 (1 cm cell) Sample, μg
Cholesterol	220
Dehydroepiandrosterone	230
Testosterone	150

Notes

1. Under the given conditions, esters do not suffer transesterification.

2. The reaction can be applied to aliphatic alcohols. A = 0.3 is given by 36 μg of ethanol.

3. The above procedure avoids the preparation of the chloride of 5-(p-sulfamoylphenylazo)salicylic acid. p-Toluenesulfonyl chloride does not intervene as a catalyst. The reaction appears to involve *in situ* generation of the acid anhydride (4):

$$(R-CO)_2O \xrightarrow{R'-OH} R-COOH + R-CO_2R'$$

(R–COOH → $(R-CO)_2O$ with C_5H_5N, Ts Cl)

D. 3,5-Dinitrobenzoyl Chloride and Piperazine

Principle and Procedure

See Chapter 2, Section III.C, colorimetric determination of primary and secondary alcohols.

Results

	A = 0.3 (1 cm cell) Sample, μg
Cholesterol	75
Dehydroepiandrosterone	65
Testosterone	60

III. 3-HYDROXYSTEROIDS

A. Mercuric Sulfoacetic Reagent

Principle (5,6)

Reaction with Deniges mercuric sulfoacetic reagent: pink or yellow-brown color.

Reagent

Triturate until dissolution 0.5 g of yellow mercuric oxide in a cooled mixture of 10 ml of water and 2 ml of concentrated sulfuric acid. Dilute 1 ml of this solution to 100 ml with a 2:1 mixture of concentrated sulfuric acid and glacial acetic acid. Allow to settle for one night, and collect the clear supernatant.

Procedure

To 1 ml of sample solution in glacial acetic acid, add 5 ml of reagent. Let stand at 20° for the given length of time (see Results), and read.

Results

	Reaction time, min	λ Max, nm	A = 0.3 (1 cm cell) Sample, μg
Cholesterol	15	500	33
7-Dehydro-cholesterol	30	500	88
Ergosterol	15	400	60

Notes

1. This reaction enables the determination of free and total cholesterol in an 0.1 ml sample of whole blood. Digitonin does not interfere (7).

2. No color or only a very slight one is developed with 19-nor steroids.

IV. KETOSTEROIDS

A. 2,4-Dinitrophenylhydrazine in Acetic-Hydrochloric Acid

Principle

See Chapter 8, Section II.B, colorimetric determination of aldehydes and ketones.

Reagents

a. An 0.1% solution of 2,4-dinitrophenylhydrazine in an 0.5% dilution of concentrated hydrochloric acid in glacial acetic acid.

b. A 1 M solution of potassium acetate in ethanol.

Procedure

To 1 ml of sample solution in glacial acetic acid, add 5 ml of reagent a. Heat at 100° for 15 min, cool to 20°, add 0.5 ml of reagent b, mix, and read at 412 nm.

Results

	A = 0.3 (1 cm cell) Sample, μg
17β-Hydroxyandrostan-3-one	190
Testosterone	78
17α-Methyltestosterone	78
Norethandrolone	129
Norethynodrel[a]	105
17α-Ethynyl-17β-hydroxy-4,9,11-estratrien-3-one[a]	25
Dehydroepiandrosterone	800
Estrone	705
3β-Hydroxypregnan-20-one	370
4-Androstene-3,17-dione	60
1,4-Androstadiene-3,17-dione	330

[a]Read at 440 nm.

	A = 0.3 (1 cm cell) Sample, μg
Pregnane-3,20-dione	165
Hydrocortisone	89
Prednisolone	345
Cortisone	115
Prednisone	260

Note

From the above table, it may be seen that the sensitivity of the reaction depends on the position of the carbonyl group: it is more sensitive with 3- than with 17- or 20-ketosteroids. With 3-ketosteroids, the order of sensitivities is the following: Δ^4 > Saturated > $\Delta^{1,4}$. Conjugated double bonds (17α-ethynyl-17β-hydroxy-4,9,11-estratrien-3-one) enhance the sensitivity.

B. p-Nitrophenylhydrazine in Acetic-Hydrochloric Acid

Principle and Procedure

See Chapter 8, Section II.C, colorimetric determination of aldehydes and ketones.

Results

	λ Max, nm	A = 0.3 (1 cm cell) Sample, μg
Testosterone	405	415
Norethynodrel	400	350
17α-Ethynyl-17β-hydroxy-4,9,11-estratrien-3-one	415	28
Hydrocortisone	405	72

C. p-Nitrophenylhydrazine and Benzyltrimethylammonium Hydroxide

Principle and Procedure

See Chapter 8, Section II.D, colorimetric determination of aldehydes and ketones.

Results

	Reaction time, min	Color	λ Max, nm	A = 0.3 (1 cm cell) Sample, μg
17β-Hydroxy-androstan-3-one	30	Red	520	33
Dehydroepi-androsterone	50	Pink	530	215
Estrone	60	Pink	530	245
Progesterone	20	Violet	545	19
Dexamethasone	30	Violet	570	34
Cortisone	20	Violet	550	21.5
Dehydrocholic acid	30	Red	510	46.5

Note

As may be seen from the table, the reaction is much more sensitive with 3- than with 17-ketosteroids.

D. 2,4-Dinitrophenylhydrazine, Nitromethane, and Benzyltrimethylammonium Hydroxide

Principle and Procedure

See Chapter 8, Section VII.A, colorimetric determination of ketones.

Results

	Temperature, °C	Reaction time, min	A = 0.3 (1 cm cell) Sample, μg
17β-Hydroxy-androstan-3-one	18 - 24	30	23.5
Testosterone	18 - 24	10	21.6
Dehydroepi-androsterone[a]	100	15	15.5
Estrone[a]	100	15	15.0
Progesterone[a]	100	15	8.4

[a]Fit the tubes with air-condensers.

V. 3- OR 17-KETOSTEROIDS

A. m-Dinitrobenzene in Ethanol-Pyridine

Principle (8)

Reaction with m-dinitrobenzene in alkaline medium to give a Meisenheimer-like σ complex, which is then oxidized by the excess reagent (Zimmermann reaction): pink-violaceous color.

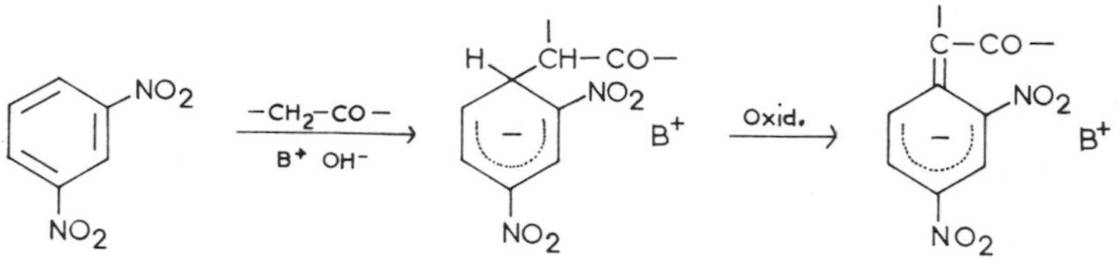

Reagent

A 2% solution of m-dinitrobenzene in pyridine.

Procedure

To 2 ml of sample solution in ethanol, add 0.5 ml of 1 N sodium hydroxide and 0.5 ml of reagent. Let stand for 1 hr in the dark, and add 7 ml of ethanol. Read at 500 nm.

Results

	A = 0.3 (1 cm cell) Sample, μg
Testosterone	2980
17α-Methyltestosterone	2460
19-Nortestosterone	330
Norethandrolone	240
Norethynodrel	215
Dehydroepiandrosterone	620
11β-Hydroxy-4-androstene-3,17-dione	825
Adrenosterone	94
19-Norprogesterone	275
Hydrocortisone	4500
Cortisone	1480

The reaction is negative with 17β-hydroxy-1,4-androstadiene-3-one (dehydrotestosterone).

Notes

1. The reaction is much more sensitive with 17- than with 3-ketosteroids. On the other hand, with 3-ketosteroids, the sensitivity is strongly enhanced when passing from the normal to the 19-nor series.

2. Janovsky and Zimmermann reactions, see Chapter 16, Section VI.

B. m-Dinitrobenzene in Dimethylformamide

Principle (9)

See preceding method: pink-violaceous color.

Reagent

A 2% solution of m-dinitrobenzene in dimethylformamide. Prepare freshly, and keep in the dark.

Procedure

To 0.5 ml of sample solution in dimethylformamide, add 0.1 ml of reagent and 0.05 ml of 1 N sodium hydroxide. Let stand for 5 min in the dark, add 4 ml of benzene, mix, let stand for 5 min in the dark, and add 1 ml of ethanol and 1 ml of pyridine. Read at 560 nm after 5 min in the dark.

Results

	A = 0.3 (1 cm cell) Sample, μg
17β-Hydroxyandrostan-3-one	55
Norethynodrel	250
17α-Ethynyl-17β-hydroxy-4,9,11-estratrien-3-one	95
Dehydroepiandrosterone	33
Progesterone	30
Prednisolone	2000
Cortisone	300

Note

Janovsky and Zimmermann reactions, see Chapter 16, Section VI.

C. m-Dinitrobenzene in Pyridine and Benzyltrimethylammonium Hydroxide

Principle

See Section V.A: pink to violet color.

Reagents

a. A 1% solution of m-dinitrobenzene in pyridine.

b. A 40% solution of benzyltrimethylammonium hydroxide (Chapter 17, Section III) in methanol.

Procedure

Operate protected against light. To 1 ml of sample solution in pyridine, add 0.5 ml of reagent a and 0.2 ml of reagent b. Let stand at room temperature for 5 min, immerse the tubes in ice water, add 3 ml of ethanol prechilled to 0°, and read immediately.

Results

	λ Max, nm	A = 0.3 (1 cm cell) Sample, μg
Testosterone	575	24
19-Nortestosterone	550	20.5
Norethandrolone	550	24
Norethynodrel	550	100
Dehydroepiandrosterone	550	22
Progesterone	500	11.5
19-Norprogesterone	500	11
Deoxycorticosterone	575	26
Adrenosterone	550	13
Cortisone	560	36

Notes

1. $\Delta^{1,4}$-3-Ketosteroids develop a bright yellow color. Determinations are possible with reading at 400 nm when

methanol is used instead of ethanol for the last dilution. A = 0.3 is given by 60 μg of prednisone or 80 μg of prednisolone.

2. Janovsky and Zimmermann reactions, see Chapter 16, Section VI.

D. 1,3,5-Trinitrobenzene

Principle (10)

See Principle, m-dinitrobenzene, Section V.A: red-orange color.

Reagents

a. Dissolve 0.05 g of 1,3,5-trinitrobenzene in 15 ml of dimethylformamide, and dilute to 50 ml with water.

b. A 10% aqueous solution of monosodium phosphate dihydrate.

Procedure

To 0.5 ml of sample solution in dimethylformamide, add 0.5 ml of water, 0.5 ml of reagent a, and 0.2 ml of 0.5 N sodium hydroxide. Heat at 60° for 15 min, protected against light, cool for 5 min in a water bath at 20°, and add 0.2 ml of reagent b and 5 ml of water. Read at 475 nm.

Results

	A = 0.3 (1 cm cell) Sample, μg
17β-Hydroxyandrostan-3-one	44
Testosterone	320
17α-Methyltestosterone	280
19-Nortestosterone	110
Norethynodrel	43
Dehydroepiandrosterone	83
Estrone	110
Androstane-3,17-dione	34

Notes

1. The addition of reagent b discharges the red color of the blank.

2. The major feature of this reaction lies in the fact that the color developed is unusually stable. The optical density lowers by less than 10% within 48 hours.

3. The following main conclusions may be drawn from the results obtained with 47 compounds (10). (Absorbances are calculated for 10^{-6} mole.)

The absorbance is about 1.9 for 3-ketosteroids with saturated ring A (17β-hydroxyandrostan-3-one and related compounds). It is slightly enhanced by a 5-10 double bond, and reduced to about 0.3 for 4-unsaturated compounds. In this latter case, however, the 19-nor derivatives give a value of about 0.75. 1,4-Unsaturated 3-ketosteroids such as 17β-hydroxy-1,4-androstadien-3-one do not react.

With 17-ketosteroids, the sensitivity depends upon the orientation of the hydroxyl group at position 3. For instance, the absorbance is 0.72 for androsterone (3α-OH) and 1.02 for epiandrosterone (3β-OH). It is about 0.8 for derivatives with an aromatic ring A, with the exception of 16α-chloromethylestrone (A = 0.23, unstable color).

20-Ketosteroids react but weakly, with the exception of 3α,11β,17α-trihydroxypregnan-20-one (A = 0.52).

7-, 11-, or 16-Ketosteroids do not react.

With 3,17-diketosteroids, the absorbance is approximately the sum of the absorbances of the corresponding 3- and 17-ketosteroids when ring A is saturated. 1-, 4-, or 1,4-Unsaturation reduces the sensitivity.

Whereas the carbonyl group at position 11 does not react, it strongly enhances the intensity of the color given by 3,17-diketosteroids: A = 1.45 for 4-androstene-3,17-dione and 3.1 for adrenosterone (4-androstene-3,11,17-trione).

The intensity of the color developed by 4-unsaturated 3,20-diketosteroids is comparable to that given by 4-unsaturated 3-ketosteroids.

4. 1,3,5-Trinitrobenzene also reacts with cardiac glycosides (11).

5. Janovsky and Zimmermann reactions, see Chapter 16, Section VI.

E. Picric Acid in Dimethylformamide and Benzyltrimethylammonium Hydroxide

Principle

See Principle, m-dinitrobenzene, Section V.A: orange color.

Reagents

a. A 1% solution of picric acid in dimethylformamide.

b. Dilute 1 ml of 40% solution of benzyltrimethylammonium hydroxide (Chapter 17, Section III) in methanol to 20 ml with dimethylformamide. Prepare freshly before use.

Procedure

To 0.5 ml of sample solution in dimethylformamide, add 0.2 ml of reagent a and 0.2 ml of reagent b. Let stand at room temperature for 2 min, and add 4 ml of water. Read at 490 nm after 5 min.

Results

	A = 0.3 (1 cm cell) Sample, μg
17β-Hydroxyandrostan-3-one	116
Testosterone	91
6-Dehydrotestosterone	40
Norethynodrel[a]	62
17α-Ethynyl-17β-hydroxy-4,9,11-estratrien-3-one[a]	35
Epiandrosterone	30
Hydrocortisone	167

[a]Read after 2 min.

	A = 0.3 (1 cm cell) Sample, μg
6-Dehydrohydrocortisone	67
Chlormadinone	28
Cortisone	105
6-Dehydrocortisone	42
Prednisone	165

Notes

1. Picric acid was also proposed for the determination of cardiac glycosides (12). See also ref. (13).

2. Janovsky and Zimmermann reactions, see Chapter 16, Section VI.

F. o-Nitrobenzaldehyde

Principle (6,14)

Reaction with o-nitrobenzaldehyde in alkaline medium: yellow color.

Reagents

a. A 1 N solution of potassium hydroxide in ethanol.

b. A 2% solution of o-nitrobenzaldehyde in ethanol. Prepare fresh immediately before use.

Procedure

To 1 ml of sample solution in ethanol, add 0.2 ml of reagent a and 0.2 ml of reagent b. Heat at 60° for the given length of time (see Results), protected against light, and cool for 2 min in ice water. Add 3 ml of water, and read.

Results

	Reaction time, min	λ Max, nm	A = 0.3 (1 cm cell) Sample, μg
17β-Hydroxyandrostan-3-one	15	450	130
Testosterone	15	410	255
19-Nortestosterone[a]	5	450	300
Norethynodrel[b]	5	460	340

	Reaction time, min	λ Max, nm	A = 0.3 (1 cm cell) Sample, μg
17α-Ethynyl-17β-hydroxy-4,9,11-estratrien-3-one	5	460	68
Dehydroepiandrosterone	15	450	104
3-Ethoxy-3,5-estradien-17-one	15	450	105
4-Estrene-3,17-dione	5	450	93
11β-Hydroxy-4-androstene-3,17-dione	5	450	145
5(10)-Estrene-3,17-dione	5	450	90
Adrenosterone	5	450	92
Progesterone	15	410	245
19-Norprogesterone[a]	5	450	270

[a]The given wavelength is not that of maximum absorption, but Beer's law is obeyed only at that wavelength and up to A = 0.4.

[b]Beer's law is obeyed up to A = 0.4.

Notes

1. The order of sensitivities is the following: 17-ketosteroids > 3-ketosteroids with saturated ring A > 3-ketosteroids with one unsaturation in ring A. The reaction is negative with 17β-hydroxy-1,4-androstadien-3-one and with 6-, 7-, 11-, 12-, or 20-ketosteroids.

2. o-Nitrobenzaldehyde was proposed for the characterization of ketones of the type —CO—CH_3 (15,16). Indigo is formed in alkaline medium according to the scheme of Baeyer and Drewsen (17), but the mechanism of the reaction seems as yet not completely known.

$$o\text{-}NO_2C_6H_4CHO + CH_3-CO-R \longrightarrow o\text{-}NO_2C_6H_4CHOH-CH_2-CO-R \longrightarrow \text{indigo}$$

3. The reagent was also applied to the characterization of cyclanones (18).

VI. 3-KETOSTEROIDS

A. 2,6-Di-tert-butyl-p-cresol and Hydrogen Peroxide

Principle (19)

Reaction with 2,6-di-tert-butyl-p-cresol in alkaline medium in the presence of hydrogen peroxide. The colors are the following: pink with normal steroids with saturated ring A and with Δ^4- or $\Delta^{5(10)}$-19-nor steroids; blue with normal steroids with unsaturated ring A and no carbonyl group at position 11; yellow-orange with steroids with unsaturated ring A and a carbonyl group at position 11.

Reagents

a. A 1% solution of 2,6-di-tert-butyl-p-cresol in ethanol.

b. An 0.03% aqueous solution of hydrogen peroxide (0.1 volume oxygen).

Procedure

To 2 ml of sample solution in ethanol, add 1 ml of reagent a, 0.2 ml of reagent b, and 2 ml of 1 N sodium hydroxide. Heat at 80° for 30 min (protected against light when operating with 19-nor steroids), cool to 18-20° in a water bath, and read.

Results

	Color	λ Max, nm	A = 0.3 (1 cm cell) Sample, μg
Dehydrocholic acid	Pink	500	180
Pregnane-3,20-dione	Pink	500	1400
19-Nortestosterone	Pink	500	335
Norethandrolone	Pink	500	485
19-Norprogesterone	Pink	500	375
Norethynodrel	Pink	390	520
Testosterone	Blue	610	31
6-Dehydrotestosterone	Blue	610	500
4-Androstene-3,17-dione	Blue	610	27
1,4-Androstadiene-3,17-dione	Blue	610	400
Hydrocortisone	Blue	610	29
Prednisolone	Blue	610	1000
Adrenosterone	Yellow-orange	460	15
Cortisone	Yellow-orange	460	26
Prednisone	Yellow-orange	460	1000

4-Estrene-3,17-dione develops a pink color, but Beer's law is not followed. Triamcinolone affords a violet color (λ Max: 550-560 nm, A = 0.3 for 190 μg), and its acetonide a green color.

Notes

1. It may be seen from the table that Δ^4-3-ketosteroids are much more reactive than the corresponding $\Delta^{1,4}$ derivatives. The reaction is negative with Δ^1-3-ketosteroids and with Δ^4-3-ketosteroids substituted at position 6. Substitutions at position 17 have no significant effect on the sensitivity.

2. The reaction is specific for 3-ketosteroids; 7-, 11-, 12-, or 17-ketosteroids do not react.

3. This procedure was derived from a paper published by Schulz and Neuss (20). It was subjected to various modifications (21), but the mechanism of the reaction is unknown. Similar results are obtained when 3,5-di-tert-butyl-4-hydroxybenzyl alcohol is used instead of 2,6-di-tert-butyl-p-cresol, but the corresponding benzaldehyde does not react (19).

4. The reagent also develops a violet color with pentacyclic triterpenes such as α- and β-amyrin and lupeol (22).

5. 2,6-di-tert-butyl-p-cresol (V) is a white solid, insoluble in water and soluble in organic solvents.

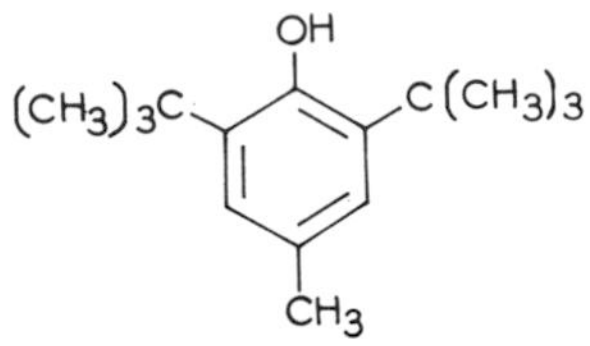

OH
$(CH_3)_3C$ $C(CH_3)_3$
CH_3
(V)

It is used as an antioxidant in fats and can be determined with testosterone propionate as reagent (23).

6. The same reaction allows the fluorimetric determination of Δ^4- or $\Delta^{1,4}$-3,11-diketosteroids (Section XVI.A).

VII. 3-KETOSTEROIDS WITH UNSATURATED RING A

A. 4-Aminoantipyrine

Principle (24)

Reaction with 4-aminoantipyrine in hydrochloric acid medium: yellow color.

Reagent

Dissolve 0.125 g of 4-aminoantipyrine in 25 ml of methanol containing 1% of concentrated hydrochloric acid.

Procedure

To 0.5 ml of sample solution in methanol, add 3 ml of reagent. Let stand at room temperature for 15 min, and read.

Results

	λ Max, nm	A = 0.3 (1 cm cell) Sample, μg
Ethisterone	360	34
Norethindrone	350	42
17α-Ethynyl-17β-hydroxy-4,9,11-estratrien-3-one	430	7.5
Norethynodrel	350	30
Progesterone	365	35
19-Norprogesterone	350	30
19-Nor-Δ^9-progesterone	380	11
Hydrocortisone	380	38
Prednisolone[a]	350	36
Fluocinolone acetonide[a]	395	52

[a]Let stand for 2 hr.

Notes

1. The reaction is also positive with oxymetholone (VI) (A = 0.3 given by 16 μg at 370 nm, after 15 min), but much less sensitive with other 3-ketosteroids with saturated ring A. For instance, A = 0.3 is read at 325 nm with 830 μg of 17β-hydroxyandrostan-3-one.

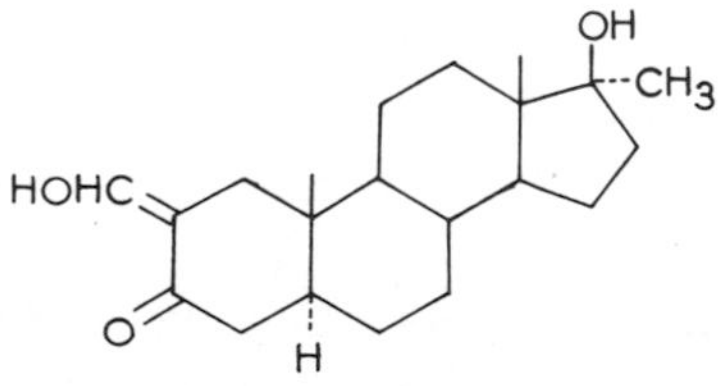

(VI)

2. 4-Aminoantipyrine also allows the colorimetric determination of phenols (Chapter 3, Section II.L).

B. Blue Tetrazolium

Principle (25)

Reduction of Blue Tetrazolium in alkaline medium into the corresponding formazan (see formulas Chapter 11, Section III.B): pink color.

Reagents

a. Dilute 5 ml of 10% aqueous solution of tetramethylammonium hydroxide to 50 ml with ethanol.

b. An 0.1% solution of Blue Tetrazolium in ethanol.

Procedure

To 1 ml of sample solution in ethanol, add 1 ml of reagent *a* and 0.5 ml of reagent *b*. Heat at 60° for 10 min in the dark, cool in ice water, add 0.5 ml of glacial acetic acid, and read at 525 nm.

Results

	A = 0.3 (1 cm cell) Sample, μg
19-Nortestosterone	12.5
Norethindrone	16
Norethindrone acetate	16
17α-Ethynyl-17β-hydroxy-4,9,11-estratrien-3-one	5.5
19-Norprogesterone	13
Testosterone	38
Progesterone	45
Norethynodrel	16
17β-Hydroxy-5(10),9(11)-estradien-3-one benzoate	15

Notes

1. Blue Tetrazolium has been proposed as reagent for the colorimetric determination of corticosteroids (Section XII.A). The reaction is based on its reduction into the

corresponding formazan by the 17-ketol grouping. However, even when other reducing substances are eliminated, the reaction is specific for this grouping only when the given operative conditions are strictly observed. Other steroids can also react on heating or when the reaction time is prolonged. Meyer and Lindberg (25) obtained positive results with a variety of 3-ketosteroids with unsaturated ring A. 3-Ketosteroids with saturated ring A and 17-ketosteroids do not react, $\Delta^{1,4}$-3-ketosteroids react but poorly. The herein-described procedure was derived from their paper.

2. As may be seen from the Results, 19-nor steroids are more reactive than those bearing a methyl group at position 10, and this fact is in agreement with the statement of the authors.

3. The mechanism of the reaction is unknown. However, from 4-androstene-3,17-dione, Meyer and Lindberg obtained the 6-hydroxy and the 6-keto derivatives.

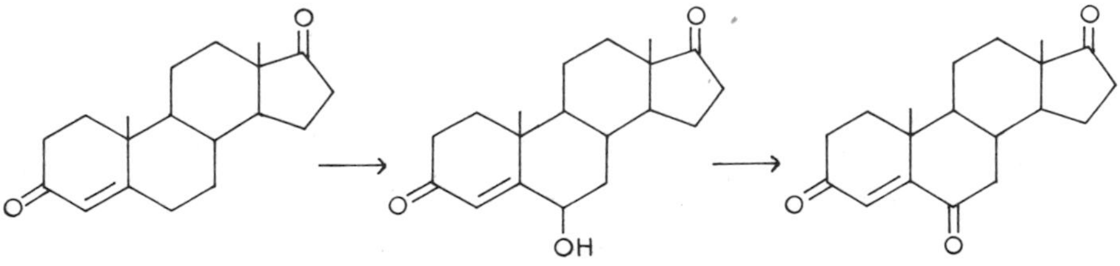

They attributed the higher reactivity of 19-norsteroids to the more complex alterations these compounds underwent. Some products showed typical aromatic properties, indicating a conversion to estrone-type steroids.

4. Tetrazolium salts and formazans, see Chapter 16, Section VIII.

VIII. Δ^1- OR Δ^4- OR $\Delta^{1,4}$- OR $\Delta^{4,6}$-3-KETOSTEROIDS

A. Dimethyl-p-phenylenediamine Oxalate

Principle (26)

Reaction with dimethyl-p-phenylenediamine in the presence of perchloric acid to give a Schiff base: yellow-orange color.

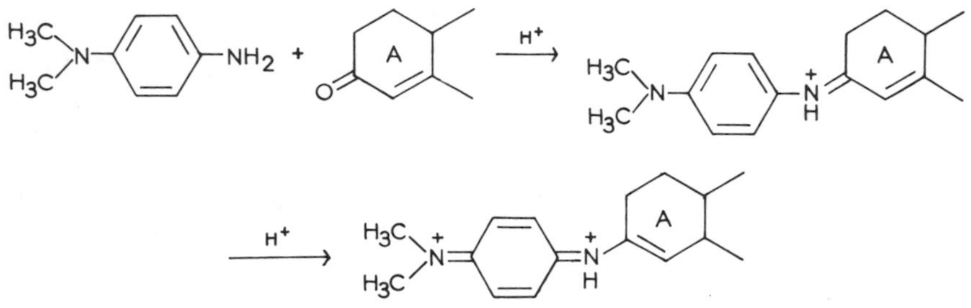

Reagents

a. An 0.01 N solution of perchloric acid obtained by diluting the suitable amount of 70% perchloric acid with methanol.

b. An 0.02 N solution of perchloric acid prepared as above.

c. An 0.1% solution of dimethyl-p-phenylenediamine oxalate in methanol.

Procedures

Δ^1- or Δ^4-3-Ketosteroids. To 0.5 ml of sample solution in reagent *a*, add 1 ml of reagent *c*. Let stand for 30 min in the dark, add 2.5 ml of methanol, and read at 420 nm.

$\Delta^{4,6}$-3-Ketosteroids. Proceed as above, the sample being dissolved in 0.5 ml of reagent *b*. Read at 450 nm.

$\Delta^{1,4}$-3-Ketosteroids. To 0.5 ml of sample solution in reagent *b*, add 1 ml of reagent *c*, and heat at 60° for 1 hr in the dark. Cool for 5 min in a water bath at room temperature, add 2.5 ml of methanol, and read at 450 nm.

Results

	A = 0.3 (1 cm cell) Sample, μg
3-Oxo-1-androsten-17-yl acetate	83
17α-Hydroxy-3,11,20-trioxo-1-pregnen-21-yl acetate	120
4-Cholesten-3-one	60
Testosterone	47
17α-Methyltestosterone	50
Norethandrolone	55
Progesterone	58
Cortisone	62
4,6-Cholestadien-3-one	50
6-Dehydrotestosterone benzoate	60
17α-Methyl-17β-hydroxy-4,6-androstadien-3-one	43
1,4-Cholestadien-3-one	76
1,4-Androstadiene-3,17-dione	60
Prednisolone	62
16α-Methylprednisolone	65

Notes

1. 3-Ketosteroids with saturated ring A, and ketosteroids with the carbonyl group located at other positions do not react.

2. Dimethyl-p-phenylenediamine also allows the colorimetric determination of aliphatic (Chapter 8, Section IV.D) and aromatic (Chapter 8, Section VI.A) aldehydes, peroxides (Chapter 10, Section III.A), and of $\Delta^{5(10)}$-3-keto-19-norsteroids (Section XI.B).

3. Stability of dimethyl-p-phenylenediamine, see Chapter 3, Section II.O, Note 2.

IX. Δ^4- OR $\Delta^{1,4}$-3-KETOSTEROIDS

A. Isoniazid

Principle (27)

Condensation in acidic medium to give the corresponding hydrazone: yellow color.

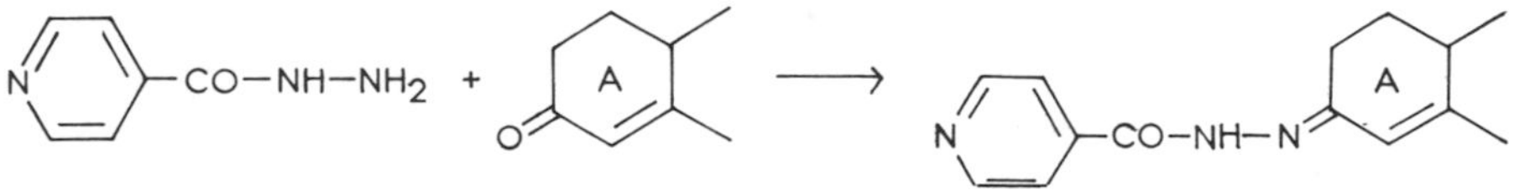

CO—NH—NH_2 + [ring A ketone] ⟶ CO—NH—N= [ring A]

Reagents

a. Dissolve 0.8 g of isoniazid in 100 ml of methanol containing 1 ml of concentrated hydrochloric acid.

b. Dilute 12.5 ml of reagent a to 100 ml with methanol.

Procedures

Δ^4-3-Ketosteroids. To 2 ml of sample solution in methanol, add 2 ml of reagent b. Let stand at room temperature for 1 hr, and read at 380 nm.

$\Delta^{1,4}$-3-Ketosteroids. To 2 ml of sample solution in methanol, add 2 ml of reagent a. Let stand at room temperature for 3 hr, and read at 405 nm.

Results

	A = 0.3 (1 cm cell) Sample, μg
Testosterone	29
17α-Methyltestosterone	30.5
19-Nortestosterone	30
Norethandrolone	31.5
17α-Ethynyl-17β-hydroxy-4,9,11-estratrien-3-one[a]	12
Progesterone	31.5
19-Norprogesterone	35.5
Cortisone	42

	A = 0.3 (1 cm cell) Sample, μg
Prednisolone	34
Dexamethasone	32.5
Prednisone	29

[a]Let stand for 3 min only, and read at 420 nm.

Notes

1. Norethynodrel reacts also, probably because of its isomerization to norethindrone in the acidic medium. After 3 hr (with reagent b) A = 0.3 is given by 63 μg (at 380 nm).

2. The reaction was applied to the determination of norethindrone, norethindrone acetate, dimethisterone [6α-methyl-17-(1-propynyl)testosterone], medroxyprogesterone acetate (6α-methyl-17α-hydroxyprogesterone acetate) and norethynodrel in oral contraceptive tablets (28).

3. 3-Ketosteroids with saturated ring A, or with the keto group located at other position, do not react (29).

4. Isoniazid was also proposed as reagent for colchicine and colchicoside (30).

X. Δ^4-3-KETOSTEROIDS

A. Sodium Nitroprusside

Principle (31)

Reaction with nitroprusside ion in the presence of sodium hydroxide: brown-orange color.

Reagents

a. A 10% aqueous solution of sodium nitroprusside.

b. A saturated aqueous solution of ammonium sulfate.

Procedure

To 3 ml of sample solution in ethanol, add 0.15 ml of reagent a and 0.15 ml of 10 N sodium hydroxide. Let stand for 15 min, add 3.3 ml of reagent b, shake vigorously, let stand for 2 min, and pipet a 3 ml aliquot of the supernatant. Add to it 0.2 ml of water, let stand for 15 min, and read at 400 nm.

Results

	A = 0.3 (1 cm cell) Sample, μg
Testosterone	950
Progesterone	660
Hydrocortisone	1400

Notes

1. Δ^1- and $\Delta^{1,4}$-3-ketosteroids, 3-ketosteroids with saturated ring A, and ketosteroids with the carbonyl group at another position do not react.

2. Nitroprusside also allows the colorimetric determination of primary (Chapter 4, Section VI.A) and secondary (Chapter 4, Section VII.A) aliphatic amines, α-methylene ketones (Chapter 8, Section VIII.A), and imidazoline derivatives (Chapter 13, Section VI.A).

3. Relations between nitroprusside and pentacyanoferrates, see Chapter 5, Section IV.A, Note 1.

B. Monoethanolamine and Potassium Hydroxide

Principle (32)

Reaction with monoethanolamine and potassium hydroxide, in dimethylformamide medium: yellow color.

Reagents

a. Pure monoethanolamine.

b. A 1 N solution of potassium hydroxide in ethanol.

Procedure

To 4 ml of sample solution in dimethylformamide, add 1 ml of reagent a and 0.4 ml of reagent b. Let stand at room temperature for 30 min, and read at 385 nm.

Results

	A = 0.3 (1 cm cell) Sample, μg
Testosterone	175
17α-Ethynyl-17β-hydroxy-4,9,11-estratrien-3-one	135
Progesterone	160
Cortisone acetate	150

Note

The reaction is much less sensitive with $\Delta^{1,4}$-3-keto-steroids. Under the above conditions, 2 mg of prednisone and of prednisolone give optical densities of 0.17 and 0.10, respectively.

C. Tetramethylammonium Hydroxide

Principle (33)

Reaction in the heat with tetramethylammonium hydroxide: yellow color.

Reagent

A 10% aqueous solution of tetramethylammonium hydroxide.

Procedure

To 1 ml of sample solution in ethanol, add 5 ml of reagent, and heat at 70° for 35 min. Allow to cool, and read at 375 nm.

Results

	A = 0.3 (1 cm cell) Sample, μg
Testosterone	105
Progesterone	110
Cortisone acetate	110

Note

The reaction is much less sensitive with $\Delta^{1,4}$-3-keto-steroids. Under the above conditions, 1 mg of prednisone and of prednisolone give optical densities of 0.17 and 0.03, respectively.

XI. $\Delta^{5(10)}$-3-KETO-19-NORSTEROIDS

A. Thionine Hydrochloride

Principle (34)

In alkaline medium, the pink color of thionine is partly discharged upon the action of the ketosteroid. Upon subsequent acidification, the remaining color turns to blue.

Reagent

An 0.0025% aqueous solution of thionine hydrochloride.

Procedure

To 1 ml of sample solution in methanol, add 1 ml of reagent, 3 ml of water, and 0.1 ml of 1 N sodium hydroxide. Let stand at room temperature for 2-5 min, and add 0.2 ml of 1 N hydrochloric acid. Read sample and blank at 600 nm against water. The difference between the readings given by the blank (A_b) and the sample (A_s) corresponds to the concentration of the steroid.

Results

	A_b - A_s = 0.3 (1 cm cell) Sample, μg
17β-Hydroxy-5(10)-estren-3-one	11.7
17α-Ethyl-17β-hydroxy-5(10)-estren-3-one	13.4
17α-Vinyl-17β-hydroxy-5(10)-estren-3-one	12.1
Norethynodrel	11.8
17β-Hydroxy-5(10),9(11)-estradien-3-one benzoate	43
19-Nor-5(10)-pregnene-3,20-dione	15
5(10)-Estrene-3,17-dione	11.8

Notes

1. The reaction is much less sensitive with Δ^4-3-ketosteroids. For instance, A_b - A_s = 0.3 is given by 2500 μg of 19-nortestosterone or 2350 μg of 19-norprogesterone.

2. This method was derived from a work of Bougault and Cattelain (35), who established that when citral is added to an alkaline solution of Methylene Blue, an immediate color change takes place from blue to pink. This phenomenon proceeding very slowly without the addition of citral, they admitted that this aldehyde acts as a catalyst. Postulating that the —CO—CH=CH— grouping was responsible for the reaction, we obtained but a poorly sensitive test with Δ^4-3-ketosteroids, whereas $\Delta^{5(10)}$-3-ketosteroids afforded very good results. It may be concluded therefrom that the reaction is almost specific for the —CO—CH_2—CH=CH— grouping, as evidenced by the action of ethyl allyl ketone (34). On the other hand, a purely catalytic effect is hardly admissible, since Beer's law is followed. The true mechanism of the reaction remains unknown.

3. Although rather satisfactory results can be obtained with Methylene Blue (VII), its slow color change in alkaline medium interferes with accurate determinations. This difficulty is overcome by substituting thionine hydrochloride (VIII) for this dye. Under the above conditions, the alkaline solution of thionine is stable, making accurate measurements possible.

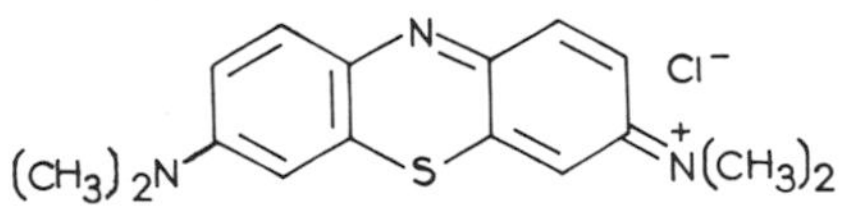

$(CH_3)_2N$ … S … N … $\overset{+}{N}(CH_3)_2$ Cl^-

(VII)

H_2N … S … N … $\overset{+}{N}H_2$ Cl^-

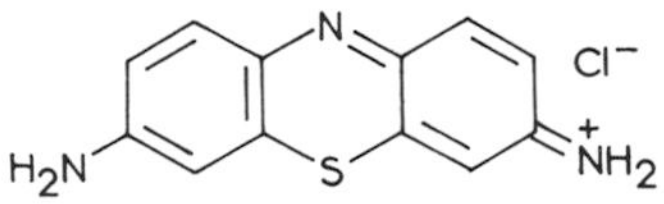

(VIII)

4. Upon the same principle, $\Delta^{5(10)}$-3-keto-19-norsteroids can be fluorimetrically determined with Pyronine G (Section XVII.A).

B. Dimethyl-p-phenylenediamine Oxalate

Principle (26)

Isomerization in acid medium into the corresponding Δ^4-3-ketosteroid, and development with dimethyl-p-phenylenediamine.

H^+ → ; $(H_3C)_2N-C_6H_4-NH_2$ → See Section VIII.A.

Reagents

a. An 0.01 N solution of perchloric acid obtained by diluting the suitable amount of 70% perchloric acid with methanol.

b. An 0.1% solution of dimethyl-p-phenylenediamine oxalate in methanol.

Procedure

Heat at 60° for 10 min 0.5 ml of sample solution in reagent a. Cool for 5 min in a water bath at room temperature, and add 1 ml of reagent b. Let stand for 30 min in the dark, add 2.5 ml of methanol, and read at 420 nm.

Results

	A = 0.3 (1 cm cell) Sample, μg
Norethynodrel	60
3-Oxo-19-nor-5(10)-androsten-17-yl acetate	65

XII. 17-KETOLSTEROIDS (FREE OR ESTERIFIED)

A. Blue Tetrazolium

Principle (36)

Reduction of Blue Tetrazolium in alkaline medium into the corresponding formazan (see formulas, Chapter 11, Section III.B): pink color.

Reagents

a. Dilute 5 ml of 10% aqueous solution of tetramethylammonium hydroxide to 50 ml with ethanol.

b. An 0.1% solution of Blue Tetrazolium in ethanol.

Procedure

To 1 ml of sample solution in ethanol, add 0.5 ml of reagent *a*, 0.5 ml of reagent *b*, and 2 ml of chloroform. Let stand at room temperature for the given length of time (see Results), exposed to subdued light, and read at 525 nm.

Results

	Reaction time, min	A = 0.3 (1 cm cell) Sample, μg
Hydrocortisone	5	18
Hydrocortisone acetate	15	21
Prednisolone m-sulfobenzoate	10	37
Triamcinolone	20	10.5
Triamcinolone acetonide	15	24
Triamcinolone-16,21-diacetate	15	12.5
Cortisone	5	17

Prednisolone-21-phosphate does not react.

Notes

1. Tetrazolium salts can be converted to formazans by other reducing compounds (see Chapter 16, Section VIII, tetrazolium salts and formazans) and by various 3-ketosteroids (Section VII.B). The above procedure is therefore not specific for 17-ketolsteroids. The presence of other reducing substances must be avoided, but under the given conditions, which are critical, other ketosteroids almost do not react. For instance, 30 μg of norethindrone gives an absorbance value of 0.04 (compare with Results, Section VII.B).

2. Studies with cortisone acetate showed that hydrolysis of the ester was prerequisite to the reaction with

Blue Tetrazolium. Differences in rates of color development were used to analyze mixtures of cortisone and hydrocortisone and of cortisone and cortisone acetate (37).

3. Triphenyltetrazolium chloride was at first proposed as reagent (38). It is still often used, although it was shown that the procedure gives poor reproducibility (39). It was established that the conditions are highly critical. Even the size of the vessel in which the color is developed must be taken into account, since the presence of air affects the assay (40). The effect of oxygen is less marked with Blue Tetrazolium, which affords much more regular calibration curves and reproducible results.

4. Esters of corticosteroids cannot be determined directly by means of conventional methods of alkaline hydrolysis. The ketol side chain thereby liberated is rapidly oxidized in the alkaline medium by atmospheric oxygen, thus affording two carboxylic acids (41):

$$\text{D-ring}{-}CO{-}CH_2OH + O_2 + 2\,OH^- \longrightarrow \text{D-ring}{-}COO^- + H{-}COO^- + 2\,H_2O$$

Even under vacuum, side reactions may occur with 17-hydroxy-17-ketolsteroids (42).

This oxidation reaction allowed studies on the stability of various 21-acyloxy corticosteroids (43), on the effect of substituents on the oxidation rate (44), and on the catalytic action of some acid salts (45). However, triamcinolone does not react under these conditions (43). It was postulated that the oxidation is completely hindered by an immediate isomerization (a) (46). It is followed by a slow release of formaldehyde according to (b), thus making some determinations possible.

(a) 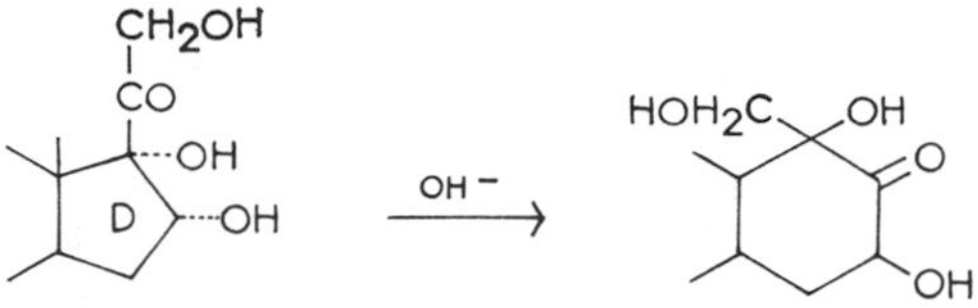 [D-ring: CH_2OH–CO–, ···OH, ···OH] $\xrightarrow{OH^-}$ [HOH_2C, OH, =O, OH]

(b) 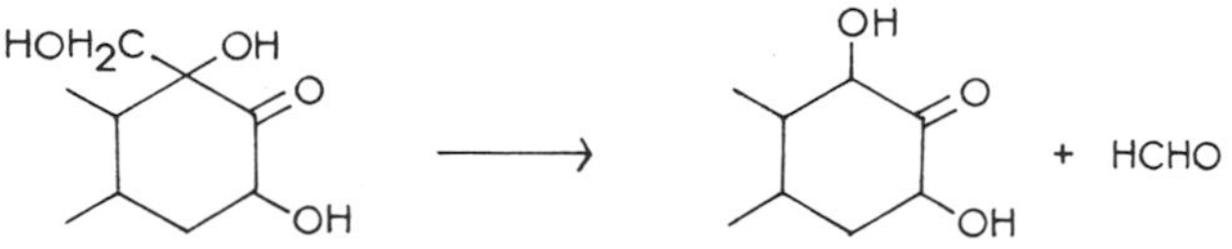 [HOH_2C, OH, =O, OH] $\longrightarrow$ [OH, =O, OH] + HCHO

Recently, Görög (47) overcame the difficulties caused by the oxidation by reducing the 20-keto group with sodium borohydride before alkaline hydrolysis, when the resulting glycol side chain is insensitive to alkali and oxygen.

It may be postulated that in the herein-described colorimetric determination as well as in methods with p-nitroso-N,N-dimethylaniline (Section XII.B) and with benzothiazole derivatives (Sections XIV.B and XIV.C) the mechanism of action of the reagent is of such a kind that the effect of oxygen becomes almost negligible, since no specific precautions against air are necessary.

B. p-Nitroso-N,N-dimethylaniline, Phenol, and Potassium Ferricyanide

Principle (48)

Reduction of p-nitroso-N,N-dimethylaniline by the ketol group, and development of the dimethyl-p-phenylenediamine formed by oxidative reaction with phenol in the presence of ferricyanide ion: green color.

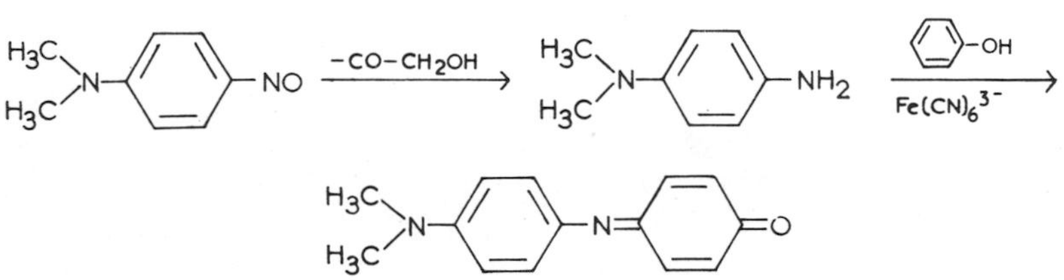

$$(H_3C)_2N-C_6H_4-NO \xrightarrow{-CO-CH_2OH} (H_3C)_2N-C_6H_4-NH_2 \xrightarrow[Fe(CN)_6^{3-}]{C_6H_5-OH} (H_3C)_2N-C_6H_4-N{=}C_6H_4{=}O$$

Reagents

a. An 0.1% solution of p-nitroso-N,N-dimethylaniline in ethanol.

b. Clark and Lubs buffer for pH 9.8, see Appendix I.C.

c. An 0.1% solution of phenol in ethanol.

d. A 1% aqueous solution of potassium ferricyanide. Prepare freshly before use.

Procedures

Free ketols and readily saponifiable esters. To 1 ml of sample solution in ethanol, add 0.5 ml of reagent a, immerse the tubes in ice water for 5 min, and add 0.5 ml of 0.1 N sodium hydroxide. Plug the tubes with cotton-wool, and let stand at 0° for the given length of time (see Results), protected against light. Add 2 ml of buffer b, 5 ml of reagent c, and 0.5 ml of reagent d, and let stand in a water bath at 20° ± 2° for 10 min. Read at 650 nm.

Hemisuccinates, cyclohexanecarboxylates, and m-sulfobenzoates. To 1 ml of sample solution in ethanol, add 1 ml of reagent a, let stand at room temperature for 5 min, and add 1 ml of 0.1 N sodium hydroxide. Plug the tubes with cotton-wool, and let stand at room temperature for the given length of time (see Results), protected against light. Add 2 ml of buffer b, 5 ml of reagent c, and 0.5 ml of reagent d, and let stand in a water bath at room temperature for 10 min. Read at 650 nm.

Results

	Reaction time, hr	A = 0.3 (1 cm cell) Sample, μg
Deoxycorticosterone acetate	5	139
Corticosterone	5	116
Hydrocortisone	2	120
Hydrocortisone acetate	2	134
Hydrocortisone hemisuccinate	2	155
Prednisolone	2	131

	Reaction time, hr	A = 0.3 (1 cm cell) Sample, μg
Prednisolone cyclohexanecarboxylate	2	174
Prednisolone m-sulfobenzoate	2	210
Dexamethasone acetate	4	144
Dexamethasone hemisuccinate	2	173
Cortisone	1	120
Cortisone acetate	1	134

Notes

1. The reaction is less sensitive with o-sulfobenzoic and pivalic esters. Phosphates and sulfates do not react.

2. Reduced p-nitroso-N,N-dimethylaniline allows the colorimetric determination of phenols (Chapter 3, Section II.O).

3. Oxidation of 17-ketolsteroids, see Section XII.A, Note 4.

XIII. 17-KETOLSTEROIDS (FREE)

A. Sodium Molybdate

Principle (49)

Reduction of the molybdate ion in acetic acid medium: blue color.

Reagent

To 0.5 ml of 25% aqueous solution of sodium molybdate, add 40 ml of glacial acetic acid.

Procedure

To 3 ml of sample solution in glacial acetic acid, add 4 ml of reagent. Let stand for 2 hr, and read at 650 nm.

Results

	A = 0.3 (1 cm cell) Sample, μg
Hydrocortisone	187
Prednisolone	172

	A = 0.3 (1 cm cell) Sample, μg
Cortisone	123
Prednisone	119

21-Esters do not react.

Note

Oxidation of 17-ketolsteroids, see Section XII.A, Note 4.

B. Sodium Metaperiodate, Phenylhydrazine Hydrochloride, and Potassium Ferricyanide

Principle

Oxidation of the ketol group with periodate ion, and development of the formaldehyde formed with phenylhydrazine and ferricyanide ion: red color.

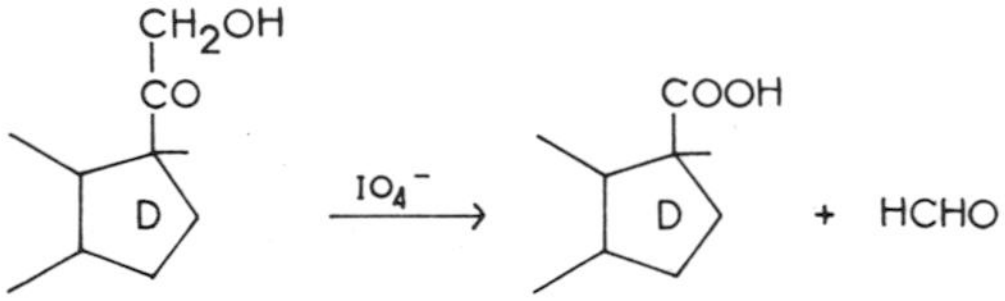

$$\text{D-ring}-CO-CH_2OH \xrightarrow{IO_4^-} \text{D-ring}-COOH + HCHO$$

$$HCHO + C_6H_5-NH-NH_2 \xrightarrow{Fe(CN)_6^{3-}} \text{Formazan dye}$$

Reagents

a. A mixture of 40 ml of 0.05 N aqueous solution of sodium metaperiodate and of 10 ml of 1.5 N hydrochloric acid.

b. A 1% aqueous solution of phenylhydrazine hydrochloride, freshly prepared.

c. A 2% aqueous solution of potassium ferricyanide, freshly prepared.

Procedure

To 1 ml of sample solution in ethanol, add 0.5 ml of reagent a, and let stand for 30 min. Add 1.5 ml of 0.1 N sodium hydroxide, 2 ml of reagent b, and 1 ml of reagent c; mix, and chill in ice water for 5 min. Add 5 ml of concentrated hydrochloric acid, mix, and dilute with 5 ml of ethanol. Let stand for 15 min, and read at 520 nm.

Results

	A = 0.3 (1 cm cell) Sample, μg
Deoxycorticosterone	215
Prednisolone	255
Cortisone	170
Prednisone	170

21-Esters do not react.

Notes

1. Oxidation of 17-ketolsteroids, see Section XII.A, Note 4.

2. Periodates and periodate oxidation, see Chapter 16, Section IV.

3. Colorimetric determination of formaldehyde, see Chapter 16, Section V.

XIV. 17-HYDROXY-17-KETOLSTEROIDS

A. Phenylhydrazine Hydrochloride in Sulfuric Acid

Principle (50,51)

Dehydration and rearrangement of the 17-hydroxy-17-ketol grouping to give a glyoxal side chain, and formation of the corresponding phenylhydrazone: yellow color.

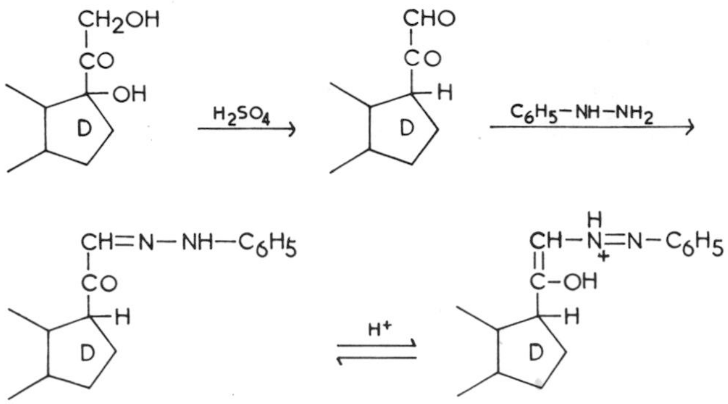

Reagent

Dissolve 0.065 g of phenylhydrazine hydrochloride in 100 ml of a cooled mixture of 310 ml of concentrated sulfuric acid and 190 ml of water.

Procedure

To 1 ml of sample solution in methanol, add 8 ml of reagent. Heat at 60° for 20 min, allow to cool, and read at 410 nm.

Results

	A = 0.3 (1 cm cell) Sample, μg
Hydrocortisone	70
Prednisolone	60
Dexamethasone	60
Cortisone	40
Prednisone	40

Notes

1. Triamcinolone and 18-norcortisone do not react (8).

2. The reaction is positive only with compounds bearing both hydroxyl and α-ketol groups at position 17. However, 17-ketolsteroids such as deoxycorticosterone can also be determined after suitable oxidation of the ketol chain to glyoxal with cupric acetate (52).

3. Under the conditions of the procedure, the 21-acetate of 21-methyl-6α,9α-difluoroprednisolone, corresponding to the side-chain structure (IX), develops a pink color.

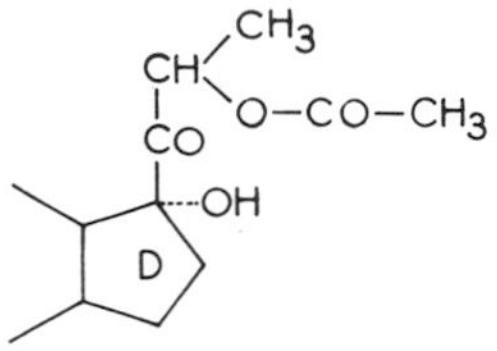

(IX)

4. The Porter-Silber reaction allowed the determination of glucuronide esters of 17-hydroxycorticosteroids in urine (53).

5. Increased sensitivity can be obtained with p-hydrazinobenzenesulfonic acid - phosphoric acid reagent (reading at 350 nm) (54).

B. 2-Hydrazinobenzothiazole and Hydrogen Peroxide

Principle (55)

Condensation with 2-hydrazinobenzothiazole in alkaline medium and development as formazan with the diazonium salt generated in the reaction medium by oxidation of the excess reagent (see mechanism, Chapter 8, Section IV.A): blue-violet color.

Since only aldehydes give rise to formazans, it may be postulated that a rearrangement occurs during the condensation, affording the hydrazone of an aldehyde (see Chapter 14, Section II.F, reaction with ketoses).

Reagents

a. Dissolve 0.040 g of 2-hydrazinobenzothiazole in 2.5 ml of 0.1 N hydrochloric acid, and dilute to 10 ml with water.

b. An 0.75% aqueous solution of hydrogen peroxide (2.5 volumes oxygen).

Procedure

Introduce 1 ml of sample solution in chloroform into the tube, and carefully evaporate to dryness on a steam bath. No trace amount of the solvent should remain in the tube, since it inhibits the reaction. Introduce 1 ml of water, add 0.5 ml of reagent *a*, and 0.5 ml of 0.1 N sodium hydroxide. Heat at 100° for 10 min, cool for 5 min in a water bath at 15°, and add 2 ml of reagent *b*. Let stand for 10 min, add 2 ml of ethanol, mix, let stand for 5 min, and read at 580 nm.

Results

	A = 0.3 (1 cm cell) Sample, μg
Hydrocortisone	39
Prednisolone	43
Dexamethasone	24.5
Triamcinolone	15
Cortisone	50
Prednisone	51

Notes

1. Readily saponifiable 21-esters can also be determined. A = 0.3 is given by 36.5 μg of dexamethasone m-sulfobenzoate or 45 μg of cortisone acetate.

2. The reaction is positive only with compounds bearing both hydroxyl and α-ketol groups at position 17. Corticosterone and 17-hydroxyprogesterone do not react.

3. 2-Hydrazinobenzothiazole also allows the colorimetric determination of aldehydes (Chapter 8, Section IV.A) and of aldoses and ketoses (Chapter 14, Section II.F).

C. 3-Methylbenzothiazolin-2-one Hydrazone and Ferric Chloride

Principle (55)

Condensation with 3-methylbenzothiazolin-2-one hydrazone in alkaline medium, and development as formazan with the diazonium salt generated in the reaction medium by oxidation of the excess reagent (see mechanism, Chapter 8, Section IV.B): blue color.

Since only aldehydes give rise to formazans, it may be postulated that a rearrangement occurs during the condensation, affording the hydrazone of an aldehyde (see Chapter 14, Section II.G, reaction with ketoses).

Reagents

a. An 0.5% aqueous solution of 3-methylbenzothiazolin-2-one hydrazone hydrochloride.

b. An 0.25% aqueous solution of ferric chloride hexahydrate.

Procedure

Introduce 1 ml of sample solution in chloroform into the tube, and carefully evaporate to dryness on a steam bath. No trace amount of the solvent should remain in the tube, since it inhibits the reaction. Introduce 1 ml of water, add 0.5 ml of reagent *a* and 0.5 ml of 0.1 N sodium hydroxide. Heat at 100° for 10 min, cool for 5 min in a water bath at 15°, add 0.5 ml of 1 N hydrochloric acid, and 2 ml of reagent *b*. Let stand for 1 hr at room temperature, and read at 630 nm.

Results

	A = 0.3 (1 cm cell) Sample, μg
11-Desoxy-17-hydroxy-corticosterone	21
Hydrocortisone	18
Prednisolone	19

	A = 0.3 (1 cm cell) Sample, μg
Cortisone	21
Prednisone	17

Note

Properties of 3-methylbenzothiazolin-2-one hydrazone, see Chapter 16, Section III.

XV. HYDROXYSTEROIDS (F)

A. Vanadium Oxinate and Magnesium Acetate

Principle and Procedure

See Chapter 2, Section XII.A, fluorimetric determination of alcohols.

Results

	Determination limits, μg	Reading 50 Sample, μg	Reading 50 Standard, μg/ml of 8-quinolinol
Cholesterol	70 - 350	137	1.28
Testosterone	40 - 200	84	1.45

XVI. Δ^4- OR $\Delta^{1,4}$-3,11-DIKETOSTEROIDS (F)

A. 2,6-Di-tert-butyl-p-cresol and Hydrogen Peroxide

Principle (56)

Reaction with 2,6-di-tert-butyl-p-cresol in alkaline medium in the presence of hydrogen peroxide. The fluorescence is not visible under ultraviolet light.

Reagents

a. A 1% solution of 2,6-di-tert-butyl-p-cresol in ethanol.

b. An 0.03% aqueous solution of hydrogen peroxide (0.1 volume oxygen).

Procedure

To 2 ml of sample solution in ethanol, add 1 ml of reagent a, 0.2 ml of reagent b, and 2 ml of 1 N sodium hydroxide. Heat at 80° for 30 min, exposed to subdued light, cool in ice water for 2 min, and read at exc: 436 nm; em: 520 nm.

Standard: A solution of 2,6-dimethyl-3,5-diacetyl-1,4-dihydropyridine in 50% ethanol.

Results

	Determination limits, μg	Reading 50 Sample, μg	Standard, μg/ml
Adrenosterone	1 - 5	1.8	0.159
Cortisone	1 - 5	2.0	0.175
Prednisone	40 - 200	82	0.185

Notes

1. The reaction is also positive with dehydrocholic acid. The determination limits are 4-20 μg, with reading at exc: 405 nm; em: 485 nm.

2. The same reagent allows the colorimetric determination of miscellaneous 3-ketosteroids (Section VI.A).

XVII. $\Delta^{5(10)}$-3-KETO-19-NORSTEROIDS (F)

A. Pyronine G

Principle (56)

In alkaline medium, the orange-yellow fluorescence of Pyronine G (also named Pyronine Y) is partly discharged upon the action of the ketosteroid.

Reagent

An 0.001% aqueous solution of Pyronine G (Y).

Procedure

During the whole procedure, operate in ice water, and protected against light. To 1 ml of sample solution in methanol, add 1 ml of reagent, 2.5 ml of water, and 0.2 ml of 1 N sodium hydroxide. Mix, let stand for 3 min, and add 0.15 ml of glacial acetic acid. Mix, and read immediately at exc: 546 nm; em: 570 nm.

This determination being based on the measurement of the decreased fluorescence of the sample solution, no standard is necessary. Adjust the spectrofluorimeter to read 100 for the blank.

Results

	Determination limits, μg	Reading for the upper limit
17β-Hydroxy-5(10)-estren-3-one	0.5 - 2.5	40
17α-Ethyl-17β-hydroxy-5(10)-estren-3-one	0.6 - 3.0	46
Norethynodrel	0.5 - 2.5	36
17β-Hydroxy-5(10),9(11)-estradien-3-one benzoate	1 - 4	40
19-Nor-5(10)-pregnene 3,20-dione	1 - 4	45

Pyronine G being very strongly adsorbed on the walls of the tubes, pipets, and cells, it is necessary to wash the glassware with concentrated sulfuric acid after each set of determinations.

Notes

1. Δ^4-3-Ketosteroids react but very weakly.

2. We have shown (Section XI.A) that the colors of thionine and Methylene Blue are partly discharged in alkaline medium under the action of $\Delta^{5(10)}$-3-keto-19-nor-steroids. We inferred from this observation that Pyronine G (X), whose structure can be compared with those of these two dyes, might lead to similar results in fluorimetric determinations.

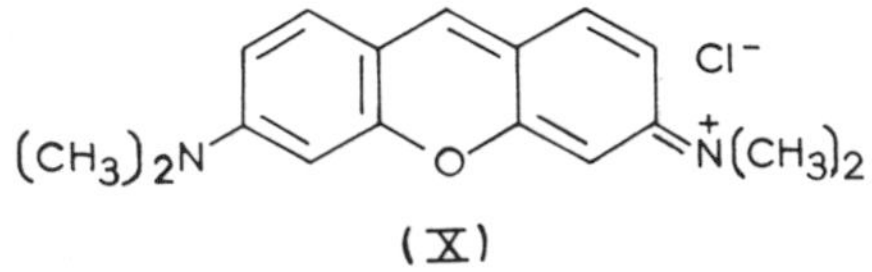

$(CH_3)_2N$ O $\overset{+}{N}(CH_3)_2$ Cl^-

(X)

XVIII. 17-KETOLSTEROIDS (FREE) (F)

A. Sodium Metaperiodate, Ethyl Acetoacetate, and Ammonium Acetate

<u>Principle</u> (<u>57</u>)

Oxidation of the ketol group with periodate ion, and development of the formaldehyde formed with ethyl acetoacetate and ammonia: blue fluorescence.

$$\text{D-ring}-CO-CH_2OH \xrightarrow{IO_4^-} \text{D-ring}-COOH + HCHO$$

$$HCHO + 2\,CH_3-CO-CH_2-CO_2C_2H_5 + NH_3 \longrightarrow \text{2,6-dimethyl-3,5-di}(CO_2C_2H_5)\text{-1,4-dihydropyridine}$$

<u>Reagents</u>

<u>a</u>. An 0.01 M aqueous solution of sodium metaperiodate.

<u>b</u>. Dissolve 0.25 g of stannous chloride dihydrate in 100 ml of 1 N hydrochloric acid. Prepare fresh before use.

<u>c</u>. A 4% solution of ethyl acetoacetate in 20% aqueous solution of ammonium acetate.

<u>Procedure</u>

To 1 ml of sample solution in 2% alcohol, add 0.2 ml of reagent <u>a</u>, and let stand at room temperature for 20 min. Add 0.8 ml of reagent <u>b</u>, 0.8 ml of 1 N sodium hydroxide, 1.2 ml of water, and 1 ml of reagent <u>c</u>. Heat at 60° for 20 min, allow to cool, and filter. Read at exc: 366 nm; em: 470 nm.

Standard: A solution of 2,6-dimethyl-3,5-dicarbethoxy-1,4-dihydropyridine in a 49:1 mixture of water and ethanol.

Results

	Determination limits, μg	Reading 50 Sample, μg	Standard, μg/ml
Deoxycorticosterone	6 - 30	12	1.75
Prednisolone	8 - 40	17	2.05
Cortisone	6 - 30	12	1.85
Prednisone	6 - 30	12	1.85

Notes

1. Periodates and periodate oxidation, see Chapter 16, Section IV.

2. Fluorimetric determination of formaldehyde, see Chapter 16, Section V.

REFERENCES

1. IUPAC Commission on the Nomenclature of Biological Chemistry, *J. Amer. Chem. Soc.*, 82, 5577 (1960).

2. IUPAC Commission on the Nomenclature of Organic Chemistry and IUPAC-IUB Commission of Biochemical Nomenclature, *Pure Appl. Chem.*, 31, 283-322 (1972).

3. M. Pesez, *Ann. Pharm. Fr.*, 14, 555 (1956).

4. J.H. Brewster and C.J. Ciotti, Jr., *J. Amer. Chem. Soc.*, 77, 6214 (1955).

5. M. Pesez, *Bull. Soc. Chim. Fr.*, 369 (1958).

6. Cf. M. Pesez and J. Robin, *Bull. Soc. Chim. Fr.*, 1930 (1962).

7. M. Herbain, *Bull. Soc. Chim. Biol.*, 41, 821 (1959).

8. M. Pesez and J. Robin, *Bull. Soc. Chim. Fr.*, 1930 (1962).

9. A. Sezerat, unpublished results.

10. M. Pesez and J. Bartos, *Ann. Pharm. Fr.*, 22, 541 (1964).

11. T. Momose, T. Matsukuma, Y. Ohkura, and Y. Nakamura, *J. Pharm. Soc. Japan*, 83, 143 (1963).

12. F.K. Bell and J.C. Krantz, Jr., *J. Amer. Pharm. Assoc., Sci. Ed.*, 37, 297 (1948); E.E. Kennedy, *J. Amer. Pharm. Assoc., Sci. Ed.*, 39, 25 (1950).

13. E.P. Schulz and M.A. Diaz, *J. Pharm. Sci.*, 53, 1115 (1964).

14. Cf. M. Pesez, *Bull. Soc. Chim. Fr*, 911 (1947); M. Pesez and M. Herbain, *ibid.*, 104 (1948).

15. J. Penzoldt, *Z. Anal. Chem.*, 24, 149 (1885).

16. F. Feigl, R. Zappert, and S. Vasquez, *Mikrochemie*, 17, 165 (1935).

17. A. v. Baeyer and V. Drewsen, *Ber.*, 15, 2856 (1882).

18. G. Zeidler and H. Kreis, *Angew. Chem.*, 54, 360 (1941).

19. J. Bartos, *Ann. Pharm. Fr.*, 17, 141 (1959).

20. E.P. Schulz and J.D. Neuss, *Anal. Chem.*, 29, 1662 (1957).

21. Cf. S. Ansari and R.A. Khan, *J. Pharm. Pharmacol.*, 12, 122 (1960); E.P. Schulz, M.A. Diaz, and L.M. Guerrero, *J. Pharm. Sci.*, 53, 1119 (1964).

22. C.H. Brieskorn and G.H. Mahran, *Naturwiss.*, 47, 107 (1960).

23. J. Jonas, *J. Pharm. Belg.*, 21, 3 (1966).

24. From E.P. Schulz, M.A. Diaz, G. Lopez, L.M. Guerrero, H. Barrera, A.L. Pereda, and A. Aguilera, *Anal. Chem.*, 36, 1624 (1964).

25. Cf. A.S. Meyer and M.C. Lindberg, *Anal. Chem.*, 27, 813 (1955).

26. M. Pesez and J. Bartos, *Talanta*, 10, 69 (1963).

27. E. Cingolani, G. Cavina, and V. Amormino, *Farmaco, Ed. Prat.*, 15, 301 (1960).

28. J.Y.P. Wu, *J. Assoc. Offic. Anal. Chemists*, 53, 831 (1970).

29. Cf. A. Ercoli, L. de Giuseppe, and P. de Ruggieri, *Farmaco*, 7, 170 (1952).

30. M. Pesez, *Ann. Pharm. Fr.*, 15, 630 (1957).

31. From H. Laubie, *Bull. Soc. Pharm. Bordeaux*, 98, 172 (1959).

32. M. Brunet, unpublished results.

33. From J.M. Cross, H. Eisen, and R.G. Kedersha, *Anal. Chem.*, 24, 1049 (1952).

34. M. Pesez and J. Bartos, *Bull. Soc. Chim. Fr.*, 912 (1964).

35. J. Bougault and E. Cattelain, *J. Pharm. Chim.*, [8]21, 437 (1935).

36. From P. Ascione and C. Fogelin, *J. Pharm. Sci.*, 52, 709 (1963).

37. D.E. Guttman, *J. Pharm. Sci.*, 55, 919 (!966).

38. Cf. W.J. Mader and R.R. Buck, *Anal. Chem.*, 24, 666 (1952).

39. F.M. Kunze and J.S. Davis, *J. Pharm. Sci.*, 53, 1259 (1964).

40. C.A. Johnson, R. King, and C. Vickers, *Analyst*, 85, 714 (1960).

41. L. Velluz, A. Petit, M. Pesez, and R. Berret, *Bull. Soc. Chim. Fr.*, 123 (1947).

42. N.L. Wendler and R.P. Graber, *Chem. Ind. (London)*, 549 (1956); D.E. Guttman and P.D. Meister, *J. Amer. Pharm. Assoc., Sci. Ed.*, 47, 773 (1958).

43. M. Pesez and J. Bartos, *Ann. Pharm. Fr.*, 20, 60 (1962).

44. M. Pesez and J. Bartos, *Bull. Soc. Chim. Fr.*, 1928 (1962).

45. J. Bartos, *Ann. Pharm. Fr.*, 21, 757 (1963).

46. L.L. Smith, M. Marx, J.J. Garbarini, T. Foell, V.E. Origoni, and J.J. Goodman, *J. Amer. Chem. Soc.*, 82, 4616 (1960).

47. S. Görög, *J. Pharm. Pharmacol.*, 21 (1969), Suppl., 46 S.

48. M. Pesez and J. Robin, *Ann. Pharm. Fr.*, 17, 624 (1959); J. Verdier, *Ann. Pharm. Fr.*, 18, 795 (1960).

49. From H. Wachsmuth and L. Van Koeckhoven, *Anal. Chim. Acta*, 22, 41 (1960).

50. C.C. Porter and R.H. Silber, *J. Biol. Chem.*, 185, 201 (1950).

51. Cf. M.L. Lewbart and V.R. Mattox, *J. Org. Chem.*, 29, 513, 521 (1964).

52. M.L. Lewbart and V.R. Mattox, *Anal. Chem.*, 33, 559 (1961).

53. B.W. Grunbaum and N. Pace, *Microchem. J.*, 16, 443 (1971).

54. A. Sanghvi, L. Taddeini, and C. Wight, *Anal. Chem.*, 45, 207 (1973).

55. J. Bartos, *Ann. Pharm. Fr.*, 20, 650 (1962).

56. M. Pesez and J. Bartos, *Talanta*, 16, 331 (1969).

57. M. Pesez and J. Bartos, *Talanta*, 14, 1097 (1967).

Part III

Chapter 16

ADDITIONAL PRACTICAL HINTS

I. DIAZO COUPLING AND DIAZONIUM SALTS

Diazo coupling and formation of diazonium salts have opened the way to a very great number of characterizations and colorimetric determinations.

The diazo coupling reaction may barely be considered as a proton-eliminating condensation of a diazonium salt with another compound possessing an active hydrogen atom.

$$Ar{-}\overset{+}{N}{\equiv}N\ X^- + H{-}Y \longrightarrow Ar{-}N{=}N{-}Y + H^+X^-$$

In some instances, further reactions can occur between the species formed and the excess reagent.

Colorimetric determinations involving diazo coupling may be subdivided into three groups:

(1) Direct diazo coupling.

(a)

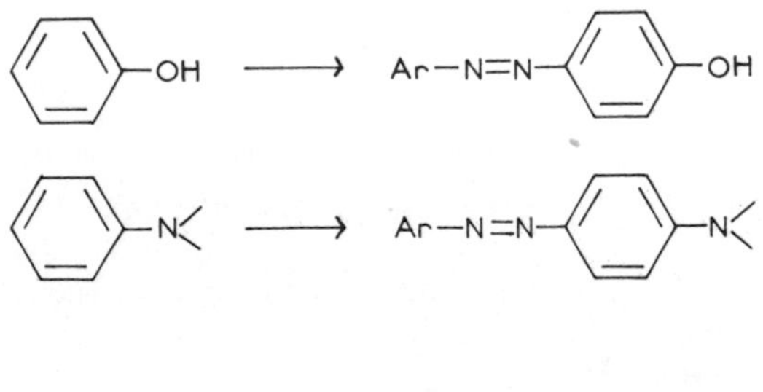

$$C_6H_5{-}OH \longrightarrow Ar{-}N{=}N{-}C_6H_4{-}OH$$

$$C_6H_5{-}N\langle \longrightarrow Ar{-}N{=}N{-}C_6H_4{-}N\langle$$

$$R{-}NH_2 \longrightarrow Ar{-}N{=}N{-}NH{-}R$$

$$R{-}CH_2{-}NO_2 \longrightarrow Ar{-}N{=}N{-}CH(R){-}NO_2$$

(b)

$$R{-}CH_2{-}CO{-}R' \longrightarrow Ar{-}N{=}N{-}CH(R){-}CO{-}R'$$

$$Ar'{-}NH{-}N{=}CH{-}R \longrightarrow (Ar{-}N{=}N)(Ar'{-}NH{-}N{=})C{-}R$$

$$R{-}CH{=}CH{-}R' \longrightarrow Ar{-}N{=}N{-}C(R){=}CH{-}R'$$

$$\text{4-R-imidazole} \longrightarrow \text{4-R-5-}(N{=}N{-}Ar)\text{-imidazole}$$

(2) Determinations through diazotization. The compound is converted into a diazonium salt which is then revealed by coupling with a suitable phenol or aromatic amine.

$$Ar{-}NH_2 \xrightarrow{HNO_2} Ar{-}\overset{+}{N}{\equiv}N$$

$$C_6H_6 \xrightarrow{HNO_3} C_6H_5{-}NO_2 \xrightarrow{H_2} C_6H_5{-}NH_2 \xrightarrow{HNO_2} C_6H_5{-}\overset{+}{N}{\equiv}N$$

$$Ar{-}NH{-}NH_2 \xrightarrow{SeO_2} Ar{-}\overset{+}{N}{\equiv}N$$

(3) Determinations through formation of nitrous acid. The nitrous acid diazotizes an aromatic amine, and the diazonium salt is then revealed as above.

$$R{-}OH \xrightarrow{HNO_2} NO_2{-}R \xrightarrow{H_2O} R{-}OH + HNO_2$$

$$R{-}SH \xrightarrow{HNO_2} R{-}S{-}NO \xrightarrow{HgCl_2} R{-}S{-}HgCl + HNO_2$$

$$H^+ \xrightarrow{NaNO_2} HNO_2$$

$$>CH-NO_2 \xrightarrow{OH^-} >C=NO_2^- \xrightarrow{H_2O_2} >C=O + HNO_2$$

It is obviously impossible to present here a comprehensive bibliography of the papers which have been published in these fields. For instance, Vejdelek and Kakác (1) quote 125 references solely for the analytical uses of p-nitrobenzenediazonium ion, and 167 references for diazotized p-sulfanilic acid, which are the most commonly used diazo reagents. More than 300 references were also mentioned in a monograph on the analytical applications of diazo coupling (2).

A. Direct Diazo Coupling

The coupling of phenols, usually carried out in alkaline medium, obeys well-known rules which are summarized in the accompanying table.

(a)

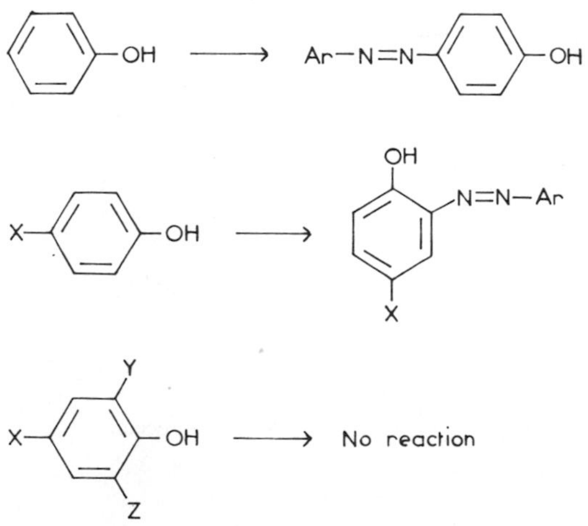

$C_6H_5-OH \longrightarrow Ar-N=N-C_6H_4-OH$

$X-C_6H_4-OH \longrightarrow$ (OH, N=N—Ar, X)

X, Y, Z —OH $\longrightarrow$ No reaction

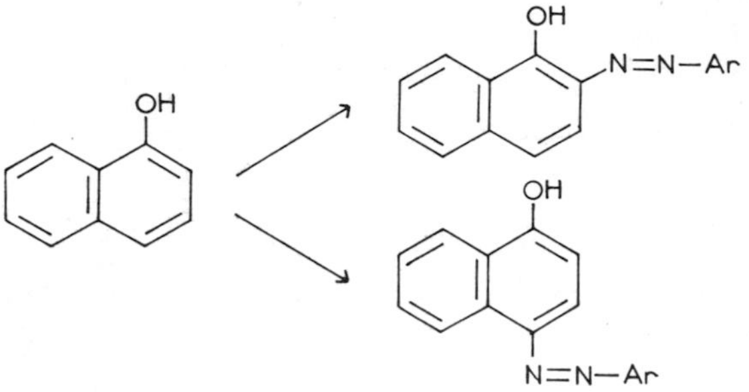

OH
N=N—Ar
OH
OH
N=N—Ar

(b)

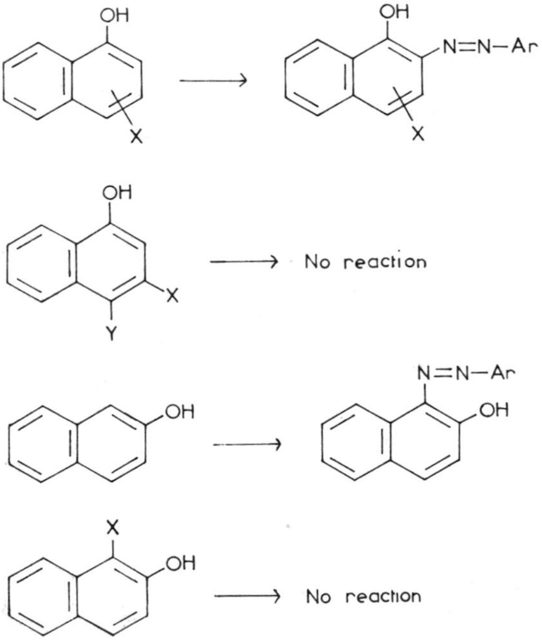

OH
OH
N=N—Ar
X
X

OH
No reaction
X
Y

N=N—Ar
OH
OH

X
OH
No reaction

Highly intense colors are generally developed, often allowing spectrophotometric determinations at the level of a few micrograms. The tints are most frequently closely linked to the structures of the phenol and of the diazonium salt, and to the pH of the reaction medium.

The capability of coupling and the reaction rate depend not only upon the location of the substituents on the phenolic ring, but also upon their kind. These facts, which have been turned to account for the selective determination of various phenols in admixture (2), become an impediment in functional group analysis, since with a given diazonium salt it is almost impossible to propose a general method applicable to all phenolic compounds. This drawback can be largely overcome with two reagents: phenitrazole, or 3-phenyl-5-nitrosamino-1,2,4-thiadiazole (I) (3), a stable primary nitrosamine which is converted into a diazonium salt in acid medium; and diazotized 2-aminobenzothiazole (II) (4).

$$C_6H_5-\underset{\text{(I)}}{C_2N_2S}-NH-NO \xrightarrow{H^+} C_6H_5-C_2N_2S-\overset{+}{N}\equiv N$$

$$C_6H_4\text{(benzothiazol-2-yl)}-\overset{+}{N}\equiv N \quad \text{(II)}$$

With each of these reagents, a single procedure permits the colorimetric determination of a great number of phenols; solely the wavelength of the absorption maximum varies with the compound tested.

As others, these reagents also couple with aromatic amines and the reaction follows the rule according to which the highest sensitivity is generally reached with N,N-dialkyl arylamines. For the same purpose, Sawicki et al. (5) introduced 4-azobenzenediazonium fluoborate. The tint developed with this reagent depends upon the acidity of the reaction mixture (see Chapter 5, Section II.D).

p-Nitrobenzenediazonium ion couples with primary aliphatic amines, which can thereby be determined in the presence of secondary and tertiary amines, and with amino acids (6).

Coupling with nitroparaffins may give rise to an equilibrium between two tautomeric forms. In the presence of excess reagent, a formazan (see Section VIII) can hence be formed. This is the case for nitromethane when it is reacted with p-iodobenzenediazonium ion (7).

$$I-C_6H_4-\overset{+}{N}\equiv N + CH_3-NO_2 \longrightarrow I-C_6H_4-N=N-CH_2-NO_2$$

$$\rightleftharpoons I-C_6H_4-NH-N=CH-NO_2 \xrightarrow{I-C_6H_4-\overset{+}{N}\equiv N} (I-C_6H_4-N=N)(I-C_6H_4-NH-N=)C-NO_2$$

Ketones react in a similar way (8). In this instance, the tautomer which leads to a formazan is the hydrazone of

$$Ar-\overset{+}{N}\equiv N + H_3C-CO-CH_3 \longrightarrow Ar-N=N-CH_2-CO-CH_3 \rightleftharpoons Ar-NH-N=CH-CO-CH_3$$

an aldehyde. It may be concluded therefrom that aldehydes can be estimated through hydrazone derivatives. For example, a colorimetric determination is based upon the reaction of their phenylhydrazones with phenitrazole (3). Sawicki and Stanley (9) proposed the use of 2-hydrazinobenzothiazole, the color being developed with p-nitrobenzenediazonium fluoborate.

The diazo species can be generated in the reaction medium by oxidizing the excess hydrazine compound. This mechanism is probably involved in the Schryver reaction, which is almost specific for formaldehyde (see Section V). Such a selectivity is not observed with 2-hydrazinobenzothiazole, although the reaction is less sensitive with other aldehydes (10) (mechanism, see Chapter 8, Section IV.A). More sensitive and less selective results are obtained with 3-methylbenzothiazolin-2-one hydrazone (11) (see Chapter 8, Section IV.B).

Other compounds with an active methylene group, such as β-diketones and β-keto esters or acids, also react, but the mechanism is often more complicated. For instance, with acetoacetic acid three consecutive reactions are involved:

$$CH_3-CO-CH_2-CO_2H + Ar-\overset{+}{N}\equiv N \longrightarrow CO_2 + H^+ + CH_3-CO-CH=N-NH-Ar$$

$$\xrightarrow{Ar-\overset{+}{N}\equiv N} CH_3-CO-C\begin{matrix}\diagup N=N-Ar \\ \diagdown\!\!\diagdown N-NH-Ar\end{matrix} \xrightarrow{Ar-\overset{+}{N}\equiv N} Ar-N=N-C\begin{matrix}\diagup N=N-Ar \\ \diagdown\!\!\diagdown N-NH-Ar\end{matrix} + CH_3-CO_2H$$

The highly colored species formed allows very sensitive measurements, and the method has been applied to the determination of acetoacetic and oxaloacetic acids in liver (12) and blood (13).

Coupling of diazonium salts with ethylenic compounds is but scarcely used in analysis. As a rule, with the exception of 2-methylpropene and 2-methyl-2-butene, only compounds with conjugated double bonds react and the yields are usually low, even with very reactive diazo compounds. Coupling occurs at the carbon atom having the highest electron density. This reaction, mainly studied by Terent'ev et al. (14), permitted the determination of some hydrocarbons such as isoprene (15).

The reaction of histidine and histamine with diazotized p-sulfanilic acid is commonly known as the Pauly reaction (16). It has been subjected to numerous variations in order to stabilize the color and to make it more sensitive. Other diazonium ions have also been proposed (17), and particular emphasis has been placed on the determination of the highly toxic histamine. In blood, it was estimated at the level of a few micrograms per liter with diazotized p-bromoaniline (18).

Other heterocyclic compounds, such as vitamin B_6 (19), 2-hydroxy-4,6-dimethylpyrimidine (20), and thymine (21) can also be determined by coupling. Uracil, cytosine, and purine derivatives from nucleic acids do not react with diazotized p-sulfanilic acid. Contrariwise, purine derivatives with a hydrogen atom at position 7 or 8 react with p-nitrobenzenediazonium ion. Theophylline, theobromine, caffeine, adenine, and guanine can be so determined, and the reaction has allowed the study of nucleic acids (22).

In alkaline medium, theophylline is decomposed to theophyllidine by opening of the pyrimidine ring, and the imidazole ring can be coupled with Fast Blue 2B Salt (zinc chloride complex of diazotized 1-amino-4-benzoylamino-2,5-diethoxybenzene). Theobromine and caffeine do not interfere (23). Diazotized 2-aminobenzothiazole allows the determination of guanine, whereas the reaction is almost negative with adenine (4).

Indole derivatives react only when the β position with respect to the nitrogen atom is unsubstituted (24).

B. Determinations through Diazotization

When coupled with a diazonium salt, primary arylamines usually afford but weakly sensitive colors. It is therefore advisable to diazotize the amine, then to develop the color by a coupling reaction. A great number of phenols and aromatic amines have been suggested for this second step. Before coupling, the excess nitrous acid must be destroyed by means of ammonium sulfamate in order to avoid side reactions. Numerous studies have dealt with the determination of local anesthetics and of sulfanilamide derivatives (2).

Diazotized p-aminosalicylic acid (PAS) is very unstable and is almost immediately converted to 2,4-dihydroxybenzoic acid (β-resorcylic acid) by loss of nitrogen (25). Primary aromatic amines can thereby be determined by diazotization in the presence of PAS. The diazotized amine couples with the β-resorcylic acid generated in the reaction medium, and prior destruction of the excess of nitrous acid is not necessary (26).

Nitro aromatic derivatives are detected and determined after the nitro group is reduced to an amine. Miscellaneous reducing agents have been proposed: zinc in acid medium, formamidine sulfinic acid (thiourea dioxide) (27), stannous chloride (28), titanous chloride (29), catalytic hydrogenation (30), etc.

Compounds with a benzene ring can be detected and determined by successive nitration, reduction, diazotization, and coupling. For instance, benzoic acid (31), cocaine (32), and phenobarbital (33) have been so estimated.

Primary arylhydrazines, on oxidation with selenium dioxide, afford diazonium salts which can be coupled with α-naphthylamine (34) or β-naphthol (35).

C. Determinations through Formation of Nitrous Acid

Methanol is determined by converting it to methyl nitrite. The ester is distilled and hydrolyzed, and the liberated nitrous acid diazotizes a primary arylamine. Development is then achieved by coupling (36).

Molecules possessing a thiol group are converted to their S-nitroso derivatives, which yield an equivalent of nitrous acid on mercuric ion-assisted hydrolysis. The acid so liberated is reacted with sulfanilamide, and the diazo compound is revealed with N-[1-naphthyl]ethylenediamine (Bratton-Marshall reagent) (37,38).

Upon the action of an inorganic or organic acid, sodium nitrite releases an equivalent amount of nitrous acid, which diazotizes sodium p-sulfanilate (39) or sulfanilamide (see Chapter 9, Section II.A). The diazo compound is then coupled with α-naphthylamine, thus allowing the colorimetric determination of free carboxylic acids.

Nitrite ion is liberated by the reaction of primary and secondary nitro aliphatic compounds with hydrogen peroxide in the presence of sodium hydroxide. The color is developed by diazotizing p-sulfanilic acid and coupling with 3-hydroxy-2-naphthoic acid (40).

D. Stability of Diazonium Salts

As a rule, diazonium salts are unstable, particularly when dissolved, and the reagent must usually be prepared immediately prior to use. When it is obtained by diazotizing a suitable amine in solution, high discrepancies may be observed between two series of runs if the operating conditions are not very precisely followed. To overcome this latter impediment, diazonium salts stable in the solid state may be used. Besides some arylsulfonates, the more stable salts are the fluoborates, generally but slightly soluble, such as the widely used p-nitrobenzenediazonium fluoborate, and double salts of diazonium chloride and a metal salt such as zinc chloride. A large variety of these complexes are commercially available (41,42). They have frequently been proposed for chromatographic detections, but also for determinations (41). Alternatively, phenitrazole has already been mentioned (Section I.A) as a highly stable diazo precursor.

Beyond the scope of this review, 2,6-dichloro-4-trimethylammoniumbenzenediazonium ion has been used analytically for the oxidation of primary alcohols to aldehydes (43), and 2,4,6-trichlorobenzenediazonium ion has been proposed for the colorimetric determination of tertiary aliphatic amines (43).

In this book, diazo coupling and diazonium salts were applied to the colorimetric determination of alcohols (Chapter 2, Section II.C), primary alcohols (Chapter 2, Section IV.B), phenols (Chapter 3, Sections II.G-II.J), primary aliphatic amines (Chapter 4, Section VI.E), tertiary aliphatic amines (Chapter 4, Section VIII.C), aromatic amines (Chapter 5, Sections II.C-II.F), primary aromatic amines (Chapter 5, Sections IV.K, IV.L), primary and secondary nitro aliphatic compounds (Chapter 7, Section III.A), primary nitro aliphatic compounds (Chapter 7, Section IV.B), nitro aromatic compounds (Chapter 7, Section VI.B), nitriles (Chapter 7, Section VIII.B), aldehydes (Chapter 8, Section III.A), carboxylic acids (Chapter 9, Section II.A), amino acids (Chapter 9, Section VI.C), thiols (Chapter 11, Section III.F), imidazole derivatives (Chapter 13, Section V.A), and hydroxysteroids (Chapter 15, Section II.B), and to the fluorimetric determination of primary alcohols (Chapter 2, Section XIII.A).

The introduction of a nitrogen-nitrogen double bond in a molecule usually has a quenching effect on fluorescence. Therefore, as a rule, the versatile diazo coupling reaction cannot, unfortunately, be extended to fluorimetric estimations. However, some primary aromatic amines were determined through their diazo derivatives by coupling with 2,6-diaminopyridine (44) (Chapter 5, Section V.B).

II. DEVELOPMENT OF AROMATIC NITRO COMPOUNDS

The Janovsky reaction of m-dinitro compounds with bases in the presence of an excess of active methylene solvent such as acetone or nitromethane is not dealt with herein, since it is the subject of another section (Section VI).

Although it may be assumed that the colors displayed by aromatic nitro compounds in alkaline media depend at times solely upon their existence as resonance hybrids, the following formulas do not account for the critical effects of the solvent and the base, and numerous experiments provide evidence that more complex mechanisms are usually involved.

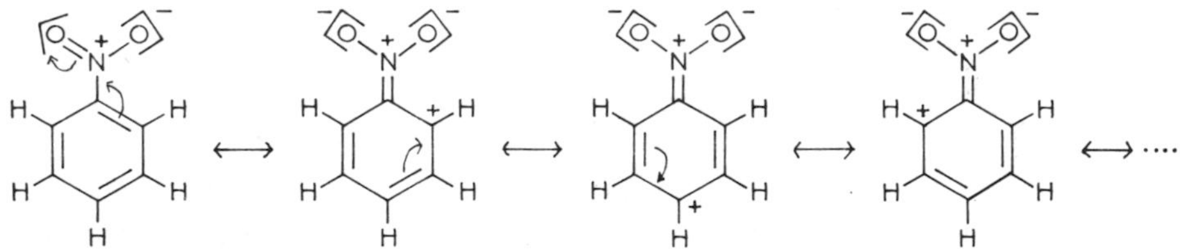

For instance, Foster (45) showed that the stoichiometry of the reaction of 1,3,5-trinitrobenzene with an amine in chloroform solution is three amine molecules to one trinitrobenzene, providing the amine molecule is small (methyl-, ethyl-, dimethyl-, or diethylamine). With more bulky amines, the stoichiometry is 1:1.

In alcoholic solution, all amines give 1:1 complexes, as established by Foster and Mackie (46), who suggested that the initial absorption spectrum is the result of covalent bond formation between the nitrogen of the amine molecule and the 2-carbon atom in the trinitrobenzene molecule, giving rise to (III). Secondary reactions then occur, and the situation may be further complicated by solvolysis with consequent attack by the ethoxide ion.

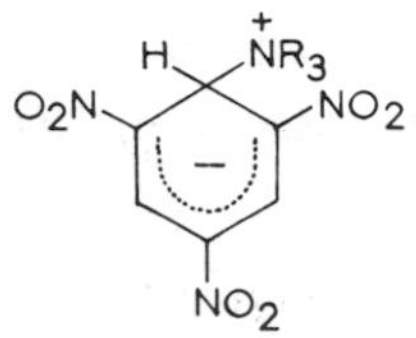

(III)

The close resemblance of the initial spectra of primary amines with trinitrobenzene in chloroform, compared with the same system in ethanol, suggests that the same species (III) is involved, irrespective of the solvent. In chloroform solutions involving small primary or secondary amines, the two extra amine molecules stabilize structure (III) by solvation.

Other reactions, whose mechanisms are almost unknown or are still being discussed, also play a role in various colorimetric determinations of aromatic nitro compounds, as emphasized by the following examples.

Porter (47) observed that certain mononitro compounds, principally derivatives of o- and p-nitroaniline and of p-nitrotoluene, develop orange, red, or purple colors in dimethylformamide upon the addition of tetraethylammonium hydroxide. None of the compounds tested gave more than a yellow color with acetone and sodium hydroxide. A mere Janovsky-like reaction can therefore hardly be admitted to explain the colors displayed. The Porter reaction was applied to the determination of chloramphenicol, but it is then worth operating with dimethylformamide containing very small amounts of acetone (48).

3,5-Dinitro-o-toluamide and related compounds form purple complexes when allowed to react with methylamine in dimethylformamide (49). Here also, the Janovsky reaction can hardly be taken into account, since it was demonstrated that the development of the color depends upon the class of the amine. Only those bases containing a primary amino group produce colored species, and more stable complexes are obtained with diamines such as 1,3-propanediamine (50).

Heotis and Cavett (51) determined some m-dinitro compounds by development with diethylamine in dimethyl sulfoxide. The authors proposed a mechanism involving a Janovsky-like reaction, giving rise to a compound of type (IV).

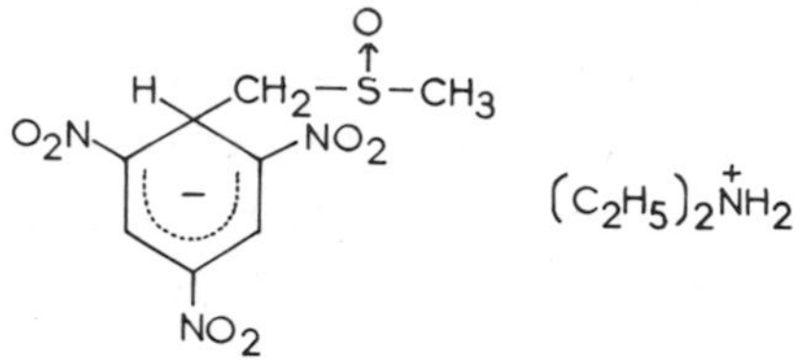

$(C_2H_5)_2\overset{+}{N}H_2$

(IV)

This statement is not in agreement with the studies of Crampton and Gold (52), who established that only primary and secondary aliphatic amines develop a color with 1,3,5-trinitrobenzene in anhydrous dimethyl sulfoxide, and NMR absorptions are consistent with structure (V).

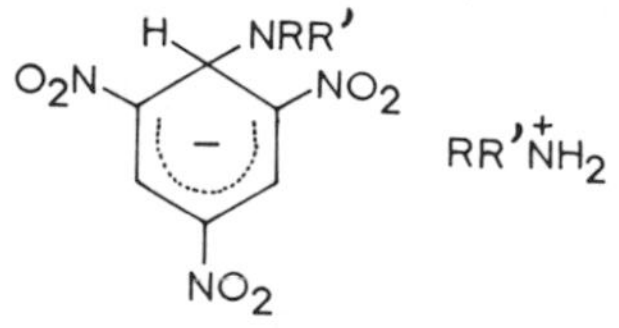

$RR'\overset{+}{N}H_2$

(V)

Tertiary amines develop only a slight color, which is attributed by the authors to trace amounts of primary or secondary amines in the reagents used.

Colorimetric determinations of various classes of compounds through the formation of aromatic m-dinitro derivatives can be bound up with the above reactions.

The 3,5-dinitrobenzoates obtained from primary and secondary alcohols are extracted into hexane and developed with dimethylformamide and propylenediamine (53).

A suitable selection of the base can avoid prior extraction of the condensate or elimination of the excess reagent, which displays no color under proper operative conditions. For instance, the esters obtained by reacting primary and secondary alcohols with 3,5-dinitrobenzoyl chloride are developed with dimethylformamide and piperazine. The N-alkyl-2,4-dinitroanilines formed by condensing primary alkylamines with 1-fluoro-2,4-dinitrobenzene are revealed with dimethylformamide and benzyltrimethylammonium hydroxide, and the same solvent and base allow the determination of aldehydes and ketones through their p-nitrophenylhydrazones (54). The esters obtained by reacting neutralized carboxylic acids with p-nitrophenacyl bromide are developed with dimethyl fulfoxide and diethylamine (55). Somewhat different mechanisms may be involved in the determination of primary and secondary aliphatic and aromatic amines with p-nitrophenylazobenzoyl chloride in dimethyl sulfoxide dioxane in the presence of benzyltrimethylammonium hydroxide (56).

In this book, development of aromatic nitro compounds in alkaline medium (with the exception of determinations through the Janovsky reaction) was applied to the colorimetric determination of primary and secondary alcohols (Chapter 2, Sections III.B, III.C), primary and secondary aliphatic amines (Chapter 4, Sections IV.B, IV.F), primary aliphatic amines (Chapter 4, Section VI.B), primary and secondary aromatic amines (Chapter 5, Section III.B), aldehydes and ketones (Chapter 8, Section II.D), p-quinones (Chapter 8, Sections X.A, X.B), carboxylic acids (Chapter 9, Section II.B), aromatic nitro compounds (Chapter 7,

Section VI.C), pentoses (Chapter 14, Section VI.E), hydroxysteroids (Chapter 15, Section II.D), and ketosteroids (Chapter 15, Section IV.C).

III. 3-METHYLBENZOTHIAZOLIN-2-ONE HYDRAZONE

Although 3-methylbenzothiazolin-2-one hydrazone (MBTH) was first prepared in 1910 (57), its analytical abilities were evidenced only in 1957 (58), and it was introduced as a reagent for colorimetric determinations in 1961 (11). It has since then become an analytical tool of considerable versatility.

MBTH can react with carbonyl derivatives through its hydrazone grouping. On the other hand, it forms a strongly electrophilic diazonium salt when acted upon by an oxidizing agent. These properties lead the way to colorimetric determinations based on the formation of formazans (Section VIII.B) and on diazo coupling (Section I).

The first procedure described by Sawicki et al. (11) allowed the determination of aldehydes, with which MBTH condenses to give a blue cation according to the following scheme:

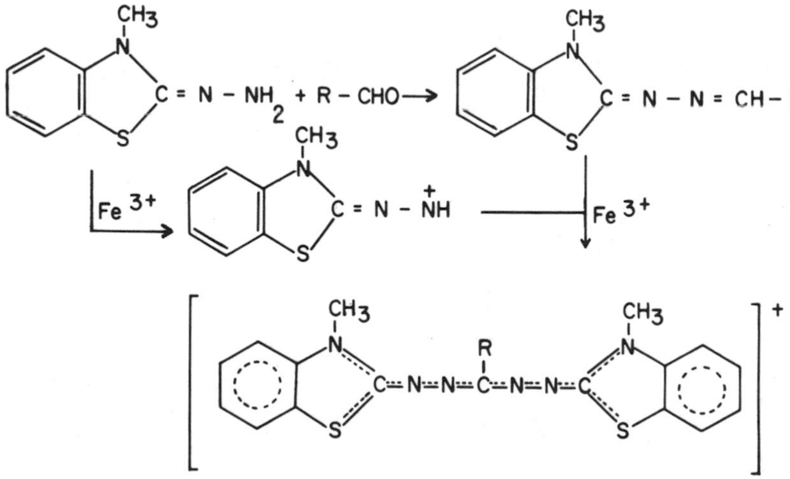

This technique was later improved, allowing more sensitive determinations (59). The reaction was applied to atmospheric analysis of aliphatic aldehydes (60) and the detection of free aldehyde groups in tissue sections and collagen (61). Very sensitive results are also obtained with glyoxylic acid (62).

Aldoses and ketoses, and 17-hydroxy-17-ketolsteroids, can be determined with a somewhat different procedure (63). It may be postulated that, at least with ketoses and ketolsteroids, the hydrazone initially formed is converted to the corresponding osazone, thus giving rise to an aldehyde-hydrazone grouping.

The method can, of course, be extended to compounds which afford aldehydes upon suitable treatment.

$$R-CH_2OH \xrightarrow{N\equiv \overset{+}{N}-C_6H_2Cl_2-\overset{+}{N}(CH_3)_3} R-CHO \qquad (43)$$

$$\left.\begin{array}{l} R-CH_2OH \\ R-CH=C< \end{array}\right] \xrightarrow{RuO_4} R-CHO \qquad (64)$$

$$\text{Glycerides} \xrightarrow{OH^-} \text{Glycerol} \xrightarrow{IO_4^-} HCHO \qquad (65)$$

$$\left.\begin{array}{l} \text{Mannitol} \\ \text{Sorbitol} \end{array}\right] \xrightarrow{IO_4^-} HCHO \qquad (66)$$

$$\underset{\text{Ephedrine}}{C_6H_5-CHOH-CH(CH_3)-NH-CH_3} \xrightarrow{IO_4^-} C_6H_5-CHO + CH_3-CHO + CH_3-NH_2 \qquad (67)$$

$$\underset{\text{Phenylephrine}}{(\underline{m}-OH)-C_6H_4-CHOH-CH_2-NH-CH_3} \xrightarrow{IO_4^-} (\underline{m}-OH)-C_6H_4-CHO + HCHO + CH_3-NH_2 \qquad (67)$$

$$R-CHOH-CHOH-CO_2H \xrightarrow{IO_4^-} R-CHO + OCH-CO_2H \qquad (68)$$

$$\text{Thiazolidine-4-carboxylic acids} \xrightarrow{I_2} \text{Aldehyde} \qquad (69)$$

$$\text{2-Amino-2-deoxyhexoses} \xrightarrow{HNO_2} \text{2,5 Anhydrohexoses} \qquad (70)$$

The colored species formed with formaldehyde can be extracted into a mixture of organic solvents (71).

It is obvious that ketones cannot react in this way. However, the colorless azine derivatives which they form with MBTH under defined conditions have characteristic spectra which permit their identification, as well as that of saturated and unsaturated aldehydes, keto acids, and many other related compounds. Carbohydrates having a pyranose structure are unreactive in this method (72).

Pyridoxal and its 5'-phosphate can be determined in the presence of each other, in the absence of oxidant (73).

Glyoxal reacting with MBTH in acetic acid medium affords a yellow diazine which allows its determination in the presence of unsubstituted monoaldehydes, when no oxidant is present (74).

It was stated that, upon the action of a suitable oxidant, MBTH forms a diazonium salt which can couple with various compounds. Phenol was so determined, the oxidant being cerium(IV) ammonium sulfate (75). The reaction was extended to miscellaneous other phenols, using various oxidants (76), and an automated method was proposed by Friestad et al. (77).

The following structure was attributed to the colored species:

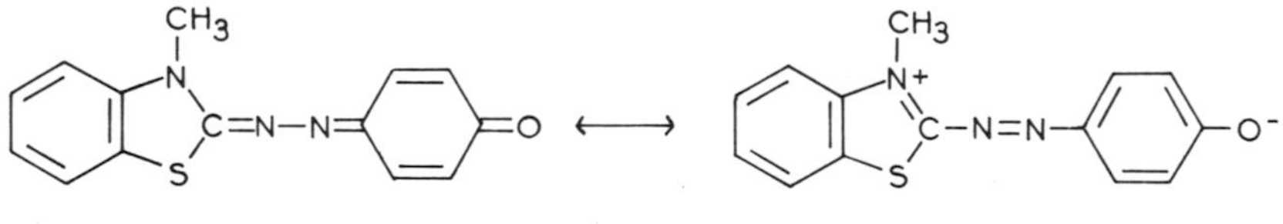

However, a general method giving good results with all the phenols tested could not be developed in our laboratory.

Contrariwise, a general procedure affording satisfactory and sensitive results is available for aromatic amines (78). Much weaker colors are obtained with aliphatic and alicyclic amines (79). More generally, this method allows the determination of compounds containing the grouping (A).

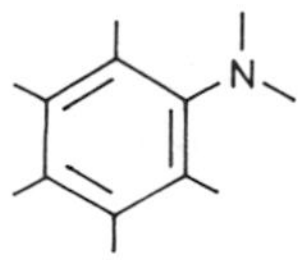

(A)

Members of this family include primary aromatic amines, aralkylamines, aryl dialkylamines, diarylamines, indoles, carbazoles, and phenothiazine-type compounds (78).

The following structure was attributed to the colored cation given by N,N-dimethylaniline:

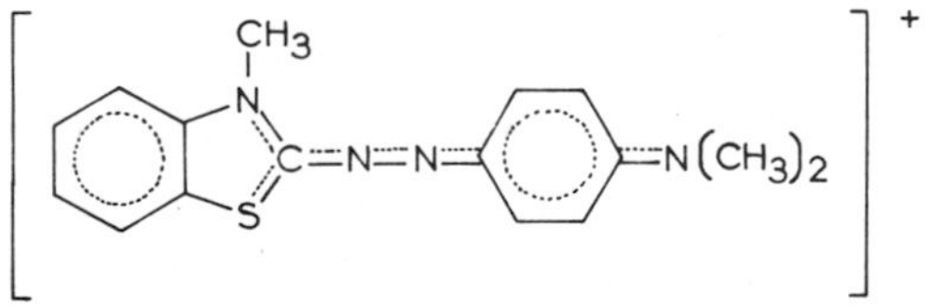

Azo dyes, stilbenes, and Schiff bases, as well as pyrrole derivatives, also react under oxidative conditions (80). This reaction was extended to the colorimetric determination of bilirubin and its oxidation products such as urobilin and biliverdin (81).

With azulene and its compounds, the mechanism probably involves the following reaction (82):

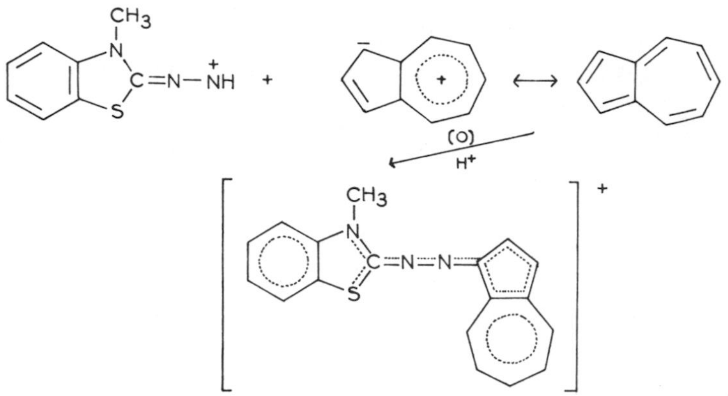

It is obvious that as a result of its versatility, this reagent can be used for specific and sensitive determinations only, provided that other species which could also give positive results are either absent or suitably eliminated from the reaction medium.

In this book, MBTH was applied to the colorimetric determination of primary alcohols (Chapter 2, Sections IV.A, IV.B), aromatic amines (Chapter 5, Section II.G), aliphatic aldehydes (Chapter 8, Section IV.B), ethylenic compounds of the type $R{-}CH{=}C\langle$ (Chapter 12, Section II.A), aldoses and ketoses (Chapter 14, Section II.G), 2-amino-2-deoxyhexoses (Chapter 14, Section X.G), and 17-hydroxy-17-ketolsteroids (Chapter 15, Section XIV.C).

IV. PERIODATES AND PERIODATE OXIDATION

The commercially available periodates are periodic acid, H_5IO_6 (mol wt 298), sodium metaperiodate, $NaIO_4$ (mol wt 214) or $NaIO_4$, $3H_2O$ (mol wt 268), sodium paraperiodate, $Na_3H_2IO_6$ (mol wt 294), and potassium metaperiodate, KIO_4 (mol wt 230).

Periodic acid is freely soluble in water. When heated in vacuum (100°, 12 mm) it is converted to the meta acid, HIO_4, an intermediate, $H_4I_2O_9$, being formed at 80°.

The solubility of sodium metaperiodate in water is dependent upon the temperature. It is about 9% at 20° for the anhydrous salt. It decreases in alkaline medium, because of the formation of $Na_2H_3IO_6$, then $Na_3H_2IO_6$, both very sparingly soluble salts.

Trisodium paraperiodate gives sodium metaperiodate by treatment with nitric acid.

Potassium metaperiodate is sparingly soluble in water (0.4% at 20°), more soluble in acid or alkaline media.

Ultraviolet and kinetic studies showed that at a pH of about 10 a dimerization of the periodate ion takes place, accompanied by a weakening of the oxidizing properties of the solution (83).

An 0.05 M solution of a metaperiodate is 0.1 N.

From the time when Malaprade discovered the periodate oxidation (84), a considerable number of papers dealing with this reaction have been published, mainly in the field of carbohydrates (elucidation of structures, determinations, etc.), and it is impossible to present here even a very rough review of all that has been done. The corresponding references may be found in several monographs (85). The table given here only summarizes the reaction schemes for various functional groups.

Simple alcohols such as methanol or ethanol, and compounds bearing two or more nonadjacent hydroxyl groups, such as 1,3-propanediol (trimethylene glycol) or pentaerythritol, are not oxidized (86). Acylation of an OH or NH_2 group often prevents oxidation. However, the sulfonate group attached to an amine is not protecting: N-Sulfonylserine consumes 2 moles of periodic acid almost as rapidly as serine (87).

The effect of the stereochemical configuration on oxidation rates has been evidenced. For instance, in the case of aldohexoses (glucose, mannose, galactose) or inositols, periodic acid oxidizes much more readily the bond between two carbon atoms bearing hydroxyl groups in the cis position than the corresponding trans configuration (88).

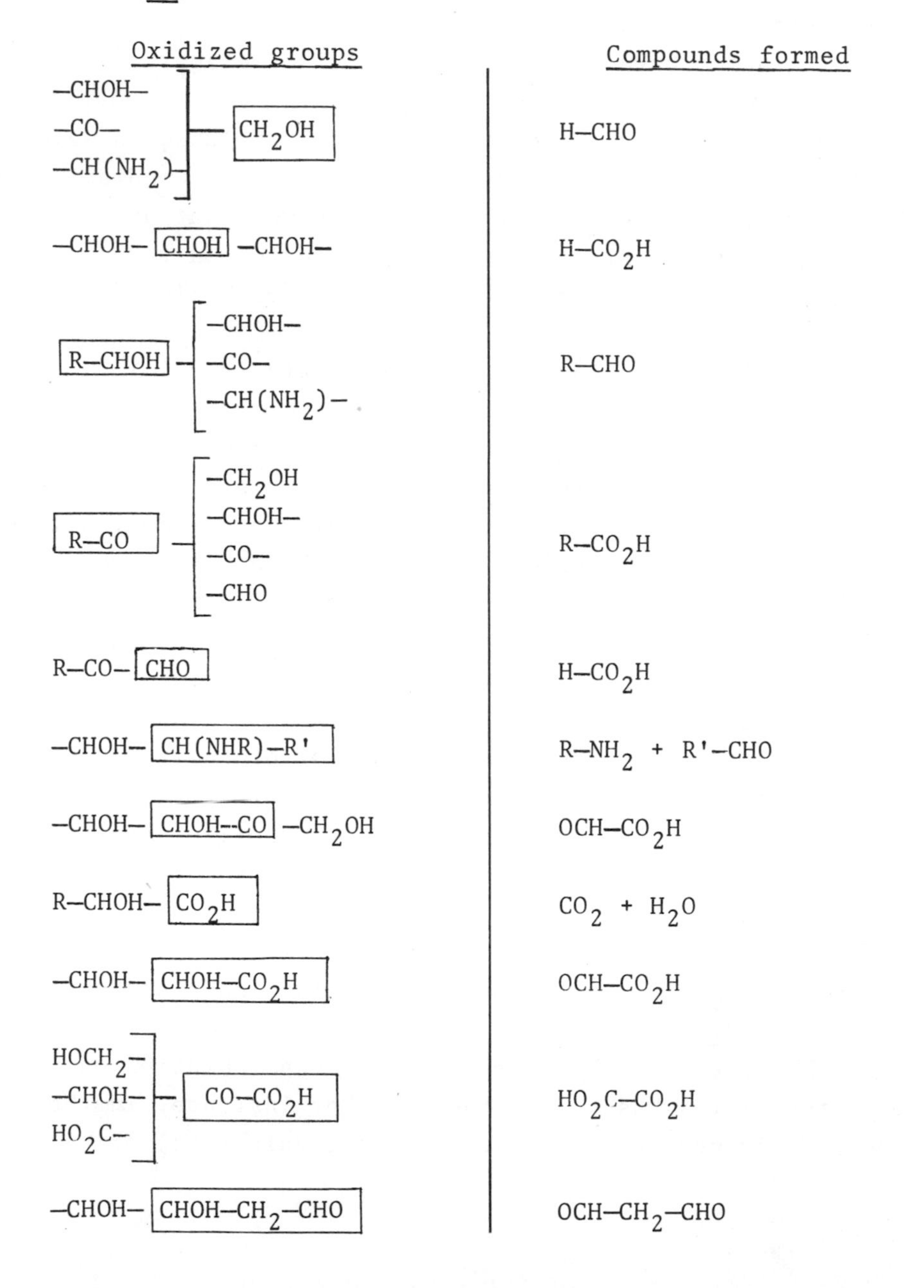

It must be emphasized that periodate oxidation is specific only when proper conditions are observed: room temperature, dilute aqueous solutions, oxidation time not exceeding a few hours. At 100°, for example, periodic acid oxidizes methanol to formaldehyde and formic acid, and ethanol to acetaldehyde.

In this book, periodate oxidation was used for the determination of 1,2-diols (Chapter 2, Sections VII.A, XIV.A), hexitols (Chapter 2, Sections VIII.A, XV.A), α-amino primary alcohols (Chapter 2, Sections IX.A, XVI.A), and 17-ketolsteroids (Chapter 15, Sections XIII.B, XVIII.A), which give formaldehyde; 2,3-dihydroxy-1-carboxylic acids (Chapter 9, Sections V.A, X.A) and hexuronic acids (Chapter 14, Sections XI.B, XVII.A), which give glyoxylic acid, and 2-deoxy sugars, which give malonic dialdehyde (Chapter 14, Sections IX.D, IX.E).

In the presence of permanganate ion, periodate ion also allows the oxidation of ethylenic compounds of the type $R{-}CH{=}CH_2$ (89). The double bond is hydroxylated by permanganate, then periodate ion oxidizes the diol or the ketol so formed, liberating formaldehyde.

$$R{-}CH{=}CH_2 \xrightarrow[IO_4^-]{MnO_4^-} \left[\begin{array}{c} CH_2OH{-}CHOH{-}R \\ \text{and} \\ CH_2OH{-}CO{-}R \end{array} \right] \longrightarrow HCHO$$

The reduced permanganate is reoxidized by the excess of periodate, so that only a catalytic amount of permanganate can be introduced into the reaction medium, and it can be replaced by an equivalent amount of a manganese(II) salt.

In this book, this reaction was applied to the colorimetric and fluorimetric determinations of compounds of the type R—CH=CH_2 (Chapter 12, Sections III.A, III.B, V.A).

V. COLORIMETRIC AND FLUORIMETRIC DETERMINATIONS OF FORMALDEHYDE

A. Colorimetry

Miscellaneous reagents have been proposed for the more or less specific and sensitive determination of formaldehyde.

Schiff reagent, obtained by passing sulfur dioxide into a solution of fuchsine until the color is just discharged, turns to red-violet upon addition of an aldehyde. This reaction, generally attributed to Schiff (90), is sometimes mentioned by the names of Caro (91), Grosse-Bohle (92) or Chautard (93). Deniges (94) pointed out that formaldehyde can be differentiated when the reaction is performed in strongly acid medium. The test depends upon the fact that the color given by formaldehyde under these conditions does not fade appreciably for several hours, whereas that which is developed by higher aldehydes fades completely within 2 hr (95,96).

(a)

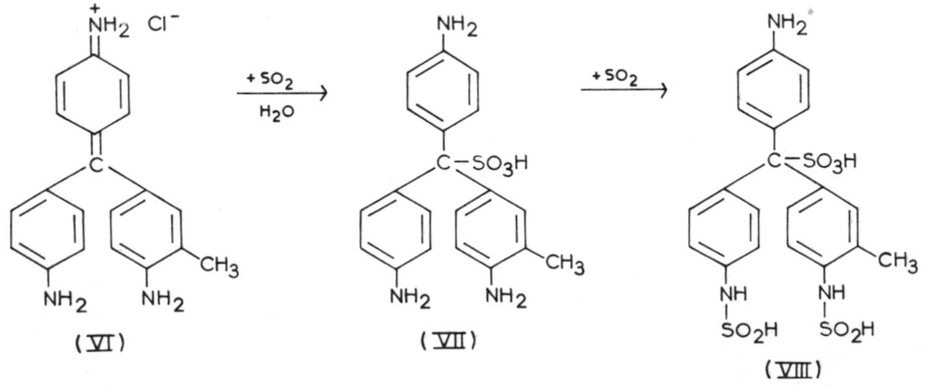

(VI) $\xrightarrow[H_2O]{+SO_2}$ (VII) $\xrightarrow{+SO_2}$ (VIII)

According to Wieland and Scheuing (97) sulfur dioxide and fuchsine (VI) form a leuco derivative (VII), then a colorless sulfinic acid (VIII). This reacts with 2 moles of a carbonyl derivative to give an unstable compound (IX), which by loss of sulfur dioxide affords the quinoidal derivative (X), responsible for the color.

(b)

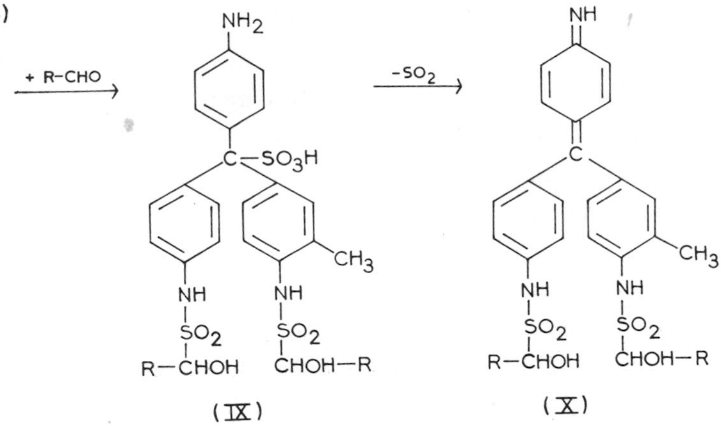

$\xrightarrow{+\ R\text{-}CHO}$ (IX) $\xrightarrow{-SO_2}$ (X)

In spite of numerous studies, the use of Schiff reagent for the determination of formaldehyde remains difficult. Beer's law is not satisfactorily followed (98), and the procedure is time-consuming.

In the reaction described by Schryver in 1910 (99), formaldehyde is at first converted into its phenylhydrazone, then a red color is developed under the action of an oxidant. It is assumed that the excess of phenylhydrazine is oxidized into the corresponding diazo compound, which then condenses with the hydrazone to give a strongly colored formazan (see Section VIII).

Numerous procedures have been proposed, particular emphasis being placed on the choice of the oxidizing agent: potassium ferricyanide, generally used in acid medium, but also works in alkaline solution (100); ferric chloride in sulfuric acid medium (101); sodium nitroprusside (101,102) or air (103), in alkaline media. Large amounts of other aldehydes may interfere (98).

In this book, the Schryver reaction was applied to the colorimetric determination of 1,2-diols (Chapter 2, Section VII.A), α-amino primary alcohols (Chapter 2, Section IX.A), ethylenic compounds of the type $R{-}CH{=}CH_2$ (Chapter 12, Section III.A), and 17-ketolsteroids (Chapter 15, Section XIII.B). All these compounds give rise to formaldehyde through suitable chemical reactions.

The reaction also allows the detection of methanol oxidized beforehand by ammonium persulfate (104).

Chromotropic acid (1,8-dihydroxynaphthalene-3,6-disulfonic acid), proposed as a reagent by Eegriwe (105), develops an intense violet color in sulfuric acid medium in the presence of formaldehyde. Kamel and Wizinger (106) admit the formation of a dinaphthylmethane derivative, which upon oxidation would give a dibenzoxanthylium-type dyestuff (XI)

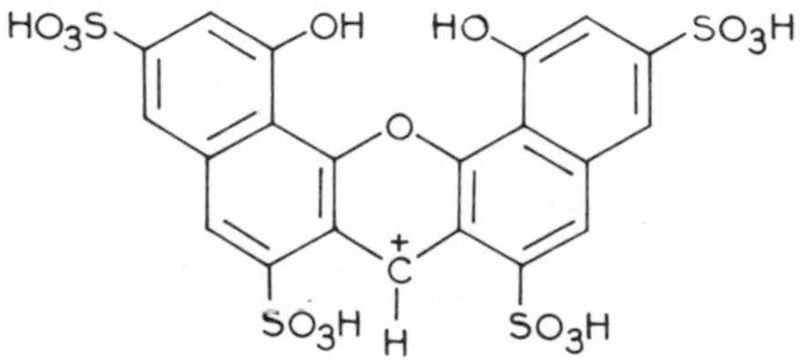

(XI)

The reaction is selective for formaldehyde, but also gives positive results with compounds which liberate formaldehyde in sulfuric acid medium, such as apiol, narceine, narcotine, piperine, piperonal, and safrole (107), glycolic acid (105,108), phenoxyacetic acid and its chloro derivatives (109), monochloroacetic acid (110), ethylenediaminetetraacetic acid (111), and hexoses through 5-hydroxymethylfurfural (112).

Chromotropic acid also allows the characterization and determination of formic acid reduced to formaldehyde by magnesium (105,113), of methanol oxidized to formaldehyde by potassium permanganate (114), and of compounds which upon suitable oxidation give rise to this aldehyde (115) (see also below, applications to determinations in this book).

Glyceraldehyde, dihydroxyacetone, and pyruvic aldehyde develop a yellow color and a green fluorescence (116). In phosphoric or sulfuric acid media, chromotropic acid affords a yellow-green fluorescence with 17-hydroxy-17-ketolsteroids, whereas the 18-nor compounds to not react (117).

In this book, chromotropic acid was applied to the colorimetric determination of hexitols (Chapter 2, Section VIII.A), aliphatic 2-nitro-1-hydroxy compounds (Chapter 7, Section V.A), ethylenic compounds of the type R—CH=CH_2 (Chapter 12, Section III.B), and hexoses (Chapter 14, Section VIII.A). All these compounds give rise to formaldehyde through suitable chemical reactions.

J-Acid (6-amino-1-naphthol-3-sulfonic acid) forms with formaldehyde a dibenzoxanthylium salt, whose structure (XII) has been established (106).

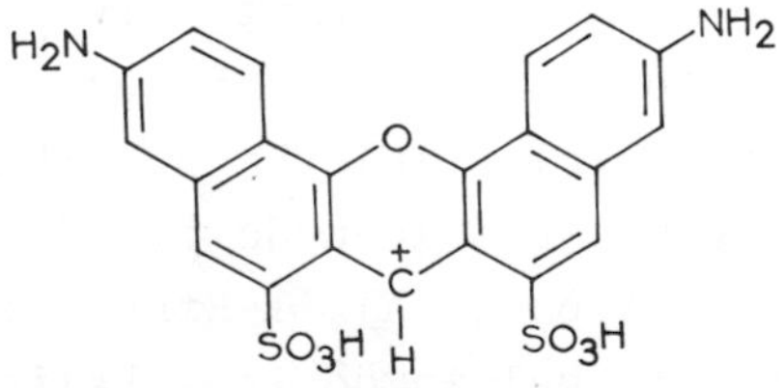

(XII)

According to the procedures described by Sawicki et al. (98), a yellow color (λ Max, 468 nm), or a blue color λ Max, 612 nm) is developed. Phenyl-J-acid (6-anilino-1-naphthol-3-sulfonic acid) can be used instead of J-acid (98), the absorption maximum of the blue color being located at 660 nm.

The sensitivities of the various colorimetric methods as compared by Sawicki et al. (98) are the following (the references correspond to the procedures used):

		Sensitivity $\left[\frac{\varepsilon \times 10^{-3}}{\text{Dilution factor}}\right]$
Schiff reagent	(96)	0.7
Schryver reaction	(100)	0.8
Chromotropic acid	(98)	1.6
J-Acid, yellow color	(98)	4.2
J-Acid, blue color	(98)	2.7
Phenyl-J-acid	(98)	4.1

The specificity of chromotropic and J-acids was attributed to the steric effect of the sulfonate group (106). Only acrolein may interfere appreciably (118).

The yellow color developed by 2,6-dimethyl-3,5-diacetyl-1,4-dihydropyridine allows a specific colorimetric determination of formaldehyde through the Hantzsch reaction (Section IX), the reagent being a solution of acetylacetone and ammonia (119).

Because of their high sensitivity, some less specific reactions have also been taken into account. Thus, with 2-hydrazinobenzothiazole (10) the absorbance value of 0.3 is read with 4 μg of formaldehyde or 150 μg of acetaldehyde (see Chapter 8, Section IV.A). 1-Ethylquinaldinium iodide gives with formaldehyde in alkaline medium the colored cation (XIII) (λ Max, 608 nm), and the reaction is 5 to 50 times less sensitive with other aldehydes (118).

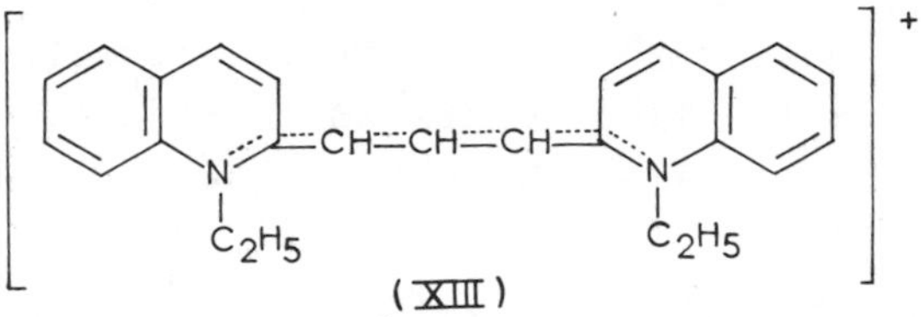

$$\left[\text{(quinoline)} \overset{\cdots}{=} CH \overset{\cdots}{=} CH \overset{\cdots}{=} CH \overset{\cdots}{=} \text{(quinoline)} \right]^+$$

N — C_2H_5 N — C_2H_5

(XIII)

Recently, Bailey and Rankin (120) developed a new method based on the catalytic effect of formaldehyde on the hydrogen peroxide oxidation of p-phenylenediamine, giving rise to the dyestuff named Bandrowski base (121):

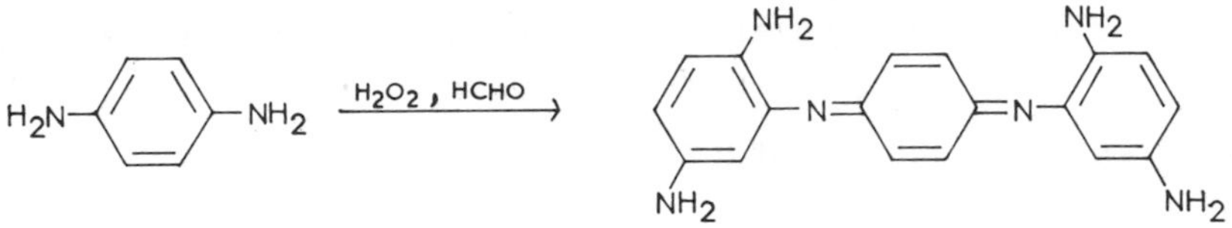

$$H_2N-C_6H_4-NH_2 \xrightarrow{H_2O_2,\ HCHO}$$

NH_2, NH_2, N=, =N, NH_2, NH_2

Although acetaldehyde be 10 times less reactive than formaldehyde, its interference should be taken into account.

B. Fluorimetry

Under suitable conditions, the dyestuff obtained by condensing formaldehyde with J-acid affords a green-yellow fluorescence (118), and this reaction was applied to the fluorimetric determination of formaldehyde formed in the oxidative degradation of miscellaneous compounds (122).

The yellow color of 2,6-dimethyl-3,5-diacetyl-1,4-dihydropyridine is accompanied by a yellow-green fluorescence, and the reaction was likewise applied to the fluorimetric determination of formaldehyde (123) and of compounds which give rise to this aldehyde upon suitable oxidation (122,124).

When, in this method, ethyl acetoacetate is substituted for acetylacetone, the 2,6-dimethyl-3,5-dicarbethoxy-1,4-dihydropyridine formed is colorless, but fluoresces intensely in the blue region. Under suitable conditions, this reaction is highly specific for formaldehyde, allowing therefore the fluorimetric determination of compounds which liberate this aldehyde by means of chemical reactions (125).

In this book, this reaction was applied to the determination of 1,2-diols (Chapter 2, Section XIV.A), hexitols (Chapter 2, Section XV.A), α-amino primary alcohols (Chapter 2, Section XVI.A), ethylenic compounds of the type $R—CH=CH_2$ (Chapter 12, Section V.A), and 17-ketolsteroids (Chapter 15, Section XVIII.A). All these compounds give rise to formaldehyde through suitable reactions.

VI. JANOVSKY AND ZIMMERMANN REACTIONS

In 1886, Janovsky (126) discovered that a solution of m-dinitrobenzene or of a substituted m-dinitrobenzene in acetone developed an intense violet color upon addition of potassium hydroxide. The reaction was extended to other ketones (127), then Reissert (128) established that it was not bound up with the carbonyl group, but with the presence of an active methylene group: $R—CH_2—NO_2$ and $R—CH_2—CN$ were shown to react also.

On the same basis, Zimmermann (129) studied the determination of 17-ketosteroids with m-dinitrobenzene in alkaline medium. Accordingly, he used an excess of the dinitro reagent and this method, applied to ketosteroids or other α-methylene ketones, has been subjected to numerous experimental variations and theoretical studies. The colored species may be extracted into an organic solvent (130); organic bases were proposed instead of potassium hydroxide (131); and 3,5-dinitrobenzoic acid (132), 1,3,5-trinitrobenzene (133), and picric acid (see Chapter 15, Section V.E) were shown to favorably replace m-dinitrobenzene. These variations were mainly aimed at stabilizing the colored species and enhancing the sensitivity. On the other hand, the mechanism of the reaction now seems well known. It is beyond the scope of this short review to present an exhaustive bibliography of all that has been done in this field; various monographs deal in detail with these matters (134).

Canbäck (135) attributed the color developed in the Janovsky reaction to the formation of a complex, and this hypothesis was later supported by a number of analytical determinations, often based on the study of IR and NMR spectra. When m-dinitro- or 1,3,5-trinitrobenzene is allowed to react with an excess of active methylene compound in alkaline medium, a complex of general formula (XIV), where R' comprises the activating group, is reversibly formed.

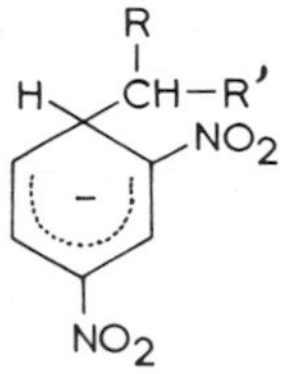

R
H CH—R'
NO_2
NO_2

(XIV)

This kind of formulation was originally introduced by Meisenheimer (136), and it may be stated that the Janovsky reaction leads to Meisenheimer-like σ complexes, as has been evidenced by studies of the derivatives obtained from m-dinitrobenzene and acetone (137,138), trinitrobenzene and miscellaneous ketones (139), m-dinitro- or trinitrobenzene, and nitromethane (138). A stable zwitterionic Meisenheimer complex was isolated by addition of an excess of triethylamine to a solution of 1,3,5-trinitrobenzene in excess of acrylonitrile (140).

In some instances, the reaction gives rise to a mixture of dyes. For example, complexes (XV) and (XVI) are simultaneously formed with methyl ethyl ketone (139).

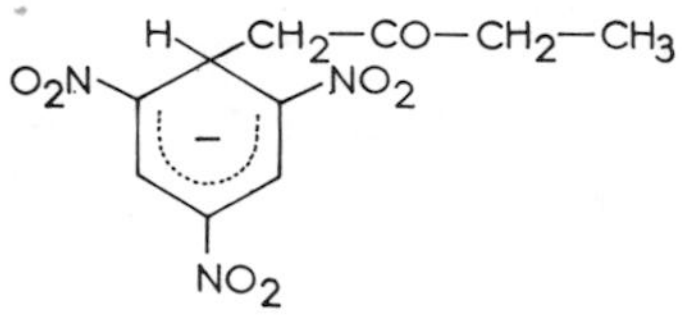

(XV) 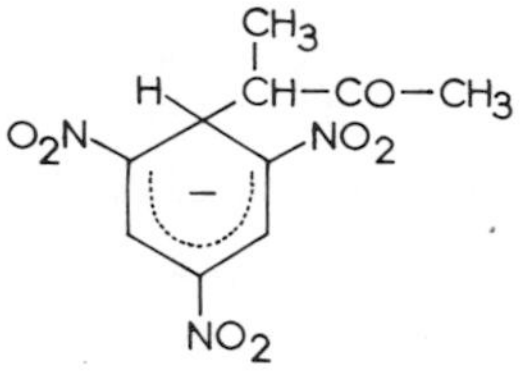 (XVI)

When operating in the presence of an excess of dinitro reagent (Zimmermann conditions), the Meisenheimer-like complex is irreversibly oxidized, thus affording the colored anion (XVII).

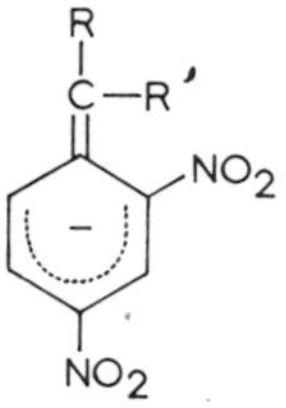

(XVII)

This oxidation step has been assumed by Canbäck (135) without experimental support, and by Ishidate and Sakaguchi (141), who postulated that the oxidant was the excess reagent, thereby reduced to m-nitroaniline. These hypotheses were later corroborated by a number of experimental data.

Kellie and Smith (142) showed that the violet product obtained from dehydroepiandrosterone and m-dinitrobenzene is in a pH-dependent equilibrium with an apparently colorless compound which they isolated as a paper-chromatographic fraction. This latter derivative was then obtained as yellow needles by Corker et al. (143), and analytical data were consistent with 3β-hydroxy-17-oxoandrost-5-en-16ξ-yl-2,4-dinitrobenzene (XVIII), the violet dye being the potassium salt of the p-quinonoid aci form (XIX).

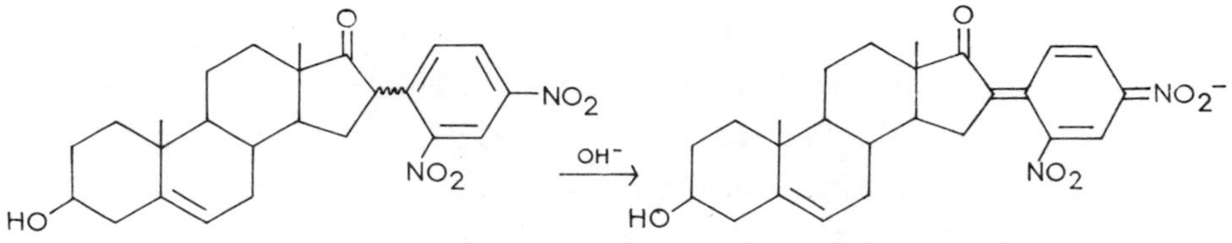

(XVIII) $\xrightarrow{OH^-}$ (XIX)

The corresponding derivative of androsterone has been characterized by Neunhoeffer et al. (144). On the other hand, King and Newall (145) obtained 2,4-dinitrobenzyl phenyl ketone from an alkaline mixture of acetophenone with an excess of m-dinitrobenzene. They also confirmed the previously assumed reduction of m-dinitrobenzene by isolating m-nitroaniline from an alkaline mixture of m-dinitrobenzene and acetone.

Foster and Mackie (146) further showed that two reactions may be involved when a m-dinitro derivative reacts with an active methylene compound. Under the Janovsky conditions, the intensity of the absorption band attains a maximum after a few minutes at a stable wavelength, whereas under the Zimmermann conditions the maximum, at 497 nm on mixing, shifts to 490 nm after 1 hr (experiments performed with m-dinitrobenzene and acetone). On the other hand, this latter wavelength is sensitive to the substituent R (formula XVII), whereas the maximum for complexes obtained under Janovsky conditions shows remarkably little sensitivity to this substituent (formula XIV).

It is therefrom well established that two reactions are involved when a m-dinitro compound reacts with an active methylene compound:

1. Janovsky conditions. With an excess of active methylene compound, a Meisenheimer-like complex (XIV) is reversibly formed, and the reaction is rapid.

2. Zimmermann conditions. With an excess of dinitro compound, the complex is irreversibly oxidized to give species (XVII). The reaction is slow.

The Janovsky reaction has been applied to the characterization and determination of m-dinitro compounds (49,51,147), and of miscellaneous benzene derivatives, including some barbiturates after suitable nitration (148). Acetone and methyl ethyl ketone have mostly been used for the development. They can be favorably replaced by nitromethane in the presence of benzyltrimethylammonium hydroxide. Beer's law is followed, thus allowing sensitive colorimetric determination of various benzene derivatives after nitration (149).

The same reagents also permit a highly sensitive determination of 2,4-dinitrophenylhydrazones of ketones. The excess 2,4-dinitrophenylhydrazine does not react under these conditions, and its prior elimination is thus avoided (150).

As a rule, the Janovsky reaction necessitates a strong base when a ketone is involved. However, with a solution of 1,3,5-trinitrobenzene in nitromethane, the alkalinity of an aliphatic amine is sufficient for the color development. The amount of complex formed being directly proportional to the amount of amine added, aliphatic amines, quaternary ammonium compounds, and guanidines can be so determined (151).

The Zimmermann reaction has been applied to the characterization and determination of diverse ketones (135), nitroaliphatic compounds (152), caprolactam (153), 2-methylchromones (154,155), terpenoids or aldehydes and ketones (156), cardiac glycosides (the reaction is due to the active hydrogen of a β,γ-lactone ring (157,158), and codeinone derivatives (159), but it was mostly used for the determination of 17-ketosteroids of biological origin

(129-134). However, a carbonyl group at other positions (particularly at position 3) can also react, and a correlation of structure and stereochemistry with reactivity has been emphasized (160,161) (see also Chapter 15, Section V.D, Note 3).

In this book, the Janovsky reaction was applied to the colorimetric determination of aliphatic amines, guanidines, and quaternary ammonium compounds (Chapter 4, Section II.A), ketones through their 2,4-dinitrophenylhydrazones (Chapter 8, Section VII.A), and benzene derivatives (Chapter 12, Section IV.A). The Zimmermann reaction was applied to the colorimetric determination of α-methylene ketones (Chapter 8, Section VIII.B), and 3- and 17-ketosteroids (Chapter 15, Sections V.A-V.E).

VII. 1,2-NAPHTHOQUINONE-4-SULFONIC ACID

The reaction of 1,2-naphthoquinone-4-sulfonic acid with primary aromatic amines was discovered by Böniger (162) as far back as 1894. This author isolated the derivatives obtained from aniline, 4-aminoazobenzene, p-phenylenediamine, p-sulfanilic acid, etc.

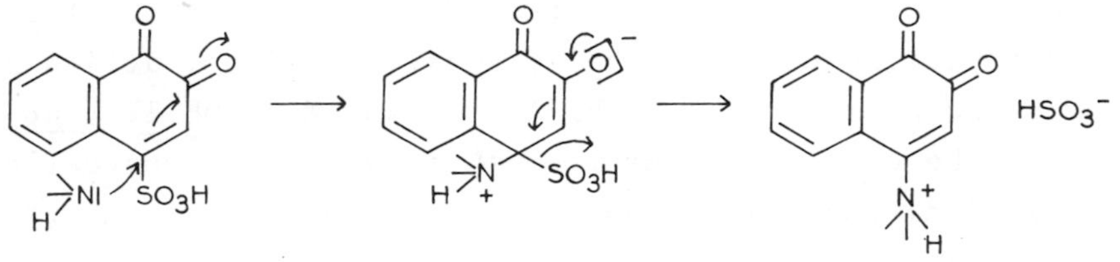

The reaction was then applied to the characterization of primary aromatic amines (163), and later formed the basis of colorimetric determinations of amino acids (164), sulfonamides (165), and primary and secondary aliphatic amines (166).

In the case of a primary amine, the adduct can be represented by either the aminoquinone structure (XX) (which was introduced in the above mechanism), the quinone-imine structure (XXI), or an equilibrium mixture thereof. Obviously, only the quinone structure can be taken into account with a secondary amine.

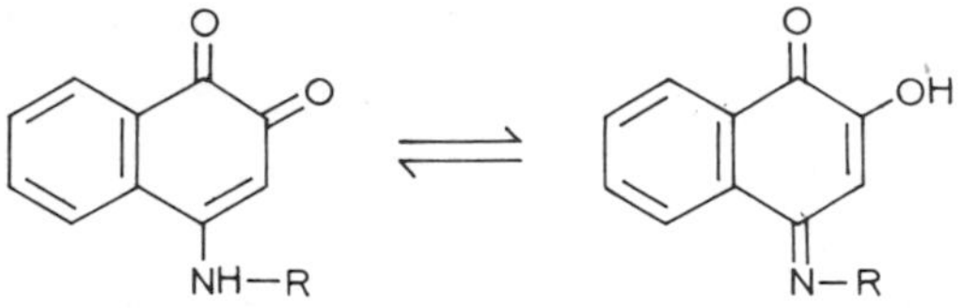

(XX) (XXI)

The absorption spectra of aniline, proline, hydroxyproline, and sarcosine derivatives were studied by Troll (167) in terms of pH. The spectra of proline and hydroxyproline compounds do not vary over the pH range 5-10, whereas with the aniline adduct the absorption maximum, at 480 nm at pH 5, shifts towards shorter wavelengths as the pH increases.

Measurements performed with the di-n-propylamine derivative showed that the absorption maximum, located at 480 nm, does not vary over the pH range 2-12. With the n-propylamine compound the absorption maximum, at 460 nm over the pH range 2-10, shifts to 440 nm at pH 12 (168). It may be concluded therefrom that, under the conventional analytical conditions, an equilibrium between forms (XX) and (XXI) may intervene for the derivatives obtained from primary amines, but the quinone structure is mostly favored.

The aminonaphthoquinones can react with 2,4-dinitrophenylhydrazine to give intensely colored species, thus enhancing the sensitivity of the colorimetric determination of primary and secondary aliphatic amines, and primary aromatic amines (169).

On the other hand, under suitable conditions, solely the N-substituted 4-amino-1,2-naphthoquinones obtained from amines can be extracted into methylene chloride, whereas the amino acid derivatives remain in the aqueous layer, thus affording a possibility of selective determinations (see Chapter 4, Section IV.D, Note 2).

N-Substituted 4-amino-1,2-naphthoquinones can also be reduced to the corresponding aminodihydroxynaphthalenes (170), which are highly fluorescent and allow the fluorimetric determination of primary and secondary aliphatic, and primary aromatic amines (171).

By means of hydrazines, naphthoquinonesulfonic acid itself is readily reduced to the fluorescent 1,2-dihydroxynaphthalene-4-sulfonic acid. The method was applied to the fluorimetric determination of hydrazines, and more particularly of benzylhydrazine in animal organs and urine (172).

Sodium naphthoquinonesulfonate also condenses with guanidine to give the compound (XXII) (173). Methylguanidine and other substituted guanidines do not react.

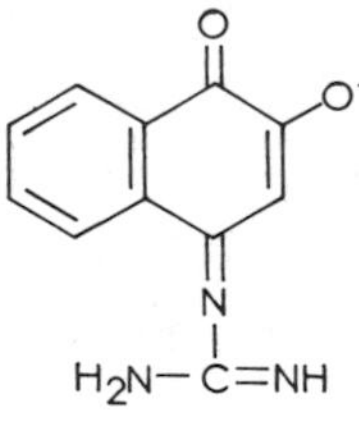

O
O⁻
N
$H_2N-C=NH$
(XXII)

Compounds with an active methylene group lead to the following type of derivatives (174,175):

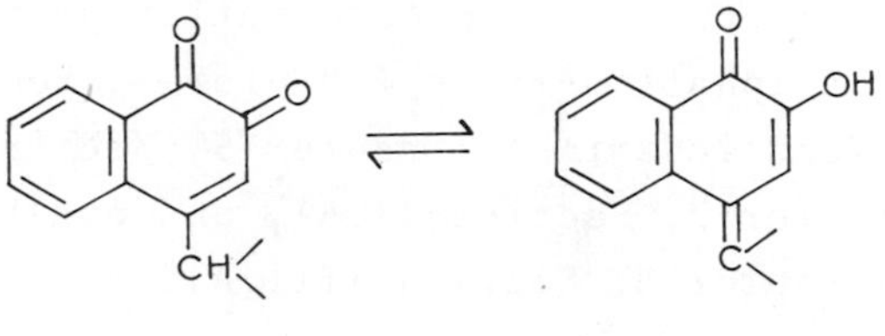

O O ⇌ O OH

CH< C<

The reaction was applied to the determination of digitoxin, which possesses such a group in the lactone ring of the aglycone (176). An active methylene group can also be formed through isomerization; pertinent examples are resorcinol, phloroglucinol, indole, pyrrole, and nitromethane, the latter affording the compound (XXIII) in alkaline medium (174,177).

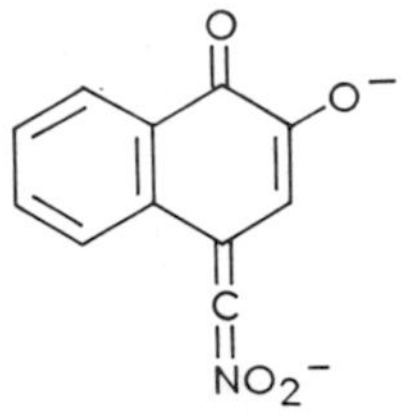

(XXIII)

The quinoid derivatives obtained from dialkyl α-methylene ketones can be reduced to the corresponding dihydroxy species, allowing the fluorimetric determination of this class of compounds (178).

Aromatic tertiary ring bases can be quaternized with methyl or ethyl iodide, thus activating the methylene group of the alkyl halide, and the resulting quaternary compounds react immediately with 1,2-naphthoquinone-4-sulfonic acid to give colored products. This reaction can be used as a test for cyclic tertiary ring bases, as well as for oxonium compounds (179).

In this book, sodium 1,2-naphthoquinone-4-sulfonate was applied to the colorimetric determination of primary and secondary aliphatic amines (Chapter 4, Sections IV.D, IV.E), primary aromatic amines (Chapter 5, Section IV.H), and to the fluorimetric determination of primary and secondary aliphatic amines (Chapter 4, Section XI.B), primary aromatic amines (Chapter 5, Section V.A), hydrazines (Chapter 7, Section IX.A), and dialkyl α-methylene ketones (Chapter 8, Section XIII.A).

VIII. TETRAZOLIUM SALTS AND FORMAZANS

Although the discovery of formazans is generally attributed to von Pechmann, he mentioned (180) that Pinner (181) was the first to obtain such a compound, namely 1,3,5-triphenylformazan, by reacting benziminoethyl ether with phenylhydrazine. Von Pechmann (182) isolated the same product by reacting the phenylhydrazone of benzaldehyde or of phenylglyoxylic acid with benzenediazonium chloride. He then extended this method to the preparation of a number of formazans (180), and it is still widely used.

It now seems well established that the reaction of the hydrazone of an aldehyde with a diazonium salt proceeds through an intermediate bisarylazo compound which isomerizes to the formazan even in the solid state, and very rapidly in a basic solution (183).

$$\mathrm{R{-}CH{=}N{-}NH{-}Ar} \;+\; \mathrm{N{\equiv}\overset{+}{N}{-}Ar} \longrightarrow \mathrm{R{-}CH(N{=}N{-}Ar)_2} + \mathrm{H^+}$$

$$\mathrm{R{-}CH(N{=}N{-}Ar)_2} \longrightarrow \mathrm{R{-}C({=}N{-}NH{-}Ar)(N{=}N{-}Ar)}$$

The first preparation of tetrazolium salts by oxidation of formazans was described by von Pechmann and Runge (184).

$$\mathrm{R{-}\overset{3}{C}({=}\overset{4}{N}{-}\overset{5}{N}H{-}Ar)(\overset{2}{N}{=}\overset{1}{N}{-}Ar)} \;\underset{\text{Red., }OH^-}{\overset{\text{Oxid.}}{\rightleftharpoons}}\; \mathrm{R{-}\overset{5}{C}\,\langle\,{=}\overset{1}{N}{-}\overset{2}{N}(Ar){-}\overset{3}{\overset{+}{N}}(Ar){=}\overset{4}{N}{-}\,\rangle}$$

Lead tetraacetate later proved to be a very suitable oxidizing agent for this purpose.

Formazans are highly colored species, whereas the corresponding tetrazolium salts are colorless. These compounds are widely used as analytical tools in chemistry, biochemistry, histology, and biology, and several dozen of them are commercially available. It is therefore obvious that a comprehensive bibliography, even limited to chemical analysis, is beyond the scope of this section, in which we present only some major reactions involving these reagents.

A. Tetrazolium Salts

It must be emphasized at first that determinations based on the reduction of colorless tetrazolium salts to highly colored formazans in alkaline media are far from specific. However, a limited selectivity can sometimes be attained, provided that the given operative conditions are strictly observed. Factors such as time, temperature, and basicity are of the highest importance. It should be also taken into account that the formazan formed can be partly reoxidized by air oxygen (185).

The colorimetric determination of reducing sugars with 2,3,5-triphenyltetrazolium chloride was devised by Mattson and Jensen (186). The method was applied by Fairbridge et al. (187) to the estimation of blood sugar, glucose, lactose, cysteine, and ascorbic acid.

Using 2,3,5-triphenyltetrazolium chloride and 2,5-diphenyl-3-(4-styrylphenyl)tetrazolium chloride, Avigad et al. (188) showed that, under suitable conditions, these salts are readily reduced by sugars manifesting enediol isomerism, but not by those in which such isomerism is impeded by the presence of a substituent in a carbinol group vicinal to the carbonyl group.

The 2,3,5-triphenyltetrazolium chloride also allows the determination of binary mixtures of reducing sugars by a simple differential rate technique developed for zero-order competitive reactions (189).

Corticosteroids can be determined through their reducing 17-ketol grouping. 2,3,5-Triphenyltetrazolium chloride was used at first (190), but it gives poorly reproducible results. More stable 2-(p-iodophenyl)-3-(p-nitrophenyl)-5-phenyltetrazolium bromide was proposed by Henly (191), but good results are also obtained with Blue Tetrazolium (192), which is now widely used. The operative conditions are, however, critical. Under somewhat different conditions and in the heat, various 3-ketosteroids with unsaturated ring A react also (193). These facts were turned to account by Smith et al. (194), who devised a colorimetric method with Blue Tetrazolium for the analysis of norethindrone, norethindrone acetate, and norgestrel. The technique proposed is selective for these progestins, showing no interference from ethynylestradiol and mestranol commonly found in oral contraceptives.

In addition to the reaction with sugars and corticosteroids, the reactivity of Blue Tetrazolium with various hydroxy ketones has been studied (195). In many cases, substantial differences were found which could be related to either the α or β position of the hydroxyl group with respect to the keto group or to whether the hydroxyl group is primary, secondary, or tertiary. Under the given conditions, the reaction is also positive with tetracycline, alloxantin, and erythromycin.

Blue Tetrazolium allows a ready determination of thiols (196). This salt is also reduced by dithiocarbamates which are obtained by reacting a primary or secondary aliphatic amine with carbon disulfide, permitting in this way a colorimetric determination of these classes of amines (197).

As already stated, it must be taken into account that these reactions can be far from specific. Sinsheimer and Salim (198) have examined the reduction of Blue Tetrazolium by polyhydric phenols, acid active hydrogen compounds, thiols, and quinones. Strongly positive results are obtained with compounds such as pyrocatechol, resorcinol, hydroquinone, diphenylacetonitrile, malononitrile, and phenyl-2-propanone. Reduction of tetrazolium salts by quinones proceeds through an oxidative cleavage of the quinone ring.

A rather peculiar reaction allows the determination of hydroperoxides and peracids: Upon the action of a hydroperoxide or a peracid in the presence of ferrous ion, ethanol is oxidized to a free radical which in turn reduces Blue Tetrazolium. Sodium fluoride, acting as a masking agent, impedes side reactions with ferric ion (199).

With sulfated and sulfonated surfactants 2,3,5-triphenyltetrazolium halides form water-insoluble salts which can be extracted into ethylene bromide. The tetrazolium cation is characterized in the organic phase by its reduction with sodium hydrosulfite to the corresponding formazan (200). This method is practically specific for detecting sulfated and sulfonated surfactants. Using benzoin as reductant (201), long-chain primary alkylsulfates can likewise be determined.

In this book, tetrazolium salts were applied to the colorimetric determination of primary and secondary aliphatic amines (Chapter 4, Section IV.G), hydroperoxides and peracids (Chapter 10, Section IV.B), thiols (Chapter 11, Section III.B), 3-ketosteroids with unsaturated ring A (Chapter 15, Section VII.B), and 17-ketolsteroids (Chapter 15, Section XII.A).

B. Formazans

The formation of formazans through reactions other than the reduction of tetrazolium salts allows miscellaneous detections and determinations. As a rule, these reactions go through the formation of hydrazone which is then condensed with a diazonium salt.

The first reaction of this kind was described as far back as 1883, at a time when formazans were unknown as definite species. Penzoldt and Fischer (202) ascertained that an alkaline solution of diazotized p-sulfanilic acid develops a red-violet color with aldehydes on standing for some time. The development of the color can be accelerated and its intensity increased by the addition of sodium amalgam. It may be inferred therefrom that the diazonium salt is partly reduced either by a fraction of the aldehyde or by the amalgam to the corresponding hydrazine, which reacts with the aldehyde, and the hydrazone so formed then condenses with the excess diazonium compound.

It is obvious that only the hydrazones of aldehydes can give rise to formazans, thus allowing selective detection and determination of this class of carbonyl compounds. Various methods have been proposed. As a rule, the intermediate hydrazone formed at first is not isolated from the reaction medium, to which the diazonium salt is then added.

Reaction with phenylhydrazine, then with diazotized p-sulfanilic acid, has been used for the detection of various aldehydes (203). Development of the phenylhydrazone with the diazonium salt generated by dehydration of 3-phenyl-5-nitrosamino-1,2,4-thiadiazole (phenitrazole) allows fairly sensitive determinations (3). Sawicki and Stanley (9) proposed 2-hydrazinobenzothiazole as a hydrazine and p-nitrobenzenediazonium fluoborate as coupling agent. Various other arylidene arylhydrazones can also be detected with this diazo compound, whereas ketone aryl-hydrazones do not give colors (204).

The diazo compound can be generated in the reaction medium by oxidizing the excess of the hydrazine. This mechanism is probably involved in the Schryver reaction, which allows a selective colorimetric estimation of formaldehyde (see Section V).

On this principle, Sawicki et al. proposed two reagents: 2-hydrazinobenzothiazole (10) which reacts according to scheme A, and 3-methylbenzothiazolin-2-one hydrazone (11), which reacts according to scheme B:

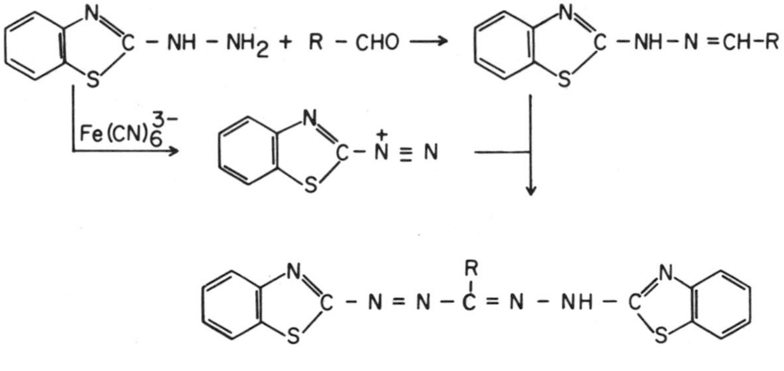

(a)

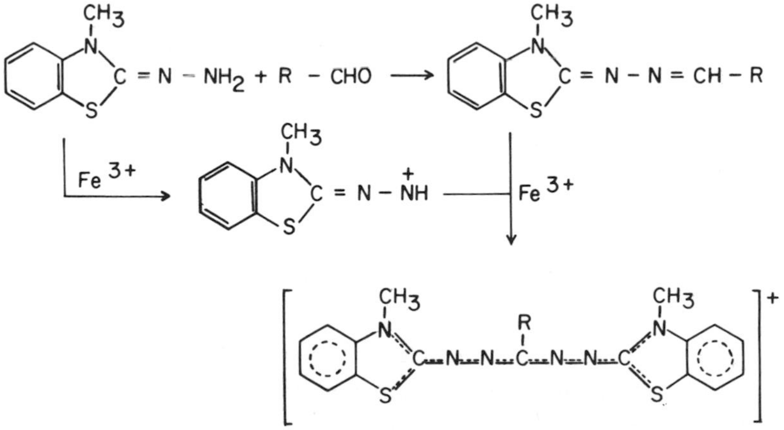

(b)

It may be assumed that similar mechanisms are involved in the colorimetric determination of aldoses and ketoses, and of 17-hydroxy-17-ketolsteroids (63) with the same reagents. The second one is very versatile, and it is the subject of a particular note in this Chapter (Section III).

A formazan can also be formed when a compound bearing an active methylene group reacts with a diazonium salt. The hydrazone formed through a diazo coupling reaction condenses with the excess reagent.

Improving the method of Rosenthal (13), Walker (205) proposed a procedure for the determination of acetoacetic acid in tissue preparations and rat liver with p-nitrobenzenediazonium ion. Under the given conditions, the reaction proceeds according to the following scheme:

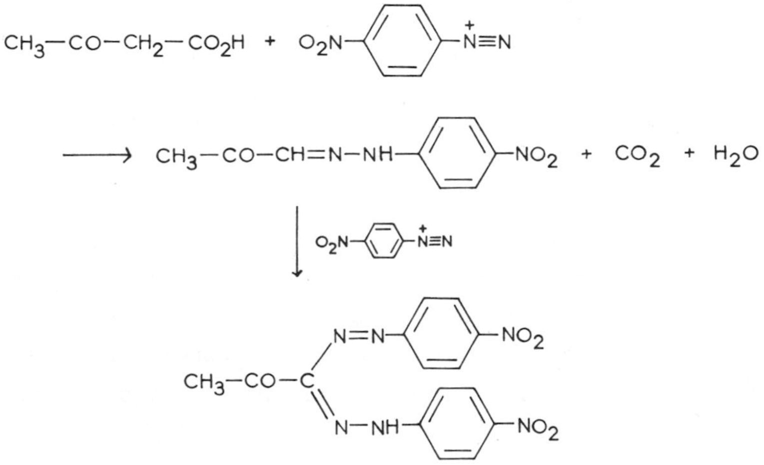

$$CH_3{-}CO{-}CH_2{-}CO_2H + O_2N{-}C_6H_4{-}\overset{+}{N}{\equiv}N$$

$$\longrightarrow CH_3{-}CO{-}CH{=}N{-}NH{-}C_6H_4{-}NO_2 + CO_2 + H_2O$$

$$\downarrow O_2N{-}C_6H_4{-}\overset{+}{N}{\equiv}N$$

$$CH_3{-}CO{-}C(N{=}N{-}C_6H_4{-}NO_2){=}N{-}NH{-}C_6H_4{-}NO_2$$

This method was later modified to measure oxaloacetic acid in the presence of moderate concentrations of acetoacetic acid (12).

In this book, beside the application of the Schryver reaction to the determination of compounds giving rise to formaldehyde under suitable conditions (Section V), the formation of formazans was turned to account for the colorimetric determination of primary alcohols oxidized to aldehydes (Chapter 2, Sections IV.A, IV.B), aliphatic aldehydes (Chapter 8, Sections III.A, IV.A, IV.B), ethylenic compounds oxidized to aldehydes (Chapter 12, Section II.A), aldoses and ketoses (Chapter 14, Sections II.F, II.G), and 17-hydroxy-17-ketolsteroids (Chapter 15, Sections XIV.B, XIV.C).

IX. DETERMINATIONS THROUGH THE HANTZSCH REACTION

According to the Hantzsch reaction (206), substituted dihydropyridines are obtained by reacting 2 moles of a β-diketone or a β-ketoester with 1 mole each of aldehyde and ammonia.

Solutions of the compound obtained from acetylacetone, formaldehyde, and ammonia (XXIV) are yellow and develop a yellow-green fluorescence, thus allowing the colorimetric and fluorimetric determinations of formaldehyde (see Section V), and the fluorimetric determination of ammonia (207).

Whereas mixtures of ethyl acetoacetate and ammonia afford dihydropyridine derivatives with numerous aldehydes under preparative conditions (208), only formaldehyde reacts satisfactorily at dilutions commonly used in analysis to give compound (XXV). Its solutions are colorless, but fluoresce strongly in the blue, thus enabling the selective determination of formaldehyde, and accordingly of compounds which give rise to this aldehyde through suitable chemical reactions (125).

Other aldehydes can be fluorimetrically determined by reacting them with mixtures of ammonia and cyclohexane-1,3-dione (125,209) or 5,5-dimethyl-1,3-cyclohexane-dione (dimedone) (209), to give, respectively, compounds (XXVI) and (XXVII).

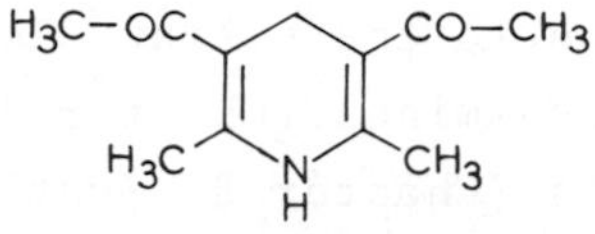

(XXIV)

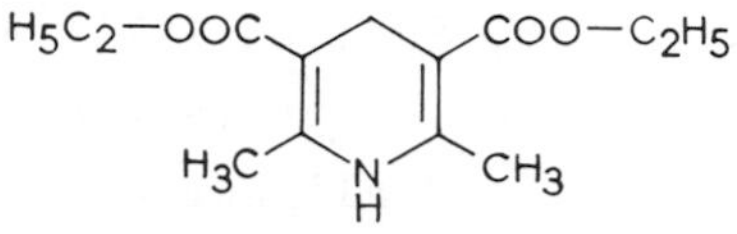

(XXV)

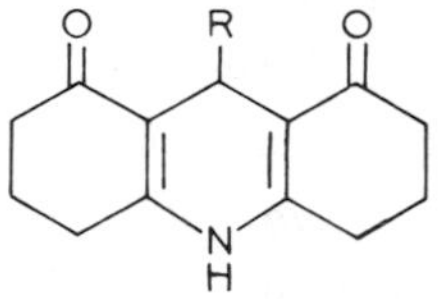

(XXVI)

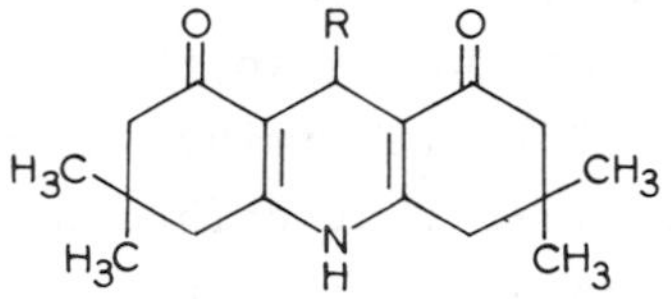

(XXVII)

A mixture of acetylacetone and formaldehyde inversely permits the colorimetric or fluorimetric determination of primary aliphatic amines and amino acids (125,210,211), and of nitriles beforehand reduced to amines (125).

A mixture of diethyl acetonedicarboxylate and formaldehyde allows the colorimetric and ultraviolet determination of primary aliphatic amines. Inversely, a mixture of the same ester with ammonium acetate permits the ultraviolet determination of aliphatic aldehydes.

In this book, the Hantzsch reaction was applied to the colorimetric determination of primary aliphatic amines (Chapter 4, Sections VI.H, VI.I), to the fluorimetric determination of 1,2-diols (Chapter 2, Section XIV.A), hexitols (Chapter 2, Section XV.A), α-amino primary alcohols (Chapter 2, Section XVI.A), primary aliphatic amines (Chapter 4, Section XII.A), aliphatic 2-nitro-1-hydroxy compounds (Chapter 7, Section X.A), nitriles (Chapter 7, Section XI.A), aliphatic aldehydes (Chapter 8, Section XI.A), α-amino acids (Chapter 9, Section XI.A), ethylenic compounds of the type $R-CH=CH_2$ (Chapter 12, Section V.A), pentoses (Chapter 14, Section XIV.A), and 17-ketolsteroids (Chapter 15, Section XVIII.A), and to the ultraviolet determination of primary aliphatic amines (Chapter 4, Section VI.I) and aliphatic aldehydes (Chapter 8, Section IV.E).

X. NINHYDRIN

Ninhydrin (XXVIII) was first prepared by Ruhemann (212), who was also the first to show that it develops a blue color with amino acids (213). Shortly afterwards, he established that carbon dioxide is evolved and aldehydes are formed in the course of this reaction, pointing to a similarity between the behaviors of ninhydrin and alloxan (214). The latter reagent had been studied by Strecker (215) who showed that, on heating aqueous solutions of a mixture of alloxan and amino acids, carbon dioxide, aldehydes, and murexide are produced, whereas alloxan is reduced to alloxantin.

Ruhemann concluded therefrom that hydrindantin (XXIX) should be formed in the reaction with ninhydrin. He prepared it by treatment of ninhydrin with hydrogen sulfide and showed that it also develops a blue color with amino acids (214). However, its formation in the course of the reaction of ninhydrin with amino acids was not demonstrated until 1938 (216)

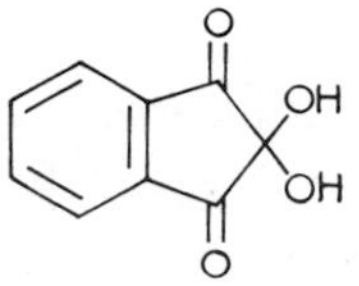

(XXVIII) 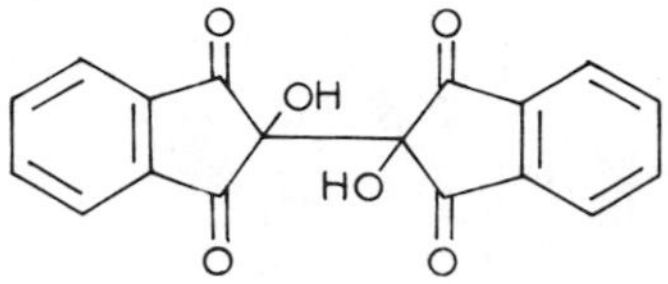 (XXIX)

Ruhemann also obtained the ammonium salt of diketohydrindylidene-diketohydrindamine (XXX) by the action of ammonium carbonate on hydrindantin, or by the oxidation of diketohydrindamine (XXXI), and he established that this compound, often named Ruhemann's Purple, is the dye formed in the reaction of ninhydrin with amino acids (217).

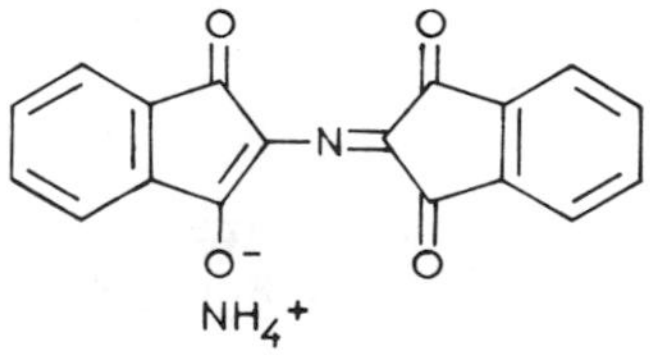

(XXX) 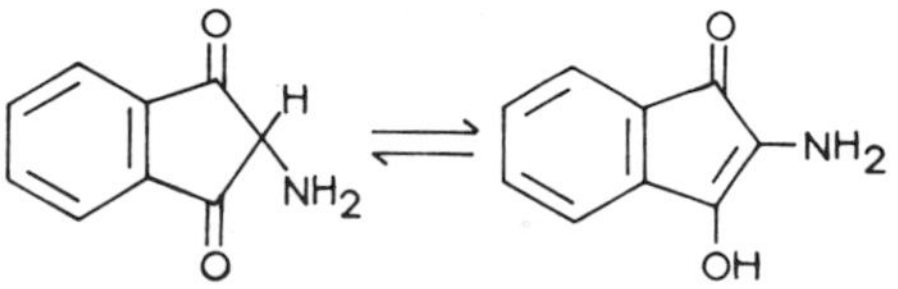 (XXXI)

A. Mechanism of the Reaction

Although Ruhemann did not propose an equation for the reaction, it may be inferred from his results that it should parallel the Strecker reaction, and should therefore be written as follows:

$$\text{ninhydrin} + R{-}CH(NH_2){-}CO_2H \longrightarrow \text{hydrindantin precursor (2-hydroxy-1,3-indandione)} + R{-}CHO + CO_2 + NH_3$$

$$\text{2-hydroxy-1,3-indandione} + \text{ninhydrin} \longrightarrow \text{hydrindantin} \xrightarrow{NH_3} \text{Ruhemann's Purple } (O^- \; NH_4^+)$$

(Scheme A)

It is now well known that this theory is wrong. For instance, it cannot account for the more rapid chromogenic reaction of α-amino acids with hydrindantin, as compared with ammonium salts.

It is beyond the scope of this Section to present the various theories which have been proposed from Ruhemann on. They are gathered in a review of McCaldin (218), who suggested a theory according to which the reactions of ninhydrin with amines, imino acids, and amino acids all proceed by the same mechanism. The interpretation is based on the mechanism of the Strecker degradation. It explains the formation of Ruhemann's Purple and hydrindantin, and why these reactions proceed more readily with compounds which contain a carboxyl group adjacent to the nitrogen atom (Scheme B).

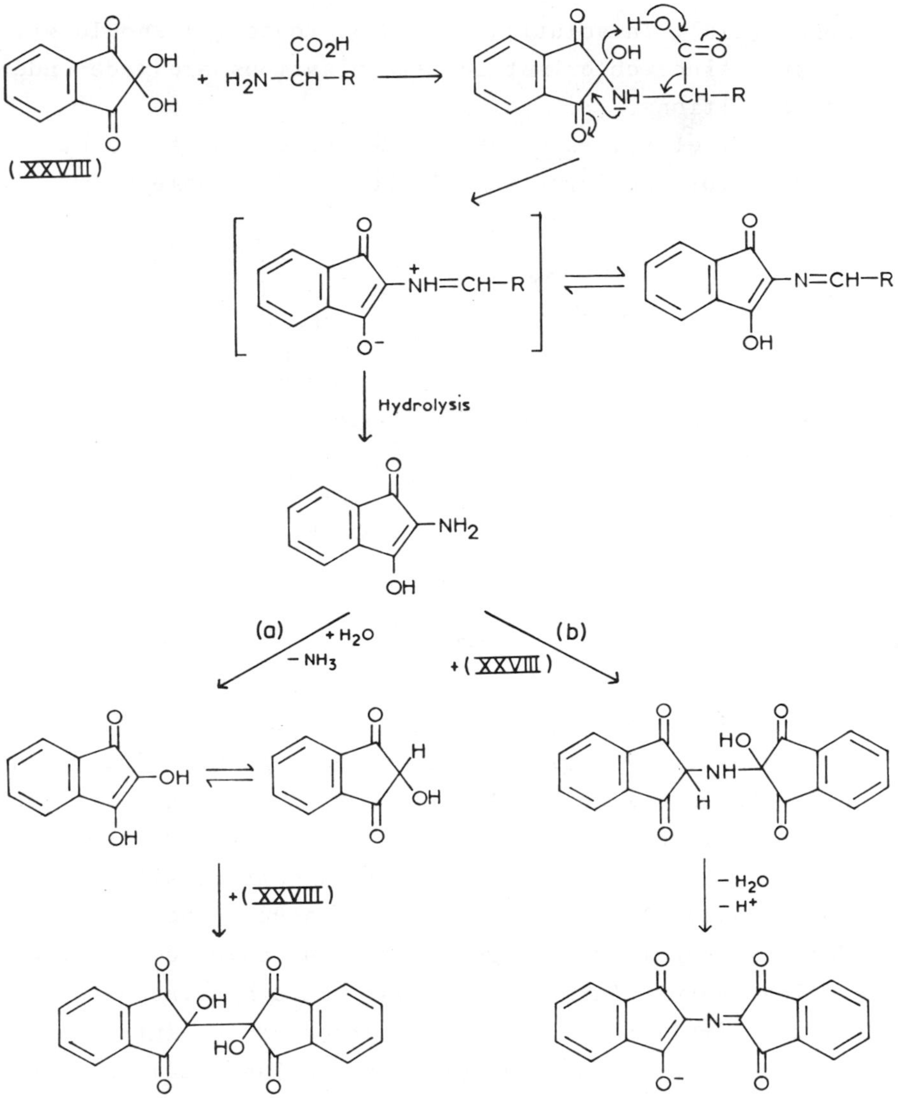

(XXVIII) + $H_2N-CH(CO_2H)-R$ ⟶

$-NH^{+}=CH-R$ ⇌ $-N=CH-R$

Hydrolysis

NH_2, OH

(a) $+H_2O$, $-NH_3$ (b) +(XXVIII)

OH ⇌ H, OH HO, NH, H

+(XXVIII) $-H_2O$, $-H^+$

OH, HO N=, O^-

(Scheme B)

At pH 1 to 2.5, the reaction proceeds chiefly by route (a), ammonia is evolved almost quantitatively, and no Ruhemann's Purple is formed. This fact was turned to account for the determination of α-amino nitrogen as ammonia (219). In solutions of pH 5, route (b) should predominate, since colorimetric determinations are made under these conditions.

Neuzil et al. (220) proposed a somewhat different mechanism for the formation of the imine (Scheme C).

(Scheme C)

Some facts are not evidenced by these schemes. Only poorly sensitive results are obtained with ninhydrin alone, and Beer's law is not followed. Moore and Stein (221) considered that these difficulties appear to result from the influence of air, since they obtained better, although still unsatisfactory, results under vacuum. They therefore added a strong reducing agent, stannous chloride, to the reaction medium. Later, they substituted hydrindantin for it (222). Troll and Cannan (223) proposed the introduction of potassium cyanide, and Yemm and Cocking (224) obtained a stoichiometric reaction for most amino acids in the presence of this salt. Although it was postulated that ascorbic acid could not be used as reductant (221), it afforded very good results in our laboratory (Chapter 9, Section VI.G).

It must be emphasized that the reducing agents added to the reaction medium do not behave like antioxidants, but all reduce ninhydrin to hydrindantin. Potassium cyanide acts as a catalyst in this reaction (225), and ascorbic acid was used for the preparation of pure hydrindantin (222).

Studying the reaction of amino acids with hydrindantin, MacFadyen and Fowler (226) demonstrated that this reagent disappears as Ruhemann's Purple is formed, mole for mole. It is known that hydrindantin dissociates in solution to give 1 mole each of ninhydrin and diketoindanol (226,227).

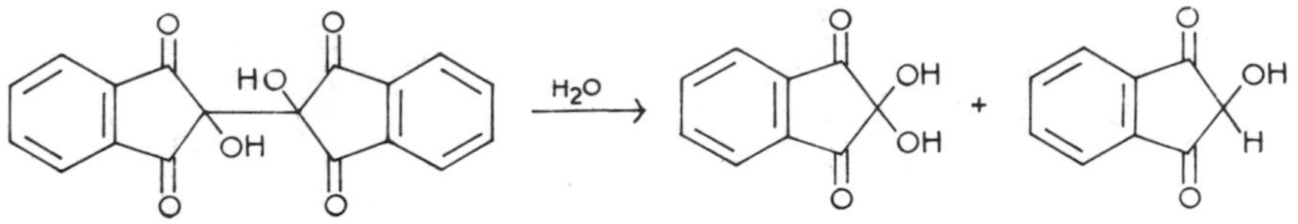

As it seems well established that the reaction of amino acids with ninhydrin is stoichiometric only in the presence of hydrindantin, it appears that this latter compound should be involved in the mechanism, a fact which is not evidenced in Scheme B, route (b).

Comparing the optical densities given by pure Ruhemann's Purple and by leucine submitted to the reaction with ninhydrin and ascorbic acid (Chapter 9, Section VI.G), we confirmed that the totality of the amino nitrogen is incorporated in the dye, hence that the reaction is quantitative.

According to our experiments, air oxygen is absorbed at room temperature by hydrindantin, but not by ninhydrin dissolved in the medium used for the determination of amino acids with ninhydrin and ascorbic acid. It may be concluded therefrom that the influence of oxygen on the reaction cannot be attributed to its action on ninhydrin. On the other hand, diketohydrindamine (XXXI) is very sensitive to air

(217), and it might be supposed that hydrindantin prevents its oxidation, thus allowing its further reaction with ninhydrin to give Ruhemann's Purple.

Very recently, Lamothe and McCormick (228) proposed a revised mechanism. Using voltammetric techniques, they showed that diketoindanol is more easily oxidized than diketohydrindamine, and they concluded therefrom that diketohydrindamine would probably not be stable in a solution containing ninhydrin. An oxidation-reduction could take place, and they suggested the following mechanism for Scheme B, route (b):

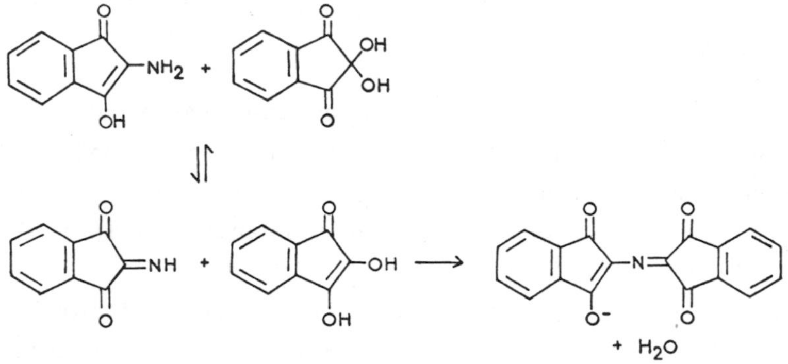

NH_2 + OH OH

OH O

=NH + —OH ⟶ —N=

O OH O^- O

+ H_2O

However, this mechanism does not clearly account for the results obtained when operating under nitrogen with ninhydrin alone (229). It appears therefore that further studies would be necessary to elucidate completely the mechanism of the reaction. In these studies, it should be taken into account that the visible spectrum of Ruhemann's Purple is strongly solvent-dependent (230).

B. Reaction of Ninhydrin with Imino Acids

Proline and hydroxyproline do not give Ruhemann's Purple. Compounds (XXXII) and (XXXIII) are formed (231), and it was admitted that the first step of the reaction follows the general mechanism proposed by McCaldin for amino acids.

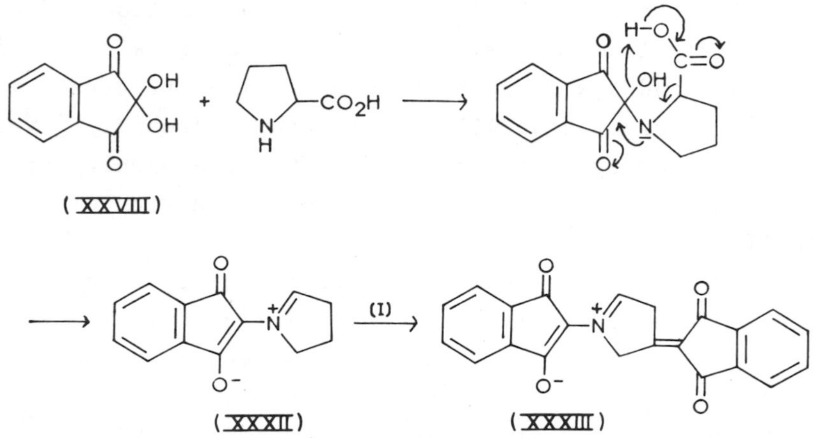

(XXVIII) (XXXII) (XXXIII)

C. Miscellaneous Reactions

In a mixture of acetic and hydrochloric acids, only cysteine develops a pink color with ninhydrin and can therefore be determined in the presence of other α-amino acids (232).

Ninhydrin has also been proposed for the characterization of ephedrine (233), the ultramicro determination of nitrogen of polypeptides and amino acids isolated from paper chromatograms (234), and for the detection of fingerprints (235).

Amines of the type $(RR')CH{-}NH_2$ react but weakly (236). However, Ruhemann's Purple forms rapidly when the mixture is kept in the dark for several hours, then exposed to sunlight (237). The method can be applied to paper chromatograms (238).

In a medium of pH 8, ninhydrin in the presence of phenylacetaldehyde develops a green fluorescence with primary aliphatic amines (239). The same reaction is given by α-amino acids, peptides, proteins, and 2-amino-2-deoxy hexoses. The structure of the fluorescent species has recently been elucidated (see Chapter 4, Section XII.B).

In alkaline medium, ninhydrin condenses with guanidine and mono- and N,N-disubstituted guanidines to give a fluorescent species (240,241). The mechanism probably involves o-carboxyphenylglyoxal proceeding from the opening of the five-membered ring of ninhydrin upon the action of the alkali.

Some nonnitrogenous compounds can also be determined. Propionaldehyde reacts with ninhydrin in concentrated sulfuric acid to produce a deep red-blue color suitable for spectrophotometric analysis, and the reaction is almost specific (242). In strong sulfuric acid, propylene glycol dehydrates and rearranges to a mixture of allyl alcohol and the enolic form of propionaldehyde, and it can be therefore likewise determined (243). The reaction was extended to the determination of hydroxypropyl groups in starch ethers (244).

A highly sensitive and remarkably specific color reaction (violet-red) is given by testosterone in sulfuric acid, allowing its determination in the presence of other steroids (245).

In this book, ninhydrin was applied to the colorimetric determination of α-amino acids (Chapter 9, Section VI.F, VI.G), and to the fluorimetric determination of primary aliphatic amines (Chapter 4, Section XII.B), guanidine and mono- and N,N-disubstituted guanidines (Chapter 6, Section IV.A), α-amino acids (Chapter 9, Section

XI.C), and 2-amino-2-deoxyhexoses (Chapter 14, Section XVI.B).

XI. DETERMINATIONS THROUGH OPENING OF THE PYRIDINE RING

In 1899, Vongerichten (246) obtained 2,4-dinitrophenylpyridinium chloride by condensing pyridine with 1-chloro-2,4-dinitrobenzene and mentioned that this compound developed a deep red color in alkaline medium. Zincke et al. (247) established that the color was due to the formation of a glutaconic dialdehyde derivative.

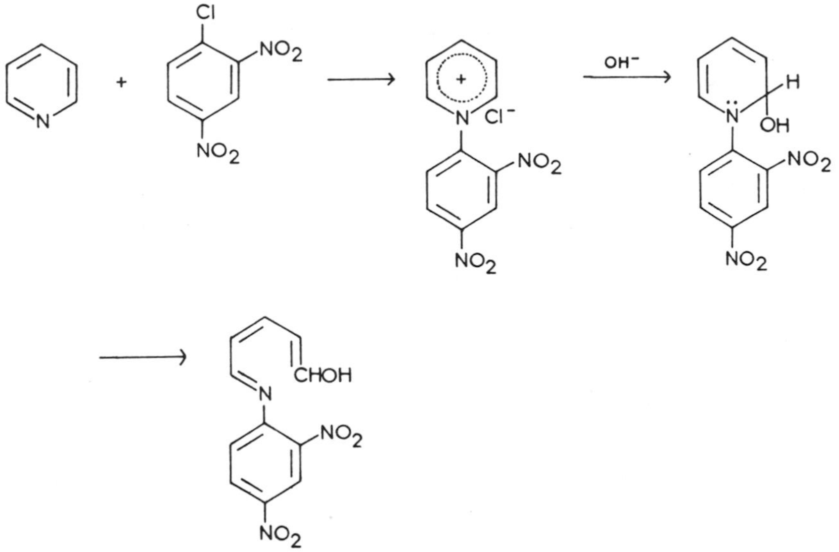

By reacting pyridine with cyanogen bromide, König (248) obtained the unstable quaternary cyano derivative (XXXIV), which on further treatment with an aromatic amine gave an imino derivative of Schiff base type (XXXV).

By reacting chloroform with pyridine in alkaline medium, Fujiwara (249) observed the development of a red color allowing the characterization of chloroform. This reaction also is bound up with the opening of the pyridine ring.

It is now well known that, as a rule, the heterocyclic ring of quaternary pyridinium compounds can undergo cleavage on alkaline hydrolysis, giving rise to an imino derivative of glutaconic dialdehyde.

Any group which acts as an electron acceptor to reduce further the electron density of the pyridine ring facilitates attack by the base. Therefore, cleavage of the pyridine ring in alkaline medium will be readily achieved by prior quaternization with compounds such as thionyl chloride, chlorosulfonic acid, phosphorus oxychloride, phosphorus pentachloride, phosgene, cyanogen halides, cyanuric acid chloride, 1-halo-2,4-dinitrobenzene, o-nitrobenzene sulfochloride, 4-chloropyridine, or gem-polyhalogen compounds. The glutaconic aldehyde derivative so obtained can be reacted with an aromatic amine to give a Schiff base type derivative.

It may be concluded therefrom that opening of the pyridine ring can give access to the colorimetric determinations of primary aromatic amines, gem-polyhalogen compounds, and pyridine derivatives.

A. Determination of Primary Aromatic Amines

In determination of primary aromatic amines, the reagent is the monoenol form of glutaconic dialdehyde obtained from pyridine in the reaction medium. The ammonium salt of this compound is formed when pyridine is exposed to ultraviolet light in the presence of air and moisture (250), and this process was proposed by Feigl and Anger (251) for the characterization of primary aromatic amines. The sodium salt of the reagent can be prepared through the König reaction with cyanogen bromide, and this method was applied to the characterization of local anesthetics (252).

The most sensitive results are, however, obtained when the reagent is prepared by decomposing 4-pyridyl-pyridinium dichloride in alkaline medium (253).

The 4-aminopyridine formed together with glutaconic dialdehyde does not behave like an aromatic amine, and therefore does not interfere. This method also allows the photometric determination of hydrazine (254).

$$\text{4-pyridylpyridinium } (N^{+}\ Cl^{-}) \cdot HCl \xrightarrow{NaOH} \underset{CHO}{CH}{=}CH{-}\underset{CHONa}{CH} + \text{4-aminopyridine } (NH_2)$$

B. Determination of gem-Polyhalogen Compounds

The Fujiwara reaction for chloroform was extended by Ross (255) to some other gem-trihalogen compounds, and has since that time been subjected to numerous studies, but its mechanism is still debated.

Working with chloroform, chloral hydrate, trichloroacetic acid, and chlorbutol (1,1,1-trichloro-2-methyl-2-propanol), Moss and Rylance (256) showed that a reaction even occurs in ice cold conditions, the absorption curve varying with the reactant. They admitted that it corresponds to the formation of the compound of general formula (XXXVI), which is very unstable. They were unable to demonstrate this complex under the normal conditions of the Fujiwara reaction except, transitorily, in the case of chlorbutol.

N^+ Cl^-, R—C—Cl, Cl (XXXVI) $\xrightarrow{NaOH}$ CH=N—C(R)(Cl)—Cl, CHONa (XXXVII) $\xrightarrow{H_2O}$ CHO CHONa (XXXVIII)

It may be assumed that, under these conditions, ring breakage immediately gives species (XXXVII). On the other hand, the authors showed that the red reaction products from the compounds tested are identical and independent of the starting material, with absorption maxima at 365 and 530 nm. To explain this similarity, it may be admitted that, under the alkaline conditions used, these compounds are at first decomposed with liberation of chloroform. This hypothesis is sustained by the fact that trichloroethanol, which does not yield chloroform in alkaline medium, gives somewhat different results, with an absorption maximum at

430-440 nm (256,257), suggesting that the whole molecule remains attached after ring breakage.

Further heating causes decomposition of (XXXVII) to glutaconic dialdehyde (XXXVIII), which absorbs at 365 nm. The authors conclude therefrom that both (XXXVII) and (XXXVIII) coexist in the Fujiwara reaction.

As a rule, only those aliphatic compounds which bear at least two halogen atoms bonded to the same carbon atom give positive results.

Under definite conditions, chloroform can be determined in carbon tetrachloride (258).

Chloralose does not react. It can, however, be determined after suitable hydrolysis (259) or oxidation (260, 261).

When a compound that contains methyl ketone or methyl carbinol groups is treated with alkali hypochlorite or hypobromite and the whole is heated with the addition of pyridine, a red or pink color immediately develops (262).

$$R{-}CO{-}CH_3 \xrightarrow{ClONa} CHCl_3 + R{-}CO_2Na$$

Methyl carbinols are oxidized beforehand by the reagent to the corresponding ketone. Positive reaction is given by ethanol, acetaldehyde, acetone, isopropyl alcohol, etc.

The sensitivity of the Fujiwara reaction can be enhanced by reacting the mixture of compounds (XXXVII) and (XXXVIII) obtained as described above with an aromatic amine, thus affording a Schiff base type imino derivative.

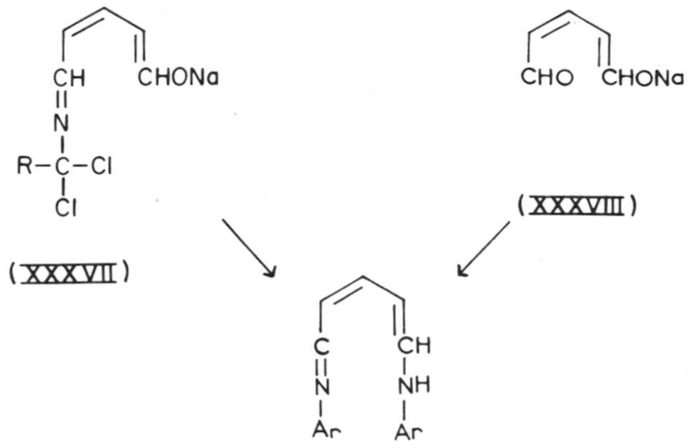

CH CHONa
N
R–C–Cl
Cl

(XXXVII)

CHO CHONa

(XXXVIII)

C CH
N NH
Ar Ar

Benzidine was proposed for colorimetric determinations (263), and p-aminobenzoic acid allows fluorimetric determinations (264).

C. Determination of Pyridine Derivatives

As a rule, only compounds unsubstituted in the position α to the heterocyclic nitrogen atom can be determined. It is impossible to present here a comprehensive bibliography of the numerous papers which have been published in this field.

The Zincke reaction was applied to the colorimetric determination of various pyridine compounds (265-267), and mainly to isoniazid, 1-fluoro-2,4-dinitrobenzene often being used instead of the chloro reagent (268).

Much more often, however, the opening of the pyridine ring is achieved through the König reaction with cyanogen bromide or chloride. The cyanogen bromide solution is usually obtained by treating potassium cyanide with a mixture of potassium bromide and potassium bromate in acidic medium. It can also be prepared by adding potassium thiocyanate to bromine water (269). Cyanogen chloride is conveniently obtained in the reaction medium by adding a solution of chloramine-T to a solution of potassium cyanide under acidic conditions.

With this latter reagent, pyridine can be determined directly in alcohol; a yellow species is formed, whose structure is as yet unknown (270). In a somewhat similar fashion, nicotinamide develops a fluorescence upon the action of cyanogen bromide in alkaline medium and can be thence specifically determined (271).

In almost all cases, however, the glutaconic dialdehyde derivative formed is developed through a further reaction. Primary arylamines affording Schiff base type species are generally used, but ammonia, some secondary arylamines, barbituric and thiobarbituric acids have also been proposed. The two latter reagents give colored species corresponding to formula (XXXIX).

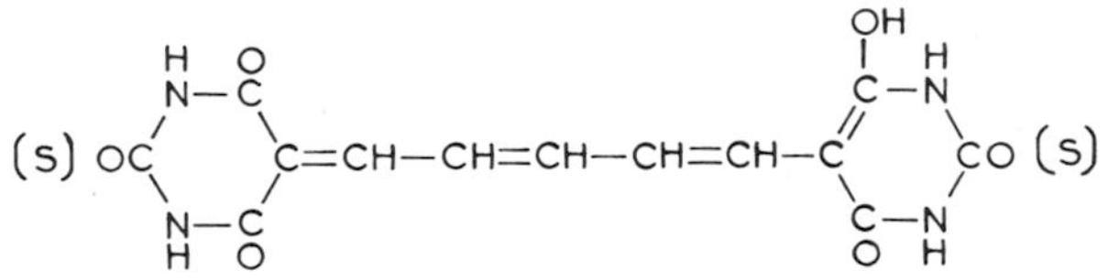

```
                                                  OH
        H  O                                      |  H
        N—C                                       C—N
       /    \                                    //   \
(S) OC       C=CH—CH=CH—CH=CH—C                  CO (S)
       \    /                                    \   /
        N—C                                       C—N
        H  O                                      O  H
```

(XXXIX)

The major reagents which have been used are gathered in the following paragraph. We only report the reference of the paper which, to the best of our knowledge, was the first dealing with the reagent, or we refer to a paper reviewing it:

5-Aminoacenaphthene (272), p-aminoacetophenone (273), p-aminobenzoic acid (274), p-aminophenol (275), p-aminopropiophenone (276), p-aminosalicylic acid (277), ammonia (278), aniline (248,272,279), barbituric acid (280), benzidine (272,281), 4,4'-diaminostilbene-2,2'-disulfonic acid (282), diphenylamine (272), p-methylaminophenol (metol) (283), N-methylaniline (284), α-naphthylamine (272), β-naphthylamine (248,272), 2-naphthylamino-1-sulfonic acid (285), orthocaine (methyl ester of 3-amino-4-hydroxybenzoic acid) (286), p-phenylenediamine (287), phthalylsulfacetamide

(277), procaine (288), sulfanilamide (289), p-sulfanilic acid (285), thiobarbituric acid (290).

Under suitable conditions, selective determinations are possible (285,291). p-Aminobenzoic acid allows also the fluorimetric determination of pyridine derivatives (292,293).

D. Miscellaneous Determinations

The glutaconic dialdehyde produced from pyridine, chloramine-T, and potassium cyanide allows the colorimetric determination of barbituric acid in the presence of barbituric acids substituted at position 5 (294). The aldehyde can also be generated from 4-pyridylpyridinium dichloride, permitting the detection of barbituric and thiobarbituric acids, and of malonic acid converted to barbituric acid by heating with urea (295).

Simazine and related chloro-s-triazines condense with pyridine in alkaline medium according to the Zincke reaction, and the condensate is developed with barbituric acid, 2-thiobarbituric acid, or ethyl cyanoacetate (296).

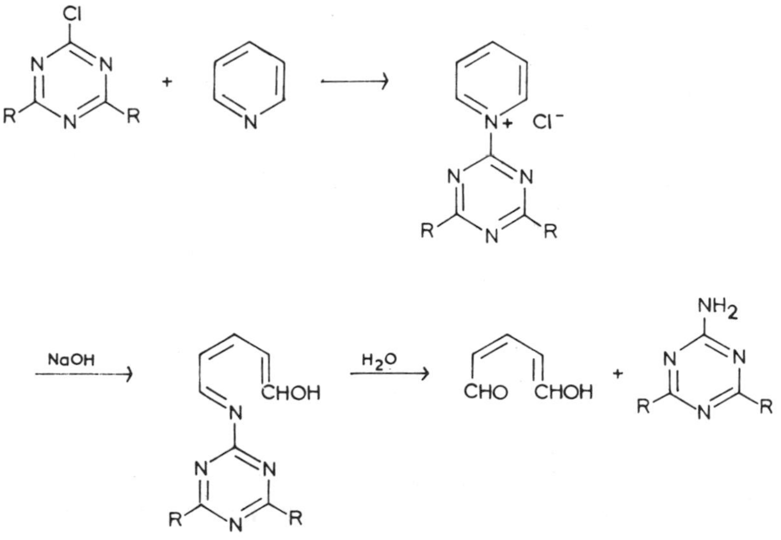

In this book, opening of the pyridine ring was applied to the colorimetric determination of primary aromatic amines (Chapter 5, Section IV.D), gem-polyhalogen compounds (Chapter 11, Sections II.A, II.B), pyridine derivatives (Chapter 13, Sections IV.B, IV.C), and the fluorimetric determination of gem-polyhalogen compounds (Chapter 11, Section IV.A) and pyridine derivatives (Chapter 13, Section XI.A).

REFERENCES

1. *F.S.A.O.V.*, vol. 1, 462 and 495.

2. J. Bartos, *Mises au Point de Chimie Analytique*, 11e série (1963), 23-61 (Masson Ed.).

3. M. Pesez, J. Bartos, and J.F. Burtin, *Talanta*, 5, 213 (1960).

4. J. Bartos, *Ann. Pharm. Fr.*, 29, 147 (1971).

5. E. Sawicki, J.L. Noe, and F.T. Fox, *Talanta*, 8, 257 (1961).

6. M. Pesez and P. Poirier, *Bull. Soc. Chim. Fr.*, 754 (1953).

7. P.S. Pelkis, R.G. Dubenko, and L.S. Pupko, *Zh. Obshch. Khim.*, 27, 2134 (1957).

8. R. Berg, *Mikrochem. Mikrochim. Acta*, 30, 137 (1941).

9. E. Sawicki and T.W. Stanley, *Mikrochim. Acta*, 510 (1960).

10. E. Sawicki and T.R. Hauser, *Anal. Chem.*, 32, 1434 (1960).

11. E. Sawicki, T.R. Hauser, T.W. Stanley, and W. Elbert, *Anal. Chem.*, 33, 93 (1961).

12. G. Kalnitsky and D.F. Tapley, *Biochem. J.*, 70, 28 (1958).

13. S.M. Rosenthal, *J. Biol. Chem.*, 179, 1235 (1949).

14. A.P. Terent'ev, *J. Gen. Chem., U.S.S.R.*, 7, 2026 (1937); A.P. Terent'ev and A.A. Demidova, *J. Gen. Chem., U.S.S.R.*, 7, 2464 (1937); A.P. Terent'ev and V. Zagorevsky, *J. Gen. Chem., U.S.S.R.*, 26, 211 (1956).

15. A.P. Altshuller and I.R. Cohen, *Anal. Chem.*, 32, 1843 (1960).

16. H. Pauly, *Z. Physiol. Chem.*, 42, 508 (1904); 44, 159 (1905).

17. S. Edlbacher, H. Baur, H.R. Staehelin, and A. Zeller, *Z. Physiol. Chem.*, 270, 158 (1941); S.M. Rosenthal and H. Tabor, *J. Pharmacol. Exptl. Therap.*, 92, 425 (1948); G. Barac, *Bull. Soc. Chim. Biol.*, 32, 287 (1950); R. Lubschez, *J. Biol. Chem.*, 183, 731 (1950); F. Sanger and A. Tuppy, *Biochem. J.*, 49, 463, 481 (1951); R.J. Block and D. Bolling, *The Amino Acid Composition of Proteins and Food* (C.C. Thomas, Springfield, Ill.), 2nd Edition (1951), 445; J. Thouvenot, N. Flavian, and R. Weber, *Bull. Soc. Chim. Biol.*, 39, 1511 (1957).

18. J. Baraud and L. Genevois, *Bull. Soc. Chim. Fr.*, 681 (1956).

19. E.B. Brown, A.F. Bina, and J.M. Thomas, *J. Biol. Chem.*, 158, 455 (1945).

20. C.R. Szalkowsky and W.J. Mader, *Anal. Chem.*, 27, 1404 (1955).

21. E.D. Day and W.A. Mosher, *J. Biol. Chem.*, 197, 227 (1952).

22. J.A. Sanchez, *J. Pharm. Chim.*, [8]29, 529 (1939); *Bol. Soc. Quim. Peru*, 9, 197 (1943).

23. E.B. Truitt, Jr., C.J. Carr, H.M. Bubert, and J.C. Krantz, Jr., *J. Pharmacol. Exptl. Therap.*, 91, 185 (1947).

24. F. Feigl and V. Anger, *Z. Anal. Chem.*, 181, 163 (1961).

25. M. Pesez, *Bull. Soc. Chim. Fr.*, 918 (1949); *Bull. Soc. Chim. Biol.*, 31, 1369 (1949).

26. M. Pesez, *Bull. Soc. Chim. Biol.*, 33, 195 (1951).

27. W.B. Koniecki and A.L. Linch, *Anal. Chem.*, 30, 1134 (1958).

28. A.G. Glazko, W.A. Dill, and M.C. Rebstock, *J. Biol. Chem.*, 183, 679 (1950).

29. P. Griess, *Justus Liebigs Ann. Chem.*, 106, 123 (1858).

30. M. Pesez and J.F. Burtin, *Bull. Soc. Chim. Fr.*, 1996 (1959).

31. F. Dickens and J. Pearson, *Biochem. J.*, 48, 216 (1951).

32. M. Guerbet, *J. Pharm. Chim.*, [7]22, 321 (1920).

33. Ranwez and Genot, quoted by M. Pesez, *Bull. des Biologistes Pharmaciens*, 41, 252 (1938).

34. F. Feigl, *Mikrochim. Acta*, 1, 127 (1937).

35. J.J. Postowski, B.P. Lugowkin, and G.T. Mandryk, *Ber.*, 69, 1913 (1936).

36. Y. Takayama, C. Kurata, and K. Enomoto, *Japan Analyst*, 5, 449 (1956).

37. A.C. Bratton and E.K. Marshall, Jr., *J. Biol. Chem.*, 128, 537 (1939).

38. B. Saville, *Analyst*, 83, 670 (1958); H.F. Liddell and B. Saville, *Analyst*, 84, 188 (1959).

39. Y. Nomura, *Bull. Chem. Soc. Japan*, 32, 536 (1959).

40. T. Meisel and L. Erdey, *Mikrochim. Acta*, 1148 (1966); J. Bartos, *Ann. Pharm. Fr.*, 27, 159 (1969).

41. A. Irwin, J.H. Pearl, and P.F. MacCoy, *Anal. Chem.*, 32, 1407 (1960).

42. M. Rigaud, M. Labadie, and J.C. Breton, *Bull. Soc. Pharm. Bordeaux*, 109, 103, 111 (1970); A.M. Brondeau, M. Labadie, and J.C. Breton, *Bull. Soc. Pharm. Bordeaux*, 110, 77 (1971).

43. M. Pesez and J. Bartos, *Bull. Soc. Chim. Fr.*, 2333 (1963).

44. L.J. Dombrowski and E.L. Pratt, *Anal. Chem.*, 43, 1042 (1971).

45. R. Foster, *J. Chem. Soc.*, 3508 (1959).

46. R. Foster and R.K. Mackie, *Tetrahedron*, 16, 119 (1961).

47. C.C. Porter, *Anal. Chem.*, 27, 805 (1955).

48. F.M. Freeman, *Analyst*, 81, 299 (1956).

49. G.N. Smith, *Anal. Chem.*, 32, 32 (1960).

50. G.N. Smith and M.G. Swank, *Anal. Chem.*, 32, 978 (1960).

51. J.P. Heotis and J.W. Cavett, *Anal. Chem.*, 31, 1977 (1959).

52. M.R. Crampton and V. Gold, *Chem. Commun.*, 549 (1965).

53. D.P. Johnson and F.E. Critchfield, *Anal. Chem.*, 32, 865 (1960).

54. M. Pesez and J. Bartos, *Talanta*, 5, 216 (1960).

55. J. Bartos, *Talanta*, 8, 556 (1961).

56. J. Bartos, *Talanta*, 8, 619 (1961).

57. E. Besthorn, *Ber.*, 43, 1519 (1910).

58. S. Hünig and K.H. Fritsch, *Justus Liebigs Ann. Chem.*, 609, 143, 172 (1957); S. Hünig and H. Balli, *Justus Liebigs Ann. Chem.*, 609, 160 (1957).

59. T.R. Hauser and R.L. Cummins, *Anal. Chem.*, 36, 679 (1964).

60. A.P. Altshuller and L.J. Leng, *Anal. Chem.*, 35, 1541 (1963); I.R. Cohen and A.P. Altshuller, *ibid.*, 38, 1418 (1966).

61. R.P. Davis and R. Janis, *Nature*, 210, 318 (1966).

62. M. Pays, P. Malangeau, and R. Bourdon, *Ann. Pharm. Fr.*, 24, 763 (1966).

63. J. Bartos, *Ann. Pharm. Fr.*, 20, 650 (1962).

64. M. Pesez and J. Bartos, *Ann. Pharm. Fr.*, 22, 609 (1964).

65. M. Pays, P. Malangeau, and R. Bourdon, *Ann. Pharm. Fr.*, 25, 29 (1967).

66. M. Pays and M. Beljean, *Ann. Pharm. Fr.*, 28, 241 (1970).

67. M. Pays and O. Danlos, *Ann. Pharm. Fr.*, 25, 533 (1967).

68. M. Pays and M. Beljean, *Ann. Pharm. Fr.*, 28, 153 (1970).

69. G.G. Guidotti, A.F. Borghetti, and L. Loreti, *Anal. Biochem.*, 17, 513 (1966).

70. A. Tsuji, T. Kinoshita, and M. Hoshino, *Chem. Pharm. Bull.*, 17, 217, 1505 (1969).

71. E. Kamata, *Bull. Chem. Soc. Japan*, 38, 2005 (1965).

72. M.A. Paz, O.O. Blumenfeld, M. Rojkind, E. Henson, C. Furfine, and P.M. Gallop, *Arch. Biochem. Biophys.*, 109, 548 (1965).

73. K. Soda, T. Yorifuji, H. Misono, and M. Moriguchi, *Biochem. J.*, 114, 629 (1969).

74. F.W. Neumann, *Anal. Chem.*, 41, 2077 (1969).

75. M. Umeda, *Yakugaku Zasshi*, 83, 951 (1963).

76. E. Kamata, *Bull. Chem. Soc. Japan*, 37, 1674 (1964); M. Pays and R. Bourdon, *Ann. Pharm. Fr.*, 26, 681 (1968).

77. H.O. Friestad, D.E. Ott, and F.A. Gunther, *Anal. Chem.*, 41, 1750 (1969).

78. E. Sawicki, T.W. Stanley, T.R. Hauser, W. Elbert, and J.L. Noe, *Anal. Chem.*, 33, 722 (1961).

79. M. Pays, R. Bourdon, and M. Beljean, *Anal. Chim. Acta*, 47, 101 (1969).

80. E. Sawicki, T.R. Hauser, T.W. Stanley, W. Elbert, and F.T. Fox, *Anal. Chem.*, 33, 1574 (1961).

81. J. Fog and E. Jellum, *Nature*, 195, 490 (1962).

82. E. Sawicki, T.W. Stanley, and W. Elbert, *Microchem. J.*, 5, 225 (1961).

83. G.J. Buist and J.W. Lewis, *Chem. Commun.*, 66 (1965).

84. L. Malaprade, *C. R. Acad. Sci., Paris*, 186, 382 (1928); *Bull. Soc. Chim. Fr.*, [4]43, 683 (1928).

85. Cf. P. Fleury and J. Courtois, *Institut International de Chimie Solvay*, 8e *Conseil de Chimie*. Rapport et Discussions (Stoop Ed., Bruxelles), 279 (1950); J.M. Bobbitt, *Advan. Carbohyd. Chem.*, 11, 1 (1956); P. Malangeau, *Mises au Point de Chimie Analytique*, 9e série (1961), 81 (Masson Ed.).

86. P. Fleury and J. Lange, *J. Pharm. Chim.*, [8]17, 313 (1933).

87. G. Nominé, R. Bucourt, and D. Bertin, *Bull. Soc. Chim. Fr.*, 561 (1961).

88. P. Fleury, J. Courtois, and A. Bieder, *Bull. Soc. Chim. Fr.*, [5]19, 118 (1952); [5]20, 543 (1953).

89. R.U. Lemieux and E. v. Rudloff, *Can. J. Chem.*, 33, 1710 (1955).

90. H. Schiff, *C. R. Acad. Sci., Paris*, 61, 45 (1865); 64, 182 (1867); *Justus Liebigs Ann. Chem.*, 140, 92 (1866).

91. Caro, from J.G. Schmidt, *Ber.*, 13, 2342 (1880).

92. H. Grosse-Bohle, *Z. Unters. Nahr. Genussm.*, 14, 78 (1907); 27, 246 (1914).

93. P. Chautard, *Bull. Soc. Chim. Fr.*, [2]45, 83 (1886).

94. G. Deniges, *C. R. Acad. Sci., Paris*, 150, 832 (1910).

95. W.J. Blaedel and F.E. Blacet, *Ind. Eng. Chem., Anal. Ed.*, 13, 449 (1941).

96. D.E. Kramm and C.L. Kolb, *Anal. Chem.*, 27, 1076 (1955).

97. H. Wieland and G. Scheuing, *Ber.*, 54, 2527 (1921).

98. E. Sawicki, T.R. Hauser, and S. McPherson, *Anal. Chem.*, 34, 1460 (1962).

99. S.B. Schryver, *Proc. Roy. Soc. (London)*, B 82, 226 (1910).

100. M. Tanenbaum and C.E. Bricker, *Anal. Chem.*, 23, 354 (1951).

101. A.B. Lyons, *J. Amer. Pharm. Assoc.*, 13, 7 (1924).

102. E. Rimini, *Bull. Soc. Chim. Fr.*, [3]20, 896 (1898).

103. R. Mari, M. Feve, and M. Dzierzinsky, *Bull. Soc. Chim. Fr.*, 1395 (1961).

104. S.B. Schryver and C.C. Wood, *Analyst*, 45, 164 (1920).

105. E. Eegriwe, *Z. Anal. Chem.*, 110, 22 (1937).

106. M. Kamel and R. Wizinger, *Helv. Chim. Acta*, 43, 594 (1960).

107. F. Feigl and L. Hainberger, *Mikrochim. Acta*, 806 (1955).

108. P. Fleury, J. Courtois, and R. Perles, *Mikrochem. Mikrochim. Acta*, 36-37, 863 (1951).

109. V.H. Freed, *Science*, 107, 98 (1948); K. Munakata, *Japan Analyst*, 1, 199 (1952).

110. F. Feigl and R. Moscovici, *Analyst*, 80, 803 (1955).

111. Y. Shain and A.M. Mayer, *Israel J. Chem.*, 1, 39 (1963).

112. B. Klein and M. Weissman, *Anal. Chem.*, 25, 771 (1953).

113. W.M. Grant, *Anal. Chem.*, 20, 267 (1948).

114. R.N. Boos, *Anal. Chem.*, 20, 964 (1948).

115. Cf. *F.S.A.O.V.*, vol. 1, 114-122.

116. B.J. Thornton and J.C. Speck, Jr., *Anal. Chem.*, 22, 899 (1950).

117. M. Pesez and J. Robin, *Bull. Soc. Chim. Fr.*, 1930 (1962).

118. E. Sawicki, T.W. Stanley, and J. Pfaff, *Anal. Chim. Acta*, 28, 156 (1963).

119. T. Nash, *Biochem. J.*, 55, 416 (1953).

120. B.W. Bailey and J.M. Rankin, *Anal. Chem.*, 43, 782 (1971).

121. E. v. Bandrowski, *Monatsh. Chemie*, 10, 123 (1889).

122. E. Sawicki and R.A. Carnes, *Mikrochim. Acta*, 602 (1968).

123. S. Belman, *Anal. Chim. Acta*, 29, 120 (1963).

124. H.C. Tun, J.F. Kennedy, M. Stacey, and R.R. Woodbury, *Carbohyd. Res.*, 11, 225 (1969).

125. M. Pesez and J. Bartos, *Talanta*, 14, 1097 (1967).

126. J.V. Janovsky and L. Erb, *Ber.*, 19, 2155 (1886); J.V. Janovsky, *Ber.*, 24, 971 (1891).

127. B. von Bitto, *Justus Liebigs Ann. Chem.*, 269, 377 (1892).

128. A. Reissert, *Ber.*, 37, 831 (1904).

129. W. Zimmermann, *Z. Physiol. Chem.*, 233, 257 (1935); 245, 47 (1937).

130. Cf. R.N. Cahen and W.T. Salter, *J. Biol. Chem.*, 152, 489 (1944); R. Henry and M. Thevenet, *Bull. Soc. Chim. Biol.*, 33, 1617 (1951); W. Zimmermann, H.V. Anton, and D. Pontius, *Z. Physiol. Chem.*, 289, 91 (1952); M. Masuda and H.C. Thuline, *J. Clin. Endocrinol. Metabolism*, 13, 581 (1953); H. Werbin and S. Ong, *Anal. Chem.*, 26, 762 (1954); O. Crepy, F. Meslin, and P. Degrez, *Ann. Biol. Clin.*, 14, 355 (1956); C.D. Migeon and J.E. Plager, *J. Clin. Endocrinol. Metabolism*, 15, 702 (1955).

131. Cf. A.M. Bongiovanni, W.R. Eberlein, and P.Z. Thomas, *J. Clin. Endocrinol. Metabolism*, 17, 331 (1957); V.T.H. James and M. De Jong, *J. Clin. Pathol.*, 14, 425 (1961); C.S. Corker, J.K. Norymberski, and R. Thow, *Biochem. J.*, 83, 585 (1962).

132. R.P. Tansey and J.M. Cross, *J. Amer. Pharm. Assoc.*, *Sci. Ed.*, 39, 660 (1950); S. Cohen and A. Kalusziner, *Anal. Chem.*, 29, 161 (1957).

133. T. Momose, Y. Ohkura, K. Kohashi, and R. Nagata, *Chem. Pharm. Bull. (Japan)*, 11, 301, 973 (1963); M. Pesez and J. Bartos, *Ann. Pharm. Fr.*, 22, 541 (1964).

134. H.L. Mason and W.W. Engstrom, *Physiol. Rev.*, 30, 321 (1950); P.L. Munson and A.D. Kenny, *Recent Prog. Hormone Res.*, 9, 135 (1954); W. Zimmermann, *Z. Physiol. Chem.*, 300, 141 (1955); see also *F.S.A.O.V.*, vol. 1, 278-291.

135. T. Canbäck, *Farm. Revy*, 48, 153, 217, 234 (1949); *Svensk. Farm. Tidskr.*, 53, 151 (1949); 54, 1 (1950).

136. J. Meisenheimer, *Justus Liebigs Ann. Chem.*, 323, 205 (1902); cf. R. Foster and C.A. Fyfe, *Rev. Pure and Appl. Chem.*, 16, 61 (1966).

137. C.A. Fyfe and R. Foster, *Chem. Commun.*, 1219 (1967).

138. C.A. Fyfe, *Can. J. Chem.*, 46, 3047 (1968).

139. R. Foster and C.A. Fyfe, *J. Chem. Soc.*, (B), 53 (1966).

140. M.J. Strauss and R.G. Johanson, *Chem. Ind. (London)*, 242 (1969).

141. M. Ishidate and T. Sakaguchi, *J. Pharm. Soc. Japan*, 70, 444 (1950).

142. A.E. Kellie and E.R. Smith, *Nature*, 178, 323 (1956).

143. C.S. Corker, J.K. Norymberski, and R. Thow, *Biochem. J.*, 83, 583 (1962).

144. O. Neunhoeffer, K. Thewalt, and W. Zimmermann, *Z. Physiol. Chem.*, 323, 116 (1961).

145. T.J. King and C.E. Newall, *J. Chem. Soc.*, 367 (1962).

146. R. Foster and R.K. Mackie, *Tetrahedron*, 18, 1131 (1962); 19, 691 (1963).

147. Cf. R.W. Bost and F. Nicholson, *Ind. Eng. Chem., Anal. Ed.*, 7, 190 (1935).

148. Cf. M. Pesez, *J. Pharm. Chim.*, 27, 120, 247 (1938); 30, 200 (1939); B.H. Dolin, *Ind. Eng. Chem., Anal. Ed.*, 15, 242 (1943); W.O. James and M. Roberts, *Quart. J. Pharm. Pharmacol.*, 18, 29 (1945); E. Rathenasinkam, *Analyst*, 75, 108, 169 (1950); R. Fabre, R. Truhaut and M. Peron, *Ann. Pharm. Fr.*, 8, 613 (1950); M.S. Schechter and I. Hornstein, *Anal. Chem.*, 24, 544 (1952); W. Hancock and E.Q. Laws, *Analyst*, 81, 37 (1956).

149. J. Bartos, *Chim. Anal.*, 53, 384 (1971).

150. J. Bartos, *Chim. Anal.*, 53, 18 (1971).

151. J. Bartos, *Talanta*, 16, 551 (1969).

152. M.R.F. Ashworth and E. Gramsch, *Mikrochim. Acta*, 358 (1967).

153. K. Czerepko, *Z. Anal. Chem.*, 182, 269 (1961).

154. A. Schönberg and M.M. Sidsky, *J. Org. Chem.*, 21, 476 (1956).

155. J.S. King, Jr., and N.H. Leake, *Analyst*, 84, 694 (1959).

156. D.H.E. Tattje, *Pharm. Weekblad*, 91, 733 (1956); 92, 729 (1957); 93, 689, 694, 1048 (1958).

157. D.H.E. Tattje, *Pharm. Weekblad*, 91, 841 (1956); *J. Pharm. Pharmacol.*, 9, 29 (1957).

158. M. Frerejacque and P. De Graeve, *Ann. Pharm. Fr.*, 21, 509 (1963).

159. T. Canbäck, *Farm. Revy*, 46, 627, 802 (1947).

160. P.E. Hall and K. Fotherby, *Experientia*, 23, 288 (1967).

161. D.N. Kirk, W. Klyne, and A. Mudd, *J. Chem. Soc.*, (C), 2269 (1968).

162. M. Böniger, *Ber.*, 27, pt. 2, 23 (1894).

163. P. Ehrlich and C.A. Herter, *Z. Physiol. Chem.*, 41, 379 (1904).

164. Cf. O. Folin, *J. Biol. Chem.*, 51, 377, 393 (1922); N.H. Furman, G.H. Morrison, and A.F. Wagner, *Anal. Chem.*, 22, 1561 (1950).

165. Cf. E.G. Schmidt, *J. Biol. Chem.*, 122, 757 (1938); F.J. Bandelin, *Science*, 106, 426 (1947); P. Mesnard and R. Crockett, *Chim. Anal.*, 42, 346, 381 (1960).

166. K. Blau and W. Robson, *Chem. Ind. (London)*, 424 (1957); D.H. Rosenblatt, P. Hlinka, and J. Epstein, *Anal. Chem.*, 27, 1290 (1955).

167. W. Troll, *J. Biol. Chem.*, 202, 479 (1953).

168. M. Pesez and J. Bartos, unpublished results.

169. J. Bartos and M. Pesez, *Bull. Soc. Chim. Fr.*, 1627 (1970).

170. H. Goldstein and G. Genton, *Helv. Chim. Acta*, 20, 1413 (1937).

171. M. Pesez and J. Bartos, *Ann. Pharm. Fr.*, 27, 161 (1969); J. Bartos and M. Pesez, *Talanta*, 19, 93 (1972).

172. M. Roth and J. Rieder, *Anal. Chim. Acta*, 27, 20 (1962).

173. M.X. Sullivan and W.C. Hess, *J. Amer. Chem. Soc.*, 58, 47 (1936).

174. F. Sachs and M. Craveri, *Ber.*, 38, 3685 (1905).

175. Cf. F. Feigl, *Spot Tests in Organic Analysis*, Elsevier (Amsterdam), 7th Ed. (1966), 153.

176. A.T. Warren, F.O. Howland, and L.W. Green, *J. Amer. Pharm. Assoc.*, *Sci. Ed.*, 37, 186 (1948).

177. F. Turba, R. Haul, and G. Uhlen, *Angew. Chem.*, 61, 74 (1949).

178. J. Bartos, *Ann. Pharm. Fr.*, 27, 691 (1969).

179. Cf. F. Feigl, *Spot Tests in Organic Analysis*, Elsevier (Amsterdam), 6th Ed. (1960), 317.

180. H. v. Pechmann, *Ber.*, 27, 1679 (1894).

181. A. Pinner, *Ber.*, 17, 182, 2002 (1884).

182. H. v. Pechmann, *Ber.*, 25, 3175 (1892).

183. A.F. Hegarty and F.L. Scott, *Chem. Commun.*, 622 (1966).

184. H. v. Pechmann and P. Runge, *Ber.*, 27, 323, 2920 (1894).

185. C.A. Johnson, R. King, and C. Vickers, *Analyst*, 85, 714 (1960).

186. A.M. Mattson and C.O. Jensen, *Anal. Chem.*, 22, 182 (1950).

187. R.A. Fairbridge, K.J. Willis, and R.G. Booth, *Biochem. J.*, 49, 423 (1951).

188. G. Avigad, R. Zelikson, and S. Hestrin, *Biochem. J.*, 80, 57 (1961).

189. H.B. Mark, Jr., L.B. Backes, D. Pinkel, and L. Papa, *Talanta*, 12, 27 (1965).

190. Cf. W.J. Mader and R.R. Buck, *Anal. Chem.*, 24, 666 (1952).

191. A.A. Henly, *Nature*, 169, 877 (1952).

192. P. Ascione and C. Fogelin, *J. Pharm. Sci.*, 52, 709 (1963).

193. A.S. Meyer and M.C. Lindberg, *Anal. Chem.*, 27, 813 (1955).

194. R.V. Smith, T.H. Hassall, and S.C. Liu, *J. Ass. Offic. Anal. Chem.*, 53, 1089 (1970).

195. P.E. Manni and J.E. Sinsheimer, *Anal. Chem.*, 33, 1900 (1961).

196. E.F. Salim, P.E. Manni, and J.E. Sinsheimer, *J. Pharm. Sci.*, 53, 391 (1964).

197. J. Bartos and M. Pesez, *Ann. Pharm. Fr.*, 28, 459 (1970).

198. J.E. Sinsheimer and E.F. Salim, *Anal. Chem.*, 37, 566 (1965).

199. J. Bartos, *Ann. Pharm. Fr.*, 30, 153 (1972).

200. J. Renault and L. Bigot, *Ann. Pharm. Fr.*, 21, 847 (1963).

201. J. Renault and L. Bigot, *Bull. Soc. Chim. Fr.*, 2093 (1964).

202. F. Penzoldt and E. Fischer, *Ber.*, 16, 657 (1883).

203. L. Rosenthaler and G. Vegezzi, *Mitt. Lebensm. Hyg.*, 45, 178 (1954).

204. E. Sawicki, T.W. Stanley, and T.R. Hauser, *Chemist-Analyst*, 47, 87 (1958).

205. P.G. Walker, *Biochem. J.*, 58, 699 (1954).

206. A. Hantzsch, *Justus Liebigs Ann. Chem.*, 215, 1, 72 (1882).

207. V.M. Sardesai and H.S. Provido, *Microchem. J.*, 14, 550 (1969).

208. Cf. A.P. Phillips, *J. Amer. Chem. Soc.*, 71, 4003 (1949).

209. E. Sawicki and R.A. Carnes, *Mikrochim. Acta*, 148 (1968).

210. M. Pesez and J. Bartos, *Talanta*, 16, 331 (1969).

211. E. Sawicki and R.A. Carnes, *Anal. Chim. Acta*, 41, 178 (1968).

212. S. Ruhemann, *J. Chem. Soc.*, 97, 1438 (1910).

213. S. Ruhemann, *J. Chem. Soc.*, 97, 2025 (1910).

214. S. Ruhemann, *J. Chem. Soc.*, 99, 792 (1911).

215. A. Strecker, *Justus Liebigs Ann. Chem.*, 123, 363 (1862).

216. R. Abderhalden, *Z. Physiol. Chem.*, 252, 81 (1938).

217. S. Ruhemann, *J. Chem. Soc.*, 99, 1486 (1911).

218. D.J. McCaldin, *Chem. Rev.*, 60, 39 (1960).

219. F.S. Schlenker, *Anal. Chem.*, 19, 471 (1947).

220. F. Neuzil, J.C. Breton, and H. Plagnol, *Bull. Mém. Ecole Nat. Méd. Pharm. Dakar*, 7, 195 (1959).

221. S. Moore and W.H. Stein, *J. Biol. Chem.*, 176, 367 (1948).

222. S. Moore and W.H. Stein, *J. Biol. Chem.*, 211, 907 (1954).

223. W. Troll and R.K. Cannan, *J. Biol. Chem.*, 200, 803 (1953).

224. E.W. Yemm and E.C. Cocking, *Analyst*, 80, 209 (1955).

225. T.C. Bruice and F.M. Richards, *J. Org. Chem.*, 23, 145 (1958).

226. D.A. MacFadyen and N. Fowler, *J. Biol. Chem.*, 186, 13 (1950).

227. M. Regitz, H. Schwall, G. Heck, B. Eistert, and G. Bock, *Justus Liebigs Ann. Chem.*, 690, 125 (1965).

228. P.J. Lamothe and P.G. McCormick, *Anal. Chem.*, 45, 1906 (1973).

229. M. Pesez and J. Bartos, unpublished results.

230. M. Friedman, *Microchem. J.*, 16, 204 (1971).

231. A.W. Johnson and D.J. McCaldin, *J. Chem. Soc.*, 817 (1958).

232. M.K. Gaitonde, *Biochem. J.*, 104, 627 (1967).

233. H. Wachsmuth, *J. Pharm. Belg.*, 4, 186 (1949).

234. P. Baudet and E. Cherbuliez, *Helv. Chim. Acta*, 40, 1612 (1957).

235. S. Oden and B. von Hofsten, *Nature*, 173, 449 (1954).

236. M. Yamagishi, *J. Pharm. Soc. Japan*, 74, 1233 (1954).

237. E. Neuzil, J. Josselin, and J.C. Breton, *C. R. Acad. Sci., Paris*, 252, 119 (1961).

238. E. Neuzil, J. Josselin, and Y. Vidal, *J. Chromatog.*, 56, 311 (1971).

239. K. Samejina, W. Dairman, J. Stone, and S. Udenfriend, *Anal. Biochem.*, 42, 237 (1971).

240. R.B. Conn, Jr., and R.B. Davis, *Nature*, 183, 1053 (1959).

241. K. Beyermann and H. Wisser, *Z. Anal. Chem.*, 245, 311 (1969).

242. L.R. Jones and J.A. Riddick, *Anal. Chem.*, 26, 1035 (1954).

243. L.R. Jones and J.A. Riddick, *Anal. Chem.*, 29, 1214 (1957).

244. D.P. Johnson, *Anal. Chem.*, 41, 859 (1969).

245. J. Labarre and E.A. Martin, *Rev. Can. Biol.*, 13, 389 (1954).

246. E. Vongerichten, *Ber.*, 32, 2571 (1899).

247. T. Zincke, *Justus Liebigs Ann. Chem.*, 330, 361 (1904); 339, 193 (1905); T. Zincke, G. Heuser, and W. Möller, *Justus Liebigs Ann. Chem.*, 333, 296 (1904).

248. W. König, *J. Prakt. Chem.*, 69, 105 (1904); 70, 19 (1904); 83, 406 (1911).

249. K. Fujiwara, *Sitzber. Abhandl. Naturforsch. Ges. Rostock*, 6, 33 (1916).

250. Cf. H. Freytag, *Ber.*, 69, 32 (1936).

251. F. Feigl and V. Anger, *J. Prakt. Chem.*, 139, 180 (1934).

252. M. Pesez, *Ann. Chim. Anal.*, 25, 37 (1943).

253. F. Feigl, V. Anger, and R. Zappert, *Mikrochemie*, 16, 67 (1934).

254. E. Asmus, J. Ganzke, and W. Schwarz, *Z. Anal. Chem.*, 253, 102 (1971).

255. J.H. Ross, *J. Biol. Chem.*, 58, 641 (1923).

256. M.S. Moss and H.J. Rylance, *Nature*, 210, 945 (1966).

257. P.J. Friedman and J.R. Cooper, *Anal. Chem.*, 30, 1674 (1958).

258. C.D. Hildebrecht, *Anal. Chem.*, 29, 1037 (1957).

259. P. Cheramy, *J. Pharm. Chim.*, [9]1, 233 (1940-1941).

260. L. Truffert, *Bull. Soc. Chim. Biol.*, 24, 195 (1942).

261. H. Griffon, *Ann. Pharm. Fr.*, 6, 165 (1948).

262. J. Adachi, *Anal. Chem.*, 23, 1491 (1951).

263. K.C. Leibman and J.D. Hindman, *Anal. Chem.*, 36, 348 (1964).

264. J. Bartos, *Ann. Pharm. Fr.*, 29, 221 (1971).

265. P. Karrer and H. Keller, *Helv. Chim. Acta*, 21, 463, 1170 (1938).

266. S.P. Vilter, T.D. Spies, and A.P. Mathews, *J. Biol. Chem.*, 125, 85 (1938).

267. E.F.G. Herington, *Analyst*, 76, 90 (1951).

268. C.W. Ballard and P.G.W. Scott, *Chem. Ind. (London)*, 715 (1952); P.G.W. Scott, *J. Pharm. Pharmacol.*, 4, 681 (1952); M. Schoog, *Münchn. Med. Wochschr.*, 94, 2135 (1952); K. Yamaguchi, *Folia Pharmacol. Japan*, 51, 631 (1955); N.F. Poole and A.E. Meyer, *Proc. Soc. Exptl. Biol. Med.*, 98, 375 (1958).

269. A.A. Shmuk and A. Borozdina, *Zh. Prikl. Khim.*, 13, 776 (1940).

270. E. Asmus, J. Kraetsch, and D. Papenfuss, *Z. Anal. Chem.*, 184, 25 (1961).

271. J.V. Scudi, *Science*, 103, 567 (1946); D.K. Schauduri and E. Kodicek, *Biochem. J.*, 44, 343 (1949).

272. M. De Clercq and R. Truhaut, *Ann. Pharm. Fr.*, 15, 529 (1957).

273. L.J. Harris and W.D. Raymond, *Biochem. J.*, 33, 2037 (1939).

274. D.K. Chaudhuri, *Indian J. Med. Res.*, 39, 491 (1951).

275. E. Stotz, *J. Lab. Clin. Med.*, 26, 1042 (1941).

276. C. Klatzkin, F.W. Norris, and F. Wokes, *Analyst*, 74, 447 (1949).

277. G.I. Luk'yanchikova, V.N. Bernshtein, and S.N. Stepanyuk, *Farm. Zh. (Kiev)*, 24, 66 (1969).

278. A. Mueller and S.H. Fox, *J. Biol. Chem.*, 167, 291 (1947).

279. E. Werle and H.W. Becker, *Biochem. Z.*, 313, 182 (1942).

280. E. Asmus and H. Garschagen, *Z. Anal. Chem.*, 139, 81 (1953).

281. A.C. Corcoran, O.M. Helmer, and I.H. Page, *J. Biol. Chem.*, 129, 89 (1939).

282. J. Fuentes-Duchemin and E. Casassas, *Anal. Chim. Acta*, 44, 462 (1969).

283. E. Bandier and J. Hald, *Biochem. J.*, 33, 264 (1939).

284. I.S. Mazel, E.A. Vasil'eva-Sokolova, and G.I. Kudryavtsev, *Vysokomol. Soedin.*, 5, 868 (1963).

285. J.P. Sweeney and W.L. Hall, *Anal. Chem.*, 23, 983 (1951).

286. R.G. Martinek, E.R. Kirch, and G.L. Webster, *J. Biol. Chem.*, 149, 245 (1943).

287. H.G. Higson, R.F. Raimondo, and E.W. Tunstall, *Anal. Chem.*, 41, 1474 (1969).

288. P.O. Dennis and H.G. Rees, *Analyst*, 74, 481 (1949).

289. Y. Raoul and O. Crepy, *Bull. Soc. Chim. Biol.*, 23, 362 (1941).

290. P. Mesnard, G. Devaux, and J. Fauquet, *Sci. Pharm. Proc.*, *25th*, 2, 77 (1965).

291. E. Asmus and D. Papenfuss, *Z. Anal. Chem.*, 185, 201 (1962).

292. J. Bartos, *Ann. Pharm. Fr.*, 29, 71 (1971).

293. R.W. Frei, A. Kunz, G. Pataki, T. Prims, and H. Zürcher, *Anal. Chim. Acta*, 49, 527 (1970).

294. E. Asmus and K. Noack, *Z. Anal. Chem.*, 249, 122 (1970).

295. V. Anger and S. Ofri, *Talanta*, 10, 1302 (1963).

296. M.T.H. Ragab and J.P. McCollum, *J. Agr. Food Chem.*, 16, 284 (1968).

Chapter 17

SYNTHESES OF REAGENTS AND STANDARDS

PURIFICATION OF SOLVENTS

I. SYNTHESES OF REAGENTS

A. 4-Azobenzenediazonium Fluoborate (1)

In a 250 ml three-necked flask equipped with a mechanical stirrer, a dropping funnel, and a thermometer, dissolve 19.7 g of 4-aminoazobenzene (C.I. Solvent Yellow 1) in the minimum amount of dimethylformamide. Chill to 0° in an ice-methanol bath, and add with stirring 49 ml of concentrated hydrochloric acid, then dropwise 7.6 g of sodium nitrite dissolved in the minimum amount of water, while the temperature is maintained at 0 - +5°. Stir for 10 min more, and, still cooling, add 1 g of urea. Filter, and to the filtered solution add 12.1 g of sodium fluoborate dissolved in the minimum amount of water. Mix well, filter the precipitate by suction on a sintered-glass funnel, wash successively with water, methanol, and ethyl ether, and dry in a vacuum desiccator over sulfuric acid. Yield: 4.8 g. Stable for several months in the refrigerator.

B. 2,6-Dichloro-4-trimethylammoniumbenzenediazonium Chloride-Fluoborate (2)

In a 100 ml three-necked flask provided with a mechanical stirrer, a dropping funnel, and a thermometer, cool a stirred suspension of 7.45 g of 2,4,6-trichlorobenzene-diazonium fluoborate (Chapter 17, Section I.H) in 20 ml of acetonitrile to -18° with a dry ice-acetone bath. Add dropwise 20 ml of a 1.26 M solution of trimethylamine in acetonitrile over a period of 1 hr, while temperature is maintained at -18° by adding dry ice to the bath when required. Stir for 15 min more, filter the precipitate with suction, and wash successively with 20 ml of acetonitrile, 20 ml of carbon tetrachloride, and twice with 20 ml portions of methylene chloride, these solvents being prechilled to about -18°. Dry in a vacuum desiccator over sulfuric acid. Yield: 5.4 g. The compound melts at about 132° with decomposition. Stable for a few days in the refrigerator.

C. 3,3-Diphenylacrolein

We obtained the sodium bisulfite derivative of this aldehyde in the course of the synthesis of 17-hydroxy-11-deoxycorticosterone, as a by-product of the oxidation of 3α-acetoxy-24,24-diphenylchola-20,23-diene (3). It can also be obtained by reacting β,β-diphenylvinylmagnesium bromide with ethyl orthoformate (4). We give here only the preparation of the aldehyde from its bisulfite derivative.

1. *Purification of the bisulfite derivative*

Reflux for 5 min 50 g of the derivative in 500 ml of ethanol, in the presence of 2 g of decolorizing carbon. Filter while hot, and allow to crystallize for 3 hr in the cold. Filter with suction, wash with 30 ml of ethanol, and dry in a vacuum desiccator over potassium hydroxide. Yield: 30 g. This compound is stable.

2. 3,3-Diphenylacrolein

In a 500 ml four-necked flask equipped with a sealed mechanical stirrer, a dropping funnel, a thermometer, and a reflux condenser connected with a sulfur dioxide absorber, suspend 30 g of the purified bisulfite derivative in 120 ml of water. Heat to 50°, and add slowly with stirring 16.5 ml of 50% sulfuric acid. Then stir for 1 hr at 60°, transfer while hot to a 200 ml separatory funnel, let stand for 1 hr, and collect the bottom layer. Add to it 30 ml of water preheated to 50°. Maintain the mixture at that temperature, and neutralize by adding sodium bicarbonate until pH 6-7. Induce to crystallize while cooling, filter with suction, and wash with water.

Without further drying, dissolve the product in 13 ml of boiling isopropyl ether, add portionwise anhydrous magnesium sulfate until all the water is eliminated, then add 2 g of decolorizing carbon, and reflux for 5 min. Filter while hot, and wash the residue with 4 ml of hot isopropyl ether. Combine the solution and the washings, induce to crystallize while cooling, and let stand at 0° for 2 hr. Filter with suction, wash with 4 ml of prechilled isopropyl ether, and dry in a vacuum desiccator over sulfuric acid. Yield: 12.4 g. Melting point: 44-45°. Stable for several years in the refrigerator.

D. Flavylium Perchlorate (5)

In a 100 ml three-necked flask equipped with a mechanical stirrer, a dropping funnel, and a thermometer, mix 6.2 ml of salicylaldehyde, 7 ml of acetophenone, 25 ml of glacial acetic acid, and 3 ml of 70% perchloric acid. Stir, maintain the temperature at about 20° with a water bath, and add dropwise 22 ml of acetic anhydride over a period of 2 hr. Let stand securely stoppered overnight, filter with suction on a sintered-glass funnel, wash with

15 ml of glacial acetic acid, then twice with 10 ml portions of ethyl ether. Dry in a vacuum desiccator over potassium hydroxide. Yield: 3.25 g. Stable for about one year in the refrigerator.

E. m-Phenylenediamine Oxalate (6)

In a 1 liter three-necked flask equipped with a reflux condenser and a dropping funnel, dissolve by warming 20 g of m-dinitrobenzene in 100 ml of ethanol, allow to cool (the compound recrystallizes as very fine crystals), and add 50 ml of 6% aqueous solution of cupric sulfate pentahydrate. Heat to reflux, and add slowly 150 ml of aqueous solution containing 25 g of potassium borohydride over a period of 1 hr 15 min, so that a regular evolution of hydrogen be obtained. During that time, the color of the foaming mixture turns from orange to gray-green, then to whitish. Maintain reflux for about 15 min more, until no more gas is evolved. Cool, filter, and extract the solution twice with 200 ml portions of ethyl ether. To the combined extracts, add 20 g of oxalic acid dissolved in 80 ml of ethanol, mix, chill to 0°, filter the crystallized oxalate with suction, and wash twice with 25 ml portions of ethyl ether. Dry in a vacuum desiccator over sulfuric acid. Yield: 21.7 g.

To recrystallize, by gentle warming dissolve the crude product together with an equal weight of oxalic acid in 140 ml of water, add 20 g of decolorizing carbon, shake, and filter quickly. Allow to cool, filter the crystallized salt with suction, and wash with a few ml of water. Dry in a vacuum desiccator over sulfuric acid. Yield: 15.5 g. Melting points: 125°, then about 270° with decomposition. Stable for several years in the refrigerator.

If the salt is prepared by mixing a solution of the commercially available free base with a solution of oxalic acid, the isolated product is always colored.

F. 3-Phenyl-5-nitrosamino-1,2,4-thiadiazole (17) (Phenitrazole)

1. Benzamidine hydrochloride

Bubble a current of dry hydrogen chloride through a mixture of 51 g of benzonitrile and 31 ml of ethanol cooled to 0° until it shows a gain in weight of 21.2 g. Allow to stand for 48 hr in a tightly stoppered flask; during this time the mixture sets to a solid mass. Crush the mass in a mortar, and introduce the crushed product into a 500 ml three-necked flask equipped with a sealed mechanical stirrer, a condenser, and a potassium hydroxide drying tube. Add 150 ml of ethanol in which 12 g of gaseous ammonia has been dissolved, and stir for 65 hr. Filter to separate ammonium chloride, and remove the solvent under reduced pressure. Induce the residual oil to crystallize, and dissolve the solid mass so obtained in 160 ml of ethanol. Add 0.3 ml of concentrated hydrochloric acid, check the pH, which should be acidic, add 4 g of decolorizing carbon, shake for a few minutes, filter, and add 300 ml of ethyl ether to the filtered solution. Induce to crystallize, let stand for 30 min, filter with suction on a sintered-glass funnel, wash with a few milliliters of ethyl ether, and dry in a vacuum desiccator over sulfuric acid. Yield: 48 g. Melting point: 165°.

2. 3-Phenyl-5-amino-1,2,4-thiadiazole

Into a 500 ml four-necked flask equipped with a mechanical stirrer, a thermometer, and two dropping funnels, introduce 20 g of benzamidine hydrochloride and 12.4 g of anhydrous potassium thiocyanate (see Note, below), and add 130 ml of methanol. The solid mixture dissolves, then potassium chloride precipitates almost immediately. Into one of the dropping funnels, introduce a solution of sodium methoxide obtained by dissolving 5.88 g of sodium in 60 ml of methanol, and into the other introduce a solution of 20.5 g (6.6 ml) of bromine in 60 ml of methanol. Cool the contents of the flask to +10° with an ice-water bath, stir, and over a period of 20-30 min add simultaneously and dropwise both the sodium methoxide and the bromine solutions, so that there is always a slight excess of bromine in the reaction medium (light yellow-brown color). The temperature should not exceed +10° during this addition. When all the bromine solution is added, proceed with the addition of the methoxide solution until the reaction mixture is colorless. Stir for 2 hr more in an ice bath, add 2 ml of glacial acetic acid, then add dropwise a saturated aqueous solution of sodium sulfide, until no color is developed when a drop of the reaction mixture is added to a few drops of an acidic solution of potassium iodide. Filter with suction, and wash the precipitate abundantly with large volumes of methanol. Evaporate the combined washings to dryness under vacuum. Take up the residue with 200 ml of water, triturate until solid, and filter with suction. Dissolve the product without further drying in 20 ml of boiling ethanol, add approximately 1 g of decolorizing carbon, reflux for a few minutes, and filter while hot. To the filtered solution, add 80 ml of water. The compound crystallizes immediately. Chill in ice, filter with suction, wash with water, and dry in a vacuum desiccator over sulfuric acid. Yield: 7.1 g. Melting point: 153°.

Note. The anhydrous potassium thiocyanate is obtained by recrystallizing a batch of the commercial product from ethanol.

3. Phenitrazole

Dissolve 7 g of 3-phenyl-5-amino-1,2,4-thiadiazole in 79 ml of formic acid with warming. Chill in ice, and add with thorough stirring 3.5 g of sodium nitrite dissolved in 20 ml of water; during the addition yellowish product precipitates. Maintain at 0° for 10 min, filter with suction, and wash with water. Without drying, dissolve the crude product in 300 ml of acetone, add approximately 1 g of decolorizing carbon, shake for a few minutes, filter, and add 500 ml of water to the filtered solution. The compound crystallizes as slightly cream-colored needles. Filter with suction, wash with water, and dry in a vacuum desiccator over potassium hydroxide. Yield: 6.6 g. Melting point: 188°, with decomposition. Stable for at least one year in the refrigerator.

G. 5-(p-Sulfamoylphenylazo)salicylic Acid (8) (Salazosulfamide)

Dissolve 17 g of sulfanilamide in a mixture of 200 ml of water and 20 ml of concentrated hydrochloric acid. Add 200 g of crushed ice, then add slowly and with stirring 35 ml of 20% aqueous solution of sodium nitrite. Maintain the temperature at 0° for 15 min more, and pour the mixture into a prechilled solution of 14 g of salicylic acid in 490 ml of 3.7% aqueous solution of potassium hydroxide. Let stand for 1 hr, and precipitate the potassium salt of salazosulfamide by bubbling a stream of carbon dioxide through the solution. Filter with suction, and dissolve without further drying in a mixture of 100 ml of water and 3.5 ml of 10 N sodium hydroxide. Add with stirring 100 ml of a 1:9 mixture of concentrated hydrochloric acid and water. Filter the precipitated acid with suction, wash with water until the washings give a negative reaction with silver nitrate, and dry at 100° in a hot-air oven. Yield: 8 g. Recrystallize from 50% ethanol. Stable for several years in the refrigerator.

H. 2,4,6-Trichlorobenzenediazonium Fluoborate (9)

In a 250 ml three-necked flask equipped with a mechanical stirrer, a dropping funnel, and a thermometer, suspend 9.8 g of 2,4,6-trichloroaniline in a mixture of 50 ml of water and 15 ml of concentrated hydrochloric acid. Stir, cool to +3° to +4° with an ice bath, and add a solution of 3.5 g of sodium nitrite in 7 ml of water over a period of 5 min. Stir for 1 hr more at the same temperature, and filter. Maintain the filtered solution in an ice bath, stir, and add portionwise 15 g of finely powdered sodium fluoborate over a period of 2 min. Stir for 10 min more, filter with suction, and wash successively with 10 ml of prechilled saturated aqueous solution of sodium fluoborate, 10 ml of ethanol, and 10 ml of ether. Dry in a vacuum desiccator over sulfuric acid. Yield: 12.1 g. Melting point: 195° with decomposition. Stable for 1 month in the refrigerator.

I. Vanadium Oxinate (Sodium Salt) (10)

Dissolve 0.27 g of ammonium vanadate(V) (ammonium metavanadate) in 7 ml of 1 N sodium hydroxide, add 3 ml of water, and boil for about 5 min until ammonia is evolved (solution A).

In a 100 ml round-bottomed flask equipped with a mechanical stirrer, dissolve by warming 1 g of 8-hydroxyquinoline in 3.5 ml of 2 N sodium hydroxide, add 11.5 ml of water, immerse the flask in a boiling water bath, stir, and add the solution A, and maintain at 100° for 10 min. Then, while the mixture is hot, add dropwise with stirring 50% acetic acid. As each drop falls into the flask, a black precipitate appears, which redissolves at once. Proceed with the addition until a drop gives a permanent precipitate (about 1 ml). Filter while hot, and chill the filtrate in ice bath for about 1 hr. The sodium salt of vanadium oxinate crystallizes as yellow crystals. Filter with suction, wash with water, and dry in a vacuum desiccator over potassium hydroxide. Yield: 0.92 g. Stable in the refrigerator.

II. SYNTHESES OF FLUORESCENCE STANDARDS

A. 2,6-Dimethyl-3,5-diacetyl-1,4-dihydropyridine (11)

Dissolve 3 g of ammonium acetate and 1 ml of 30% aqueous solution of formaldehyde in 50 ml of water, and add 2 g of acetylacetone. Reflux for 10 min, and let stand for 2 hr in an ice bath. Filter with suction, and dry in a vacuum desiccator over sulfuric acid. Yield: 1.2 g.

Recrystallize from 20 ml of ethanol per gram of crude solid, with treatment with decolorizing carbon. Yield: 0.74 g. Melting point: 230°. Kept protected from light, this compound is stable.

B. 2,6-Dimethyl-3,5-dicarbethoxy-1,4-dihydropyridine (11)

Dissolve 3 g of ammonium acetate and 1 ml of 30% aqueous solution of formaldehyde in 50 ml of water, and add 2.6 g of ethyl acetoacetate. Reflux for 10 min, and let stand for 2 hr in an ice bath. Filter with suction, and dry in a vacuum desiccator over sulfuric acid. Yield: 1.15 g.

Recrystallize from 14 ml of ethanol per gram of crude solid, with treatment with decolorizing carbon. Yield: 0.68 g. Melting point: 192-193°. Kept protected from light, this compound is stable.

C. 1,8-Dioxodecahydroacridine (11)

Dissolve 3 g of ammonium acetate and 1 ml of 30% aqueous solution of formaldehyde in 50 ml of water, and add 2.24 g of cyclohexane-1,3-dione (dihydroresorcinol). Reflux for 30 min, and let stand for 2 hr in an ice bath. Filter with suction, and dry in a vacuum desiccator over sulfuric acid. Yield: 0.65 g.

Suspend the crude product in 8 ml of ethanol, add 17 ml of a 1 N solution of potassium hydroxide in ethanol, treat with decolorizing carbon, filter, and acidify with glacial acetic acid until the solution turns from orange to light yellow. Let stand for 2 hr at 0°, filter, wash with a few milliliters of ethanol, and dry in a vacuum desiccator over sulfuric acid. Yield: 0.4 g. The compound can also be recrystallized from 30 ml of ethanol per gram. It is stable when protected from light and does not melt below 300°.

D. 2-Hydroxy-3-methylquinoxaline (12)

Mix a solution of 0.22 g of sodium pyruvate in 10 ml of 1 N hydrochloric acid with a solution of 0.27 g of o-phenylenediamine in 12 ml of 1 N hydrochloric acid. Let stand for 1 hr at room temperature, filter the crystallized quinoxaline with suction, wash with a few drops of 1 N hydrochloric acid, and dry in a vacuum desiccator over potassium hydroxide. Yield: 0.18 g.

Recrystallize from the minimum amount of ethanol, with treatment with decolorizing carbon. Yield: 0.09 g. Melting point: 250-252°. Stable if protected from light.

E. 2-Methyl-5-carboxy-7-aminoquinoline (13)

To a mixture of 10 ml of paraldehyde and 10 ml of ethanol chilled to 0°, slowly add 50 ml of a 50% aqueous solution of 3,5-diaminobenzoic acid dihydrochloride pre-chilled to 0°. Let stand for a few hours at 15-17°, filter the crystallized compound, wash with water, and dry in a vacuum desiccator over sulfuric acid. Yield: 4 g.

Dissolve the crude product in 5 ml of 1 N sodium hydroxide, treat with decolorizing carbon, filter, and precipitate the compound by bubbling a stream of carbon dioxide through the solution. Filter with suction, wash with water, and dry at 100° under vacuum. The compound crystallizes with 1 mole of water. It can also be recrystallized from 60 ml of glacial acetic acid per gram. The compound dried at 130° is then anhydrous. Both recrystallization procedures afford low yield. This compound has no definite melting point, and it is stable if protected from light.

III. NOTE ON BENZYLTRIMETHYLAMMONIUM HYDROXIDE

When we first used this quaternary ammonium base (14), it was commercially available only as a 35-40% aqueous solution, and we prepared it on a laboratory scale by reacting benzyltrimethylammonium iodide with silver oxide. Unfortunately, these solutions were rather unstable.

Since then, the base has become available mainly, if not only, as a 40% solution in methanol, which has proved very stable when kept in the refrigerator. (It is also known by the trade name Triton B.)

All the published procedures in the course of which we had used the aqueous solution were rechecked with the methanolic one, and eventually suitably modified for publication in this book.

As a matter of fact, the base in methanol probably is an equilibrium mixture of hydroxide and methoxide, but since the suppliers still use the term hydroxide for this solution, we kept it up.

IV. PURIFICATION OF SOLVENTS

When a procedure, which had previously proved satisfactory, yields erratic results, it is worth checking up at first if the solvent is not to be incriminated.

In spite of numerous studies which have been carried out on the purification of solvents for analytical purposes, no general answer to this problem can as yet be proposed. A trace of a given impurity may be of the highest importance and can escape detection, whatever be the sensitivity of the method used. On the other hand, it so happened that in some instances we obtained better results with a laboratory reagent grade solvent than with the corresponding analytical grade. Methods such as washing with water, an acid or a base, fractional distillation, eventually from sodium, calcium hydride, calcium chloride, etc., or fractional crystallization can sometimes afford a batch giving still worse results. The behavior of a given batch of solvent can also depend upon the procedure in which it is involved.

Since we tried to present only methods which do not necessitate extreme precautions, and which are not particularly sensitive to trace amounts of impurities in the reagents and solvents, we used as a rule solvents of laboratory reagent grade without further purification for both colorimetric and fluorimetric determinations. Only four solvents were in some instances responsible for wrong results: pyridine (in the reactions of gem-polyhalogen compounds), dimethylformamide, dimethyl sulfoxide, and glacial acetic acid.

When slightly hydrated and exposed to light, pyridine can suffer decomposition through opening of the ring (15). We therefore recommend the selection of a batch which gives a but faintly colored reagent blank, and fractional distillation, protected against light, collecting only the heart-cut (about 80% of the volume), and keeping it protected against light.

Purification of organic solvents by adsorption of the impurities often gives good results (16). For instance, dimethylformamide sometimes afforded highly colored blanks when used for the determination of carbonyl compounds (Chapter 8, Section II.D). Satisfactory results were obtained by shaking the solvent with alumina, or merely with decolorizing carbon (0.1 g of adsorbent per milliliter of solvent) for a few minutes, and filtering.

Dimethyl sulfoxide may be the most troublesome solvent. Miscellaneous physical and chemical methods have been proposed for its purification (17). The difficulties we encountered in the determination of carboxylic acids (Chapter 9, Section II.B) could not be overcome by treatment with alumina or decolorizing carbon. Contrariwise, good results were obtained with the recrystallized solvent (melting point: 18°). A batch of dimethyl sulfoxide was cooled to about +5° until complete crystallization, then allowed to stand until about half of the whole bulk was melted. It was then quickly filtered with suction on a sintered-glass funnel which was precooled in the refrigerator, protected against moisture. The residue on the funnel was immediately collected into a beaker, and kept protected against moisture in a desiccator over sulfuric acid until melted, then stored in a tightly stoppered flask. No other precautions were taken against moisture.

We also once obtained a highly colored blank with a batch of acetic acid when determining carbonyl compounds with 2,4-dinitrophenylhydrazine. We ascertained that this was due to the presence of a trace amount of aldehyde in the solvent. Purification of this acid proved to be a tremendous work. It is therefore advisable merely to select another batch when wrong results are obtained.

None of the solvents used in fluorimetric determinations afforded troubles. However, it may be worth mentioning that Fischer and Cooper (18) showed that when a solvent is unduly fluorescent, a suitable purification can readily be achieved by distilling it in the presence of a small amount (0.5-1% w/v) of a long-chain hydrocarbon, or ordinary paraffin wax. Cetyl alcohol was used for methanol.

REFERENCES

1. E. Sawicki, T.W. Stanley, and T.R. Hauser, *Chemist Analyst*, 48, 30 (1959).

2. H. Meerwein, K. Wunderlich, and K.F. Zenner, *Angew. Chem.*, 74, 807 (1962); M. Pesez and J. Bartos, *Bull. Soc. Chim. Fr.*, 2333 (1963).

3. L. Velluz, *Substances Naturelles de Synthèse*, Masson Ed. Paris, vol. 6, 56 (1953).

4. E.P. Kohler and R.G. Larsen, *J. Amer. Chem. Soc.*, 57, 1452 (1935).

5. R. Wizinger and H.V. Tobel, *Helv. Chim. Acta*, 40, 1305 (1957); J. Bartos, *Pharm. Weekblad*, 93, 594 (1958).

6. From M. Pesez and J.F. Burtin, *Bull. Soc. Chim. Fr.*, 1996 (1959).

7. J. Goerdeler, *Chem. Ber.*, 87, 57 (1954); J. Goerdeler and K. Deselaers, *Chem. Ber.*, 91, 1025 (1958).

8. M. Pesez, *Ann. Pharm. Fr.*, 14, 555 (1956).

9. K.B. Whetsel, G.F. Hawkins, and F.E. Johnson, *J. Amer. Chem. Soc.*, 78, 3360 (1956); M. Pesez and J. Bartos, *Bull. Soc. Chim. Fr.*, 2333 (1963).

10. M. Pesez and J. Bartos, *Bull. Soc. Chim. Fr.*, 1930 (1961).

11. M. Pesez and J. Bartos, *Talanta*, 14, 1097 (1967).

12. From J.E. Spikner and J.C. Towne, *Anal. Chem.*, 34, 1468 (1962).

13. L. Velluz, G. Amiard, and M. Pesez, *Bull. Soc. Chim. Fr.*, 678 (1948).

14. M. Pesez and J. Bartos, *Talanta*, 5, 216 (1960).

15. H. Freytag, *Ber.*, 67, 1995 (1934); 69, 32 (1936).

16. Cf. G. Hesse, B.P. Engenbrecht, H. Engelhardt, and S. Nitsch, *Z. Anal. Chem.*, 241, 91 (1968).

17. Cf. T. Chaudron and A. Sekera, *Chim. Anal.*, 53, 310 (1971).

18. N. Fischer and R.M. Cooper, *Chem. Ind. (London)*, 619 (1968).

Part IV

APPENDIX

MISCELLANEOUS TABLES FOR PRACTICAL USERS

I. BUFFERS

A. Clark and Lubs Buffer, pH 2.2-6.2 (1) (Potassium Biphthalate, and Hydrochloric Acid or Sodium Hydroxide)

Stock solutions

a. An 0.2 M aqueous solution of potassium biphthalate (potassium hydrogen phthalate): 1 liter contains 40.844 g of the salt dried at 110°.

b. An 0.2 M solution of hydrochloric acid.

c. An 0.2 M solution of sodium hydroxide, carbon dioxide-free.

To 50 ml of solution a, add the volume of solution b or solution c given in the following table, and dilute to 200 ml with carbon dioxide-free water.

pH	Solution b, ml	pH	Solution c, ml
2.2	46.70	4.0	0.40
2.4	39.60	4.2	3.70
2.6	32.95	4.4	7.50
2.8	26.42	4.6	12.15
3.0	20.32	4.8	17.70
3.2	14.70	5.0	23.85
3.4	9.90	5.2	29.95
3.6	5.97	5.4	35.45
3.8	2.63	5.6	39.85
		5.8	43.00
		6.0	45.45
		6.2	47.00

B. Clark and Lubs Buffer, pH 5.8-8.0 (1) (Potassium Dihydrogen Phosphate and Sodium Hydroxide)

Stock solutions

a. An 0.2 M aqueous solution of potassium dihydrogen phosphate: 1 liter contains 27.218 g of the salt dried at 110°.

b. An 0.2 M solution of sodium hydroxide, carbon dioxide-free.

To 50 ml of solution a, add the volume of solution b given in the following table, and dilute to 200 ml with carbon dioxide-free water.

pH	Solution b, ml	pH	Solution b, ml
5.8	3.72	7.0	29.63
6.0	5.70	7.2	35.00
6.2	8.60	7.4	39.50
6.4	12.60	7.6	42.80
6.6	17.80	7.8	45.20
6.8	23.65	8.0	46.80

C. Clark and Lubs Buffer, pH 7.8-10.0 (1) (Boric Acid, Potassium Chloride, and Sodium Hydroxide)

Stock solutions

a. An 0.2 M aqueous solution of both boric acid and potassium chloride: 1 liter contains 12.368 g of boric acid recrystallized from water and dried at room temperature in the air, and 14.90 g of potassium chloride.

b. An 0.2 M solution of sodium hydroxide, carbon dioxide-free.

To 50 ml of solution a, add the volume of solution b given in the following table, and dilute to 200 ml with carbon dioxide-free water.

pH	Solution b, ml	pH	Solution b, ml
7.8	2.61	9.0	21.30
8.0	3.97	9.2	26.70
8.2	5.90	9.4	32.00
8.4	8.50	9.6	36.85
8.6	12.00	9.8	40.80
8.8	16.30	10.0	43.90

D. McIlvaine Buffer, pH 2.2-8.0 (2) (Disodium Hydrogen Phosphate and Citric Acid)

Stock solutions

a. An 0.2 M aqueous solution of disodium hydrogen phosphate dihydrate (Sorensen's phosphate): 1 liter contains 35.60 g of the salt.

b. An 0.1 M aqueous solution of citric acid monohydrate: 1 liter contains 21.00 g of the acid.

Mix the volumes of solutions a and b given in the following table.

pH	Solution a, ml	Solution b, ml
2.2	4.0	196.0
2.4	12.4	187.6
2.6	21.8	178.2
2.8	31.7	168.3
3.0	41.1	158.9
3.2	49.4	150.6
3.4	57.0	143.0
3.6	64.4	135.6
3.8	71.0	129.0
4.0	77.1	122.9
4.2	82.8	117.2
4.4	88.2	111.8
4.6	93.5	106.5
4.8	98.6	101.4
5.0	103.0	97.0
5.2	107.2	92.8
5.4	111.5	88.5
5.6	116.0	84.0
5.8	120.9	79.1
6.0	126.3	73.7
6.2	132.2	67.8
6.4	138.5	61.5
6.6	145.5	54.5
6.8	154.5	45.5
7.0	164.7	35.3
7.2	173.9	26.1
7.4	181.7	18.3
7.6	187.3	12.7
7.8	191.5	8.5
8.0	194.5	5.5

E. Kolthoff and Vleeschhouwer Buffer, pH 9.2-11.0 (3)
(Sodium Carbonate and Sodium Tetraborate)

Stock solutions

a. An 0.05 M aqueous solution of anhydrous sodium carbonate: 1 liter contains 5.30 g of the salt dried at 160°.

b. An 0.05 M aqueous solution of sodium tetraborate decahydrate: 1 liter contains 19.02 g of the salt recrystallized from water and dried at room temperature in a desiccator over potassium bromide dihydrate.

Mix the volumes of solutions a and b given in the following table.

pH	Solution a, ml	Solution b, ml
9.2	0.00	100.0
9.4	35.70	64.30
9.6	55.50	44.50
9.8	66.70	33.30
10.0	75.40	24.60
10.2	82.15	17.85
10.4	86.90	13.10
10.6	91.50	8.50
10.8	94.75	5.25
11.0	97.30	2.70

F. Kolthoff and Vleeschhouwer Buffer, pH 11.0-12.0 (3)
(Disodium Hydrogen Phosphate and Sodium Hydroxide)

Stock solutions

a. An 0.1 M aqueous solution of disodium hydrogen phosphate dihydrate (Sorensen's phosphate): 1 liter contains 17.80 g of the salt.

b. An 0.1 M solution of sodium hydroxide, carbon dioxide-free.

To 25 ml of solution a, add the volume of solution b given in the following table, and dilute to 50 ml with carbon dioxide-free water.

pH	Solution b, ml
11.0	4.13
11.2	6.00
11.4	8.67
11.6	12.25
11.8	16.65
12.0	21.60

G. Teorell and Stenhagen Buffer, pH 2.0-12.0 (4) (Phosphoric, Citric, Boric, and Hydrochloric Acids, and Sodium Hydroxide)

a. Prepare an aqueous solution of phosphoric acid by diluting 35 ml of 85% phosphoric acid (d = 1.69) to 1 liter, and standardizing against 1 M sodium hydroxide (solution c) in the presence of phenolphthalein.

b. Prepare an aqueous solution of citric acid: 1 liter contains 70.00 g of citric acid monohydrate. Standardize against 1 M sodium hydroxide as above.

c. A 1 M solution of sodium hydroxide, carbon dioxide-free.

d. An 0.1 M solution of hydrochloric acid.

Stock solution

Prepare this solution by measuring into a 1 liter flask the amounts of solutions a and b both equivalent to 100 ml of solution c, adding 343.0 ml of solution c, and diluting to 1 liter with carbon dioxide-free water.

To 20 ml of this stock solution, add the volume of solution d given in the following table, and dilute to 100 ml with carbon dioxide-free water.

pH	Solution d, ml	pH	Solution d, ml
2.0	73.30	5.0	45.18
2.1	70.35	5.1	44.60
2.2	67.85	5.2	44.05
2.3	65.70	5.3	43.50
2.4	63.85	5.4	42.94
2.5	62.25	5.5	42.36
2.6	60.80	5.6	41.80
2.7	59.55	5.7	41.23
2.8	58.45	5.8	40.61
2.9	57.40	5.9	40.00
3.0	56.50	6.0	39.42
3.1	55.70	6.1	38.74
3.2	54.95	6.2	38.09
3.3	54.30	6.3	37.45
3.4	53.70	6.4	36.74
3.5	53.20	6.5	36.06
3.6	52.65	6.6	35.36
3.7	52.10	6.7	34.65
3.8	51.55	6.8	33.92
3.9	51.02	6.9	33.25
4.0	50.50	7.0	32.65
4.1	49.97	7.1	31.98
4.2	49.45	7.2	31.45
4.3	48.90	7.3	30.83
4.4	48.35	7.4	30.35
4.5	47.80	7.5	29.87
4.6	47.26	7.6	29.43
4.7	46.75	7.7	29.05
4.8	46.22	7.8	28.68
4.9	45.68	7.9	28.33

pH	Solution d, ml	pH	Solution d, ml
8.0	28.02	10.0	17.92
8.1	27.69	10.1	17.43
8.2	27.45	10.2	16.97
8.3	27.25	10.3	16.64
8.4	26.90	10.4	16.36
8.5	26.60	10.5	16.15
8.6	26.10	10.6	15.95
8.7	25.63	10.7	15.70
8.8	24.90	10.8	15.40
8.9	24.33	10.9	15.02
9.0	23.75	11.0	14.52
9.1	23.05	11.1	13.93
9.2	22.38	11.2	13.20
9.3	21.72	11.3	12.30
9.4	21.12	11.4	11.23
9.5	20.52	11.5	10.00
9.6	19.94	11.6	8.40
9.7	19.37	11.7	6.60
9.8	18.81	11.8	4.70
9.9	18.35	11.9	2.60
		12.0	0.40

H. Britton and Robinson Buffer, pH 2.40-12.02 (5) (Citric, Boric, and Hydrochloric Acids, Potassium Dihydrogen Phosphate, Barbital Sodium, and Sodium Hydroxide)

Stock solutions

a. Prepare 1 liter of aqueous solution containing 6.01 g of citric acid monohydrate, 3.89 g of potassium dihydrogen phosphate dried at 110°, 1.77 g of boric acid, 5.90 g of barbital sodium (veronal sodium), and 28.6 ml of 1 M hydrochloric acid.

b. An 0.2 M solution of sodium hydroxide.

To 100 ml of solution *a*, add the volume of solution *b* given in the following table.

pH	Solution *b*, ml	pH	Solution *b*, ml
2.40	0	7.12	52
2.55	2	7.30	54
2.73	4	7.45	56
2.92	6	7.62	58
3.12	8	7.79	60
3.35	10	7.98	62
3.57	12	8.15	64
3.80	14	8.35	66
4.02	16	8.55	68
4.21	18	8.76	70
4.40	20	8.97	72
4.57	22	9.20	74
4.75	24	9.41	76
4.91	26	9.65	78
5.08	28	9.88	80
5.25	30	10.21	82
5.40	32	10.63	84
5.57	34	11.00	86
5.70	36	11.23	88
5.91	38	11 44	90
6.10	40	11.60	92
6.28	42	11.75	94
6.45	44	11.85	96
6.62	46	11.94	98
6.79	48	12.02	100
6.94	50		

II. FLUORESCENCE STANDARDS COMPARISONS

Some of the compounds used as standards in our fluorimetric procedures are commercially available: erythrosine sodium, esculin, fluorescein sodium, protocatechuic acid, and quinine sulfate. The syntheses of the other ones are described in Chapter 17, Section II, and we explained in Chapter 1, Section V.C why we referred to these primary standards. However, three commercially available compounds, namely Acridine Yellow (3,6-diamino-2,7-dimethylacridine monohydrochloride), fluorescein sodium, and α-naphthol can eventually be substituted for them, and the following tables will permit comparisons for those who do not intend to synthesize the primary standards. But it must be recalled that the use of these secondary standards may lead to only approximate comparisons.

On the other hand, since the linear relationship of concentration versus reading holds in the range of concentrations given in the tables, and since, unless otherwise stated, the solvent blank is negligible, extrapolations can be readily made for reading 50.

The solutions of these secondary standards are prepared as follows:

Acridine Yellow. Dissolve by warming 0.020 g of the dye in 5 ml of dimethyl sulfoxide, cool, and dilute to 100 ml with ethanol. This solution, which contains 200 μg/ml of Acridine Yellow, is suitably diluted either with ethanol *(Solution A)* or with a 1:1 mixture of ethanol and 1 N sulfuric acid *(Solution B)*.

Fluorescein sodium. Dissolve in water *(Solution C)*.

α-*Naphthol.* Dissolve in a 1:4 mixture of 0.2 N sodium hydroxide and ethanol *(Solution D)*.

A. 2,6-Dimethyl-3,5-diacetyl-1,4-dihydropyridine

Exc, nm	Em, nm	Concentration, primary standard, μg/ml	Corresponds to Solution	Corresponds to μg/ml	References Chapter	References Section
Primary standard in ethanol						
395	485	1.00	A	0.041	4	XII.B
					9	XI.C
					14	XVI.B
405	485	1.00	A	0.0318	4	XII.B
					9	XI.C
					14	XVI.B
405	490	3.00	A	0.10	7	XI.A
405	505	0.50	C	0.106	6	IV.A
405	510	0.50	C	0.097	4	XII.A
					6	IV.A
436	505	0.70	C	0.0325	13	XI.A
436	510	0.70	C	0.034	13	XI.A
436	515	0.70	C	0.0365	13	XI.A
436	550	1.00	B	0.0135	11	IV.A
Primary standard in 50% ethanol						
366	510	5.00	C	0.49	4	XI.C
436	510	1.00	C	0.050	14	XII.A
436	520	0.15	C	0.0080	15	XVI.A
Primary standard in 1:49 ethanol-water						
436	540	1.00	B	0.086	13	X.A

B. 2,6-Dimethyl-3,5-dicarbethoxy-1,4-dihydropyridine

Exc, nm	Em, nm	Concentration, primary standard, μg/ml	Corresponds to Solution	Corresponds to μg/ml	References Chapter	References Section
Primary standard in 3:97 ethanol-water						
340	470	2.50	D	0.096	7	IX.A
366	470	2.50	D	0.96	12	V.A
Primary standard in 1:49 ethanol-water						
366	470	2.50	D	1.2	2	XIV.A
					2	XV.A
					2	XVI.A
					5	V.A
					15	XVIII.A
Primary standard in 1:99 ethanol-water						
366	470	1.00	D	0.46	7	X.A
Primary standard in 0.7:99.3 ethanol-water						
366	480	0.62	D	0.35	13	XI.A

C. 1,8-Dioxodecahydroacridine

Exc, nm	Em, nm	Concentration, primary standard, μg/ml	Corresponds to Solution	Corresponds to μg/ml	References Chapter	References Section
Primary standard in 1:49 ethanol-water						
366	470	0.50	D	2.55	8	XI.A
Primary standard in 1:49 ethanol-2 N sodium hydroxide						
405	520	0.32	B	0.076	8	XIII.A
436	510	0.20	C	0.105	2	XIII.A
436	515	0.20	C	0.112	2	XIII.A
436	520	0.08	B	0.0142	8	XIII.A
436	520	0.08	C	0.048	8	XIII.A
436	520	0.70	C	0.42	8	XI.A
436	525	0.15	C	0.090	2	XIII.A
436	530	0.18	B	0.0295	8	XIII.A

D. 2-Hydroxy-3-methylquinoxaline

No satisfactory secondary standard was found. Acridine Yellow, solution A, can be used, but at such dilutions that the solvent blank also fluoresces, affording a reading of about 10-15. In the following table, the data correspond to the effective readings, i.e., to the sum: fluorescence of the dye + fluorescence of the solvent. They must therefore be considered as but indicatory and very roughly approximate.

Exc, nm	Em, nm	Concentration, primary standard, μg/ml	Corresponds to		References	
			Solution	μg/ml	Chapter	Section
Primary standard in 50% sulfuric acid						
366	490	0.30	A	0.065	9	XII.A
366	500	0.56	A	0.14	9	XII.A
366	505	0.36	A	0.07	9	XII.A
366	520	0.27	A	0.062	9	XII.A

E. 2-Methyl-5-carboxy-7-aminoquinoline

Exc, nm	Em, nm	Concentration, primary standard, μg/ml	Corresponds to		References	
			Solution	μg/ml	Chapter	Section
Primary standard in glacial acetic acid						
405	485	0.030	A	0.044	4 9	XIII.A XI.A
405	490	0.044	A	0.060	9	XI.A
405	500	0.040	A	0.0485	9	XI.A
436	495	0.35	A	0.049	14	XIV.A
Primary standard in 10% acetic acid						
405	495	0.25	A	0.335	8	XII.A
405	505	0.030	A	0.036	14	XV.A

REFERENCES

1. W.M. Clark and H.A. Lubs, *J. Biol. Chem.*, 25, 479 (1916).

2. T.C. McIlvaine, *J. Biol. Chem.*, 49, 183 (1921).

3. I.M. Kolthoff and J.J. Vleeschhouwer, *Biochem. Z.*, 189, 191 (1927).

4. T. Teorell and E. Stenhagen, *Biochem. Z.*, 299, 416 (1938).

5. H.T.S. Britton and R.A. Robinson, *J. Chem. Soc.*, 1456 (1931).

AUTHOR INDEX

Numbers in parentheses are reference numbers and indicate that an author's work is referred to although his name is not cited in the text. Underlined numbers give the page on which the complete reference is listed.

I

J

M

N

O

SUBJECT INDEX

This Index includes:

1. The names of all compounds for which an absorbance or a fluorescence value is given in the book. These names and the corresponding pages are not underlined. Numbers labeled F refer to fluorimetric determinations. Salified acidic or basic compounds are referred to as free acids or bases.

2. The names of compounds which can be either characterized or determined with reagents used in the Procedures described in the book, but which are only mentioned and were not tested in our laboratories, and the names of compounds tested, but with which Beer's law does not hold. The corresponding pages are underlined.

3. The reagents used in the methods described in the book. Their names are underlined. In this case, an underlined number corresponds to the page where the synthesis of the reagent is described (either under Reagents or in Chapter 17). Numbers labeled F refer to fluorimetric reagents. As a rule, mineral and organic acids and bases, alkaline carbonates, chlorides, nitrates, phosphates, sulfates, salts used only as buffers, and solvents are not mentioned. The only exception to this rule concerns the compounds of these classes which are incorporated in the final colored or fluorescent species through a covalent bond.

4. Fluorescence standards. Only the pages where the synthesis of these compounds are given are mentioned. The corresponding numbers are labeled S.

A

O

Q

R